D1570017

Functional and Numerical Methods in Viscoplasticity

Functional and Numerical Methods in Viscoplasticity

IOAN R. IONESCU
Institute of Mathematics
Romanian Academy
Bucharest, Romania

and

MIRCEA SOFONEA
Department of Mathematics
Blaise Pascal University
Clermont-Ferrand, France

OXFORD NEW YORK TOKYO
OXFORD UNIVERSITY PRESS
1993

Oxford University Press, Walton Street, Oxford OX2 6DP
Oxford New York Toronto
Delhi Bombay Calcutta Madras Karachi
Kuala Lumpur Singapore Hong Kong Tokyo
Nairobi Dar es Salaam Cape Town
Melbourne Auckland Madrid
and associated companies in
Berlin Ibadan

Oxford is a trade mark of Oxford University Press

Published in the United States
by Oxford University Press Inc., New York

© Ioan R. Ionescu and Mircea Sofonea, 1993

All rights reserved. No part of this publication may be reproduced, stored in a retrieval system, or transmitted, in any form or by any means, without the prior permission in writing of Oxford University Press. Within the UK, exceptions are allowed in respect of any fair dealing for the purpose of research or private study, or criticism or review, as permitted under the Copyright, Designs and Patents Act, 1988, or in the case of reprographic reproduction in accordance with the terms of licences issued by the Copyright Licensing Agency. Enquiries concerning reproduction outside those terms and in other countries should be sent to the Rights Department, Oxford University Press, at the address above.

A catalogue record for this book is available from the British Library

Library of Congress Cataloging in Publication Data
Ionescu, Ioan R.
Functional and numerical methods in viscoplasticity / Ioan R. Ionescu and Mircea Sofonea.
(Oxford mathematical monographs)
Includes bibliographical references and index.
1. Viscoplasticity. I. Sofonea, Mircea. II. Title. III. Series.
QA931.I55 1993 532'.0533—dc20 92-35315

ISBN 0–19–853590–2

Typeset by Integral Typesetting, Great Yarmouth, Norfolk
Printed in Great Britain by
Biddles Ltd, Guildford & King's Lynn

Cstack
SCIMON
8-11-93

To

Rodica and Răzvan

and

Carmen and Mircea

PREFACE

The development of interest in viscoplasticity was determined by the investigation of the mechanical properties of materials such as pastes, paint oils, and dough. Later on, elastic/viscoplastic models were applied to various materials, such as metals, soils, polymers, rubbers, and rocks.

The large variety of constitutive equations used in viscoplasticity and their applications in engineering have encouraged new mathematical techniques in the study of the existence, behaviour, and numerical approximation of the solutions for a large class of mixed problems. So, the use of *functional and numerical methods* in the study of viscoplasticity problems has rapidly developed in various directions over the last few decades.

The purpose of this book is to give the reader some mathematical methods that can be applied in viscoplasticity. It represents an extension of the original research work of the authors, and is aimed at students and researchers interested in the mathematical and numerical problems of non-linear mechanics and their application.

The contents of the book have been compiled to cover the path starting from the mechanical model and ending with the numerical solution of the associated mathematical formulation. Our intention was to give a complete treatment of the studied problems, so theoretical results (concerning the existence, uniqueness, and behaviour of the solution) as well as numerical approaches and some engineering applications are presented. In this way the authors try to indicate how practical problems can be compiled as problems of functional analysis, and how this mathematical approach can be used for the analysis and numerical treatment of these engineering problems.

As *functional methods* in the study of the mechanical problems presented in this book, elliptic variational inequalities, ordinary differential equations in Banach spaces, linear semigroups of continuous operators, monotony, and fixed points methods are used. These functional tools allow us to obtain the following types of results:

- existence and uniqueness results of local or global solutions
- continuous dependence of solutions with respect to the data and parameters
- evolution of perturbations
- stability, long-term behaviour, and blow-up of the solutions.

The study of the behaviour of the solutions with respect to the parameters allows us to point out:

- the link between different mechanical models (elasticity as a limit case of viscoelasticity, perfect plasticity as an approach of viscoplasticity)
- the blocking property of the solution as a function of the yield limit and

of the friction coefficient in the study of the flow of a Bingham fluid with friction
- the study of strain localization phenomena for rate-type viscoplastic models.

The theoretical results obtained from different functional methods are enhanced by relevant examples, counter-examples, and 'pathological' examples.

The *numerical approach* of the solution follows from the use of the following techniques: penalization and regularization of elliptic variational inequalities, finite element discretization, Euler's method for ordinary differential equations, and Newton's method for non-linear equations. These techniques allow us to obtain results of the following type:

- convergence of the numerical solutions
- estimates of the error on finite or infinite time intervals.

The numerical methods presented here are completed by some relevant examples and applications in concrete problems of:

- wire drawing
- mining engineering.

The book is intended to be self-contained and accessible to a large number of readers. To this end, Chapter 1 presents some preliminaries on continuous mechanics, and Chapter 2 summarizes some basic results on functional spaces used in viscoplasticity. Since the reading of this book supposes some knowledge of *functional* and *numerical analysis*, an Appendix has been added summarizing some of the necessary prerequisites.

Chapters 3, 4, and 5 represent the main part of the book. So, in Chapters 3 and 4 some mixed problems for semilinear rate-type constitutive equations arising from small-strain viscoplasticity are studied in both the quasistatic and dynamic cases. In Chapter 5, boundary-value problems describing the flow of a Bingham fluid with friction are considered.

The list of references given at the end of the book is not intended to be a complete bibliography on the subject. It refers only to papers which were used for or are directly connected with the subjects treated in this book. Each chapter is concluded with a detailed section entitled 'Bibliographical notes', containing references to the principal results treated, as well as information on important topics related to, but not included in, the body of the text. The main points of this book such as formulae, definitions, theorems, and remarks are numbered consecutively using a decimal system within each section. References to these points in the sections are made by decimal notations with explicit indication of the chapter concerned if it is different from that in which the reference is made. Thus, for example (3.5) means the fifth formula of the third section of the chapter in which it occurs, in the absence of a specified chapter number. References to points in the Appendix are preceded by the capital letter A.

The authors would like to thank Professors N. Cristescu, I. Suliciu, and E. Soos for the interesting discussions which generated their interest in the field.

Aubière I.R.I.
December 1992 M.S.

CONTENTS

NOTATION

$\mathbb{N}$	the set of natural numbers;
$\mathbb{R}$	the real line $(-\infty, +\infty)$;
$\mathbb{R}_+$	the set $[0, +\infty)$;
C, C_i	strictly positive generic constants ($i \in \mathbb{N}$);
$x \to f(x)$	function correspondence;
f_x	the (partial) derivative with respect to x;
∇f	the gradient of the function f;
$\varepsilon(f), D(f)$	the symmetric part of the gradient of f, i.e. $\frac{1}{2}(\nabla f + \nabla^T f)$;
∂f	the subdifferential of the function f;
$\dot{f}, \ddot{f}$	first and respectively second time derivative of f, sometimes interpreted in the sense of distributions;
$(x_1; x_2; \ldots; x_N)$	an element of the product space $X_1 \times \cdots \times X_N$;
X^N	the space given by $X^N = \{x = (x_i) \mid x_i \in X, i = \overline{1, N}\}$;
$X_s^{N\times N}$	the space given by $X_s^{N\times N} = \{x = (x_{ij}) \mid x_{ij} = x_{ji} \in X, i, j = \overline{1, N}\}$;
$\lVert \cdot \rVert_X$	the norm on the space X;
$\langle , \rangle_X$	the inner product on the space X;
X'	the strong dual of the normed space X;
$B(x, r)$	the set of elements y in X such that $\lVert y - x \rVert_X \leq r$;
$B(X, Y)$	the space of linear continuous operators from X to Y; in particular, $B(X) = B(X, X)$;
$\langle , \rangle_{Y,X}$	duality pairing between Y and X;
θ_X	the zero element of X;
I_X	the identity map on X;
2^X	the set of all subsets of X;
$\mathscr{S}_N$	the space of second-order symmetrical tensors on $\mathbb{R}^N$, i.e. $\mathscr{S}_N = \mathbb{R}_s^{N\times N}$;
I_N	the identity map on $\mathbb{R}^N$, $\mathscr{S}_N$;
0_N	the zero element of $\mathbb{R}^N$, $\mathscr{S}_N$;
$\cdot, \lvert\cdot\rvert$	inner product and norm on $\mathbb{R}^N$ and $\mathscr{S}_N$;
σ'	the deviator of the tensor $\sigma \in \mathscr{S}_N$;
$\sigma_{\Vert}$	the positive quantity defined by $\sigma_{\Vert} = \frac{1}{2}\sigma' \cdot \sigma'$ $\forall \sigma \in \mathscr{S}_N$;
Ω	domain in $\mathbb{R}^N$, i.e. an open connected set with the boundary denoted by Γ and the closure $\bar{\Omega}$;
a.e.	almost everywhere;
Γ_i	open subsets of Γ with the closure in Γ denoted by $\bar{\Gamma}_i$ ($i \in N$);
meas Γ_i	the $(N-1)$-dimensional Lebesgue measure of Γ_i;
ν	the exterior unit normal at Γ.

If Ω is a bounded domain in $\mathbb{R}^N$ with smooth boundary, $m \in \mathbb{N}$ and $1 \leq p \leq \infty$, the following classical notations are used:

$C^m(\Omega)$	the space of real-valued and m times continuously differentiable functions on Ω; we write $C(\Omega)$ instead of $C^0(\Omega)$;
$C_0^m(\Omega)$	the space of functions in $C^m(\Omega)$ with compact support in Ω; in particular $C_0(\Omega) = C_0^0(\Omega)$;
$C^m(\bar{\Omega})$	the space of functions which, together with their derivatives through order m are bounded and uniformly continuous;
$C^\infty(\Omega)$, $C^\infty(\bar{\Omega})$	the spaces defined by $$C^\infty(\Omega) = \bigcap_{m \in \mathbb{N}} C^m(\Omega),\ C^\infty(\bar{\Omega}) = \bigcap_{m \in \mathbb{N}} C^m(\bar{\Omega});$$
$C_0^\infty(\Omega)$ or $\mathscr{D}(\Omega)$	the space of functions in $C^\infty(\Omega)$ with compact support in Ω;
$\mathscr{D}(\Omega)$	the dual of $\mathscr{D}(\Omega)$, i.e. the space of all scalar distributions on Ω;
$L^p(\Omega)$	the space of p-summable real-valued functions on Ω if $1 \leq p < \infty$, with the usual modification if $p = \infty$; $L^p(\Gamma)$ is defined in a similar way for Γ;
$W^{m,p}(\Omega)$	the Sobolev space of all functions in $L^p(\Omega)$ having the property that all distributional derivatives up to order m belong to $L^p(\Omega)$; in particular $H^m(\Omega) = W^{m,2}(\Omega)$;
$W_0^{m,p}(\Omega)$	the closure of $\mathscr{D}(\Omega)$ in $W^{m,p}(\Omega)$; in particular $H_0^m(\Omega) = W_0^{m,2}(\Omega)$;
$H^{1/2}(\Gamma)$	the Sobolev space of order $\frac{1}{2}$ on Γ.

If $N \in \mathbb{N}$, we also use the following notation:

$$
\begin{aligned}
\boldsymbol{D} &= [\mathscr{D}(\Omega)]^N \\
\mathscr{D} &= [\mathscr{D}(\Omega)]_s^{N \times N} \\
H &= L^2(\Omega) \\
\boldsymbol{H} &= [L^2(\Omega)]^N \\
\mathscr{H} &= [L^2(\Omega)]_s^{N \times N} \\
H_1 &= H^1(\Omega) \\
\boldsymbol{H}_1 &= [\boldsymbol{H}^1(\Omega)]^N \\
\mathscr{H}_2 &= \{\sigma \in \mathscr{H} \mid \operatorname{Div} \sigma \in \boldsymbol{H}\} \\
\boldsymbol{H}_\Gamma &= [H^{1/2}(\Gamma)]^N \\
\boldsymbol{H}'_\Gamma &= [H^{-1/2}(\Gamma)]^N \\
\boldsymbol{V}_1 &= \{v \in \boldsymbol{H}_1 \mid v = 0 \text{ on } \Gamma_1\} \\
\mathscr{V}_1 &= \{\sigma \in \mathscr{H}_2 \mid \sigma v = 0 \text{ on } \Gamma_2\} \\
\mathscr{V}_2 &= \{\sigma \in \mathscr{H}_2 \mid \operatorname{Div} \sigma = 0 \text{ in } \Omega,\ \sigma v = 0 \text{ on } \Gamma_2\} \\
\boldsymbol{V}_{\Gamma_1} &= \{\xi \in \boldsymbol{H}_\Gamma \mid \xi = 0 \text{ on } \Gamma_1\}
\end{aligned}
$$

$\gamma_1: \boldsymbol{H}_1 \to \boldsymbol{H}_\Gamma$	the trace map for vectorial functions
$\gamma_2: \mathscr{H}_2 \to \boldsymbol{H}'_\Gamma$	the trace map for tensorial functions
$z_1: \boldsymbol{H}_\Gamma \to \boldsymbol{H}_1$	the continuous section of the inverse of γ_1
$z_2: \boldsymbol{H}'_\Gamma \to \mathscr{H}_2$	the continuous section of the inverse of γ_2

If X is a Banach space, $T > 0$, $1 \leq p \leq \infty$ and $k \in \mathbb{N}$, the following notations are also used:

$L^p(0, T, X)$	the space of strongly measurable functions on $(0, T)$ to X such that $\int_0^T \|(x(t)\|_X^p \, dt < +\infty$ with the usual modification if $p = \infty$;
$W^{k,p}(0, T, X)$	the space of functions in $L^p(0, T, X)$ having the property that all vectorial distributional derivatives up to order k belong to $L^p(0, T, X)$;
$C^k(I, X)$	the space of k times continuously differentiable functions on the real interval $I \subset \mathbb{R}$ to X; we write $C(I, X) = C^0(I, X)$;
$\|\cdot\|_{k,T,X}$	the norm on the space $C^k([0, T], X)$ or in $W^{k,\infty}(0, T, X)$;
$\|\cdot\|_X^\infty$	the norm on the space of bounded functions belonging to $C(\mathbb{R}_+, X)$, i.e. $\|x\|_X^\infty = \sup_{t \in \mathbb{R}_+} \|x(t)\|_X$.

1

PRELIMINARIES ON MECHANICS OF CONTINUOUS MEDIA

The object of this chapter is to summarize some basic concepts and general principles of continuous mechanics. To this end, elements of kinematics, the balance laws, the stress tensors, and the constitutive laws in the study of continuous media are presented. All the functions involved here are assumed sufficiently smooth and the classical notations ∇, div, Div for the gradient operator, the vectorial divergence operator, and the tensorial divergence operator are used. The notations $\mathcal{M}_3$ and $\mathcal{S}_3$ represent the space of tensors and symmetric tensors on $\mathbb{R}^3$, while I_3, 0_3 represent the unit and the zero element of $\mathcal{M}_3$ and $\mathcal{S}_3$. Moreover, for every tensor S, we denote its transpose by S^{T}, the trace of S by $\operatorname{tr} S$ and finally, the summation rule upon the repeated indices is used everywhere in this section.

1 Kinematics of continuous media

We start this section by presenting some basic ideas on deformation theory. After this, we introduce the finite strain tensor and the small strain tensor, and present their physical interpretation.

1.1 Material and spatial description

As usual in the literature of the field, we understand the term *continuous medium* to mean a body which completely fills the space that it occupies, leaving no pores or empty spaces, so that its properties are describable by continuous functions. A continuous medium is made up of *particles*.

In the following we assume that an origin O and an orthonormal basis (e_1, e_2, e_3) have been chosen in three-dimensional Euclidean space, which will therefore be identified with the space $\mathbb{R}^3$. From the notational viewpoint, we identify the point x with the vector Ox.

Let us consider a continuous medium which at time $t = 0$ occupies a subset Ω of $\mathbb{R}^3$ and and the time $t > 0$ occupies a subset Ω_t of $\mathbb{R}^3$. We suppose that Ω and Ω_t are bounded, open, and connected subsets of $\mathbb{R}^3$. Ω is called the *reference configuration* or the *non-deformed configuration* of the body, and for every $t > 0$, Ω_t is called the *actual configuration* or the *deformed configuration* of the body. Let P be an arbitrary particle of the body, let $X = (X_1, X_2, X_3)$ be the position of the particle in the basis (e_1, e_2, e_3) at $t = 0$ and let $x = (x_1, x_2, x_3)$ be the position of the same particle P at $t > 0$.

The components X_1, X_2, and X_3 are called the *Lagrangian coordinates* of the particle P and the description which uses the components X_1, X_2, and X_3 as independent variables is called the *Lagrangian description.* We may think of this description as a *material description* by using the reference position X as a label for a material particle, and call X the *material coordinates* of the particle P. In the same way the components x_1, x_2, and x_3 are called the *Eulerian coordinates* or *spatial coordinates* of the particle P, and the description which uses the components x_1, x_2, and x_3 and the time variable t as independent variables is called the *Eulerian description* or the *spatial description.*

The *motion* of a body is determined by the position x of the material points in space as a function of the reference position X and the time t, and hence is a family of functions $\chi(\cdot, t)\colon \Omega \to \chi(\Omega, t) = \Omega_t$ defined by

$$x = \chi(X, t) \tag{1.1}$$

for every $t > 0$. We suppose that for every $t > 0$, $\chi(\cdot, t)$ is a continuous injective function, and denote by $\chi^{-1}(\cdot, t)\colon \Omega_t \to \Omega$ the inverse of χ with respect to its first argument, that is

$$X = \chi^{-1}(x, t). \tag{1.2}$$

Using (1.1) and (1.2), every quantity defined on the body can be regarded either as a function of X and t (the case of the material description), or as a function of x and t (the case of the spatial description). So, the material description fixes attention on a given particle or part of the body, while the spatial description fixes attention on a given point or region of the space.

Now let ϕ be a scalar or vector function defined on $\Omega \times \mathbb{R}_+$, that is, on the Lagrangian frame. The derivative of ϕ with respect to time with X constant is called the *material* or *Lagrangian derivative* of ϕ and will be denoted by $\mathrm{d}\phi/\mathrm{d}t$ or by $\dot{\phi}$:

$$\dot{\phi}(X, t) = \frac{\mathrm{d}\phi}{\mathrm{d}t}(X, t) \qquad \forall X \in \Omega, \quad t > 0. \tag{1.3}$$

In particular, the *velocity* and the *acceleration* of the particle P at the time $t > 0$ are defined using the material derivatives as follows:

$$v = \dot{x} = \frac{\mathrm{d}}{\mathrm{d}t}\chi(X, t) \tag{1.4}$$

$$a = \dot{v} = \ddot{x} = \frac{\mathrm{d}^2}{\mathrm{d}t^2}\chi(X, t). \tag{1.5}$$

Now let ϕ be a scalar or vector function defined in the Eulerian frame, that is, ϕ depends on $x \in \Omega_t$ and $t > 0$. The material derivative of ϕ is the derivative of the function $t \to \phi(\chi(X, t), t)$, for $X \in \Omega$ fixed, and generally it is different from the time derivative of the function $t \to \phi(x, t)$ for a fixed $x \in \Omega_t$, called the *Eulerian* or *spatial derivative* of ϕ and denoted by $\partial\phi/\partial t$.

More precisely, using the chain rule we have

$$\dot{\phi}(x,t) = \frac{\mathrm{d}}{\mathrm{d}t}\phi(x,t) = \frac{\mathrm{d}}{\mathrm{d}t}[\phi(\chi(X,t),t)] = \frac{\partial\phi}{\partial t}(x,t) + \frac{\partial\phi}{\partial x_i}(x,t)\frac{\partial\chi_i}{\partial t}(x,t)$$

and, using the notation ∇_x for the spatial gradient of ϕ, by (1.4) we have

$$\dot{\phi} = \frac{\partial\phi}{\partial t} + (v\cdot\nabla_x)\phi. \tag{1.6}$$

In particular, from (1.6) and (1.5) it follows that

$$a = \frac{\partial v}{\partial t} + (v\cdot\nabla_x)v. \tag{1.7}$$

1.2 Deformation and strain tensors

A *deformation* of the reference configuration Ω is the function χ defined by (1.1) for a fixed $t > 0$. Given a deformation we define at each point of Ω the matrix

$$F = (F_{ij}), \qquad F_{ij} = \frac{\partial\chi_i}{\partial X_j} \tag{1.8}$$

called the *deformation gradient*. The determinant of F denoted $J = J(X,t)$ measures the local change of volume. Indeed, if we denote by $\mathrm{d}X$ the *volume element* at the point X of the reference configuration and by $\mathrm{d}x$ the volume element at the point $x = \chi(X,t)$ of the deformed configuration we have

$$\mathrm{d}x = |J(X,t)|\,\mathrm{d}X.$$

In order to preclude compression to zero volume we shall assume that $J > 0$, i.e. the motion χ satisfies the *orientation-preserving condition*, and by the previous equality it follows that

$$\mathrm{d}x = J\,\mathrm{d}X. \tag{1.9}$$

If A denotes a measurable subset of the reference configuration Ω, the volume of A is given by

$$\mathrm{vol}\,A = \int_A \mathrm{d}X. \tag{1.10}$$

If we use the notation $A_t = \chi(A,t)$, then we have $\mathrm{vol}\,A_t = \int_{A_t}\mathrm{d}x$ and using (1.10) and the formula for changes of variables in multiple integrals we get

$$\mathrm{vol}\,A_t = \int_A J\,\mathrm{d}X. \tag{1.11}$$

By (1.10) and (1.11) it follows that if $J = 1$ the motion χ has the volume-preserving property. Conversely if for every measurable subset A of Ω we have $\mathrm{vol}\,A = \mathrm{vol}\,A_t$ it follows that $J = 1$. Hence the equality

$$J(X,t) = 1, \qquad \forall X\in\Omega, \quad t > 0 \tag{1.12}$$

characterizes the motions having the volume-preserving property. Such motions are called *isochoric motions*. A body is called *incompressible* if it may have only isochoric motions.

Let us denote by ∇_X and ∇_x the gradient operators on the reference and actual configuration respectively. Using (1.8) we have

$$F = \nabla_X \chi \tag{1.13}$$

and

$$\nabla_X \phi = \nabla_x \phi \cdot F \tag{1.14}$$

for all scalar or vector functions $\phi(x, t) = \phi(\chi(X, t), t)$.

Together with a deformation $\chi(\cdot, t)$ it is often convenient to introduce the *displacement* u which is the vector field $u(\cdot, t)\colon \Omega \to \mathbb{R}^3$ defined by the relation

$$x = \chi(X, t) = X + u(X, t), \qquad \forall X \in \Omega \quad \text{and} \quad t > 0. \tag{1.15}$$

The displacement field u estimates the difference between the material points X and their corresponding points x. Moreover, using (1.4) and (1.5) it follows that

$$v = \dot{u}, \qquad a = \ddot{u}. \tag{1.16}$$

Let H be the *displacement gradient* defined by

$$H = \nabla_X u \qquad \left(H_{ij} = \frac{\partial u_i}{\partial X_j} \right). \tag{1.17}$$

Using (1.15) it follows that the displacement gradient and the deformation gradient are related by the equation

$$F = I_3 + H \qquad (F_{ij} = \delta_{ij} + H_{ij}). \tag{1.18}$$

Let us now introduce the symmetric tensor

$$C = F^{\mathrm{T}} F \qquad (C_{ij} = F_{ki} F_{kj}) \tag{1.19}$$

called the *right Cauchy–Green strain tensor*. In order to deduce the significance of this tensor let us differentiate (1.1) for $t =$ constant. Using (1.8) it follows that

$$\mathrm{d}x = F\,\mathrm{d}X \qquad (\mathrm{d}x_i = F_{ij}\,\mathrm{d}X_j) \tag{1.20}$$

hence

$$|\mathrm{d}x|^2 = \mathrm{d}x \cdot \mathrm{d}x = (F\,\mathrm{d}X) \cdot (F\,\mathrm{d}X) = (F_{ki}\,\mathrm{d}X_i)(F_{kj}\,\mathrm{d}X_j) = F_{ki} F_{kj}\,\mathrm{d}X_i\,\mathrm{d}X_j$$

and by (1.19) we get

$$|\mathrm{d}x|^2 = C_{ij}\,\mathrm{d}X_i\,\mathrm{d}X_j = \mathrm{d}X \cdot C\,\mathrm{d}X. \tag{1.21}$$

It follows that the knowledge of the tensor C allows us to compute the lengths of the infinitesimal vector $\mathrm{d}x$ in the deformed configuration.

With the intention of showing that the tensor C is indeed a good measure of 'strain', understood here in its intuitive sense of 'change in *form* or *size*', let us first consider a class of motions that induce no 'strain': a motion is called a *rigid motion* if it is of the form

$$\chi(X, t) = a(t) + Q(t)X \qquad \forall X \in \Omega, \quad t > 0 \tag{1.22}$$

where $a(t) \in \mathbb{R}^3$, $Q(t) \in Q^3_+$, Q^3_+ being the set of *rotations* in $\mathbb{R}^3$, i.e. the set of orthogonal matrices of order 3 whose determinant is $+1$. In other words, for every $t > 0$ the corresponding deformed configuration is obtained by rotating the reference configuration around the origin by the rotation $Q(t)$ and by translating it by the vector $a(t)$. This indeed corresponds to the idea of rigid motion where the reference configuration is 'moved' but without strain.

If χ is the rigid motion given by (1.22) then $F(X, t) = Q(t)$ for all $X \in \Omega$ and $t > 0$ and therefore by (1.19) it follows that

$$C = I_3 \qquad \forall X \in \Omega, \quad t > 0. \tag{1.23}$$

It is remarkable that conversely, if (1.23) is satisfied then the corresponding motion is a rigid motion, i.e. is of the form (1.22) with $a(t) \in \mathbb{R}^3$, $Q(t) \in Q^3_+$. As a consequence, the condition (1.23) characterizes rigid motion, i.e. an arbitrary motion is a rigid one iff (1.23) is satisfied. In order to compare the deformed element with its original shape we introduce the relative measure of the deformation, E, by

$$E = \tfrac{1}{2}(C - I_3) = \tfrac{1}{2}(F^{\mathrm{T}}F - I_3) \tag{1.24}$$

called the *finite strain tensor* or the *Green–Saint Venant tensor*. Using (1.21) we have

$$|\mathrm{d}x|^2 - |\mathrm{d}X|^2 = C_{ij}\,\mathrm{d}X_i\,\mathrm{d}X_j - \delta_{ij}\,\mathrm{d}X_i\,\mathrm{d}X_j = (C_{ij} - \delta_{ij})\,\mathrm{d}X_i\,\mathrm{d}X_j.$$

Hence by (1.24) it follows that

$$|\mathrm{d}x|^2 - |\mathrm{d}X|^2 = 2E_{ij}\,\mathrm{d}X_i\,\mathrm{d}X_j, \tag{1.25}$$

i.e. the tensor E characterizes the variation of the square of length of the material element. Using the definition of the strain tensor E we can rephrase the previously mentioned result:

$$\chi \text{ is a rigid motion iff } E = 0_3. \tag{1.26}$$

In what follows we are interested in expressing the tensors C and E in terms of the displacement gradient H. Using (1.18) and (1.19) it follows that

$$C = I_3 + H + H^{\mathrm{T}} + H^{\mathrm{T}}H \tag{1.27}$$

and by (1.24) we get

$$E = \tfrac{1}{2}(H + H^{\mathrm{T}} + H^{\mathrm{T}}H). \tag{1.28}$$

The equality (1.28) shows that the finite strain tensor E is a non-linear function on u (recall that $H = \nabla_X u$). In order to linearize the kinematics relations we shall suppose that $\eta = |H| \ll 1$ and we neglect all the quantities of order 2 or more in η. So we consider the tensor ε defined by

$$\varepsilon = \tfrac{1}{2}(H + H^{\mathrm{T}}) = \tfrac{1}{2}(\nabla_X u + \nabla_X^{\mathrm{T}} u), \tag{1.29}$$

called the *small strain tensor*. Using the above approximation we have $\varepsilon \approx E$ hence ε will have the same physical meaning as E. Sometimes we shall use the notation $\varepsilon(u)$ instead of ε in order to underline the dependence of the small strain tensor on the displacement field u.

Let us introduce the set of displacements $\mathscr{R}$ defined by

$$\mathscr{R} = \{u\colon \Omega \to \mathbb{R}^3 \,|\, u(X) = a + AX, a \in \mathbb{R}^3, A = -A^{\mathrm{T}} \in \mathscr{M}_3\} \tag{1.30}$$

We have the following result

$$\varepsilon(u) = 0_3 \qquad \text{iff} \qquad u(\cdot, t) \in \mathscr{R} \quad \forall t > 0. \tag{1.31}$$

In the following we shall call $\mathscr{R}$ the set of *rigid displacements*. This definition is justified by the fact that in the linearized theory $\varepsilon \approx E$, hence by (1.31) and (1.26) we have that if $u(\cdot, t) \in \mathscr{R}$ then the corresponding motion χ is of an 'almost' rigid type.

We are interested now in characterizing isochoric motion in the linearized theory. Using (1.18) we have $J = \det F = \det(I_3 + H)$ and neglecting the terms of order 2 and 3 in η it follows that

$$J = 1 + H_{kk} = 1 + \operatorname{tr} H = 1 + \operatorname{div}_X u. \tag{1.32}$$

It follows that a motion χ is isochoric iff the corresponding displacement field satisfies the equality $\operatorname{div}_X u = 0$.

We finish this section by remarking that in the linearized theory if the spatial gradient $\nabla_x \phi$ of a scalar or vector function ϕ is of order η then the material gradient $\nabla_X \phi$ of the same function is of order η too and conversely. Moreover, in this case we have $\nabla_X \phi \approx \nabla_x \phi$.

Indeed, using (1.14) and (1.18) we get

$$\nabla_X \phi = \nabla_x \phi \cdot F = \nabla_x \phi (I_3 + H) = \nabla_x \phi + \nabla_x \phi \cdot H$$

hence if $\nabla_x \phi = O(\eta)$ we have $\nabla_x \phi \cdot H = O(\eta^2)$, which implies $\nabla_X \phi = \nabla_x \phi$. Conversely, let $\nabla_X \phi = O(\eta)$. Using again (1.14) and (1.18) we have $\nabla_x \phi = \nabla_X \phi (I_3 + H)^{-1}$, and since in the linearized theory $(I_3 + H)^{-1} = I_3 - H$, we get $\nabla_x \phi = \nabla_X \phi (I_3 - H) = \nabla_X \phi$.

As a consequence, in this case we make no distinction between the material and spatial coordinates in the calculus of the partial derivatives of ϕ. In particular the partial derivatives in (1.29) may be understood with respect to the variables (x_i):

$$\varepsilon = \tfrac{1}{2}(\nabla_x u + \nabla_x^{\mathrm{T}} u). \tag{1.33}$$

For consistency of notation we shall write (1.29) and (1.33) on the form

$$\varepsilon(u) = (\varepsilon_{ij}(u)) \qquad \varepsilon_{ij}(u) = \tfrac{1}{2}(\partial_j u_i + \partial_i u_j) \tag{1.34}$$

where ∂_j represents the partial derivative operator with respect to the X_j or x_j variables.

1.3 The rate of deformation tensor

We denote by L the *spatial gradient of the velocity*, i.e.

$$L = \nabla_x v \qquad \left(L_{ij} = \frac{\partial v_i}{\partial x_j}\right). \tag{1.35}$$

The relation between the tensor L and the gradient of deformation F follows from the chain rule:

$$L_{ij} = \frac{\partial v_i}{\partial x_j} = \frac{\partial v_i}{\partial X_k} \cdot \frac{\partial \chi_k^{-1}}{\partial x_j}. \tag{1.36}$$

Using (1.4) and (1.8) we get

$$L = \dot{F}F^{-1} \qquad (L_{ij} = \dot{F}_{ik}F_{kj}^{-1}). \tag{1.37}$$

The symmetric part of L is called the *rate of deformation tensor* and is denoted by D:

$$D = \tfrac{1}{2}(L + L^{\mathrm{T}}) = \tfrac{1}{2}(\nabla_x v + \nabla_X^{\mathrm{T}} v) \qquad \left(D_{ij} = \frac{1}{2}\left(\frac{\partial v_i}{\partial x_j} + \frac{\partial v_j}{\partial x_i}\right)\right). \tag{1.38}$$

The significance of this definition is given in the following: from (1.20) by differentiation with respect to the time variable it follows that $\dot{\overline{\mathrm{d}x}} = \dot{F}\,\mathrm{d}X$ and $\mathrm{d}X = F^{-1}\,\mathrm{d}x$; hence finally we get $\dot{\overline{\mathrm{d}x}} = \dot{F}F^{-1}\,\mathrm{d}x$; using now (1.37) we obtain

$$\dot{\overline{\mathrm{d}x}} = L\,\mathrm{d}x \qquad (\dot{\overline{\mathrm{d}x_i}} = L_{ij}\,\mathrm{d}x_j). \tag{1.39}$$

It follows that

$$\frac{\mathrm{d}}{\mathrm{d}t}|\mathrm{d}x|^2 = |\dot{\overline{\mathrm{d}x}}|^2 = 2\dot{\overline{\mathrm{d}x}}\cdot\mathrm{d}x = 2L\,\mathrm{d}x\cdot\mathrm{d}x = (L + L^{\mathrm{T}})\,\mathrm{d}x\cdot\mathrm{d}x = 2D\,\mathrm{d}x\cdot\mathrm{d}x.$$

We have thus obtained

$$|\dot{\overline{\mathrm{d}x}}|^2 = 2D_{ij}\,\mathrm{d}x_i\,\mathrm{d}x_j. \tag{1.40}$$

Hence the rate of deformation tensor characterizes the rate of variation of the distances between adjoint material points in a Eulerian frame.

As in the case of small strain tensor we shall use sometimes the notation $D(v)$ instead of D in order to underline the dependence of the rate deformation tensor on the velocity field v. Moreover, the following result holds: $D(v) = 0$ iff

$$v(x, t) = a(t) + A(t)x \qquad \forall x \in \Omega_t, \quad t > 0, \tag{1.41}$$

where $a(t) \in \mathbb{R}^3$ and $A(t)$ is a skew-symmetric tensor. In order to characterize isochoric motion we start from the equalitites $\dot{J} = J \operatorname{tr} L = J \operatorname{tr} D = J \operatorname{div}_x v$, and we conclude that a motion is isochoric iff

$$\operatorname{div}_x v = \operatorname{tr} D = \operatorname{tr} L = 0. \tag{1.42}$$

Moreover, an incompressible body may only have motions for which (1.42) holds.

To complete this section we denote by W the skew-symmetric part of L, i.e.

$$W = \tfrac{1}{2}(L - L^{\mathrm{T}}) \qquad \left(W_{ij} = \frac{1}{2}\left(\frac{\partial v_i}{\partial x_j} - \frac{\partial v_i}{\partial x_i}\right)\right). \tag{1.43}$$

W is called the *spin tensor* and it characterizes the instantaneous rotation velocity at the spatial point x at the current moment t.

2 Balance laws and stress tensors

In this section we present the balance laws of continuous mechanics and we deduce some of their consequences.

2.1 The balance law of mass

In mechanics of continuous media it is supposed that there exists a function $\rho_0 \colon \Omega \to \mathbb{R}_+$ called the *density of mass* (regarded as a property of a body) such that the mass of a part $\mathscr{P}$, which in the reference configuration occupies the open subset $\omega \subset \Omega$, is given by

$$m(\mathscr{P}) = \int_\omega \rho_0(X)\,\mathrm{d}x.$$

Let $t > 0$. In the deformed configuration the part $\mathscr{P}$ occupies the open set $\omega_t = \chi(\omega, t) \subset \Omega_t$, and if we denote by $\rho(\cdot, t) \colon \Omega_t \to \mathbb{R}_+$ the density of mass in the Euler frame the mass of $\mathscr{P}$ is given by

$$m_t(\mathscr{P}) = \int_{\omega_t} \rho(x, t)\,\mathrm{d}x.$$

Assuming the principle of conservation of mass we have $m_t(\mathscr{P}) = m(\mathscr{P})$ $\forall t > 0$, and hence for every part $\mathscr{P}$,

$$\int_{\omega_t} \rho(x, t)\,\mathrm{d}x = \int_\omega \rho_0(X)\,\mathrm{d}X \qquad \forall \omega \subset \Omega, \quad t > 0.$$

Having in mind that $\omega_t = \chi(\omega, t)$ we get $\int_{\omega_t} \rho\,\mathrm{d}x = \int_\omega \rho J\,\mathrm{d}X$, and hence we deduce that $\int_\omega \rho J\,\mathrm{d}x = \int_\omega \rho_0\,\mathrm{d}X$, and because ω is an arbitrary open subset of Ω it follows that

$$\rho_0 = \rho J \qquad (\rho_0(X) = \rho(\chi(X, t), t)J(X, t) \qquad \forall X \in \Omega, \quad t > 0). \quad (2.1)$$

This equation represents the *balance law of mass in the material description.* Taking the derivative with respect to the time variable and using again the equality $\dot{J} = J \operatorname{div}_x v$ we get

$$\dot{\rho} + \rho \operatorname{div}_x v = 0 \qquad \forall x \in \Omega_t, \quad t > 0. \quad (2.2)$$

which represents the *balance law of mass in the spatial description.* Alternatively, using the formula (1.6) for the material derivative, (2.2) becomes

$$\frac{\partial \rho}{\partial t} + \operatorname{div}(\rho v) = 0 \qquad \forall x \in \Omega_t, \quad t > 0. \quad (2.3)$$

From (1.12) and (2.1) one can deduce that a motion is isochoric iff the density of every particle is constant in time (i.e. $\rho(\chi(X, t), t) = \rho_0(X)\ \forall X \in \Omega, t > 0$).

2.2 The balance laws of momentum

Consider a continuous medium which at the instant $t > 0$ occupies the deformed configuration Ω_t. The action of thc outside world on the body may be described by the *applied body forces*, defined by a vector field $b(\cdot, t)\colon \Omega_t \to \mathbb{R}^3$ and called the *density of the applied body forces per unit mass.* An elementary force $b(x, t)$ is exerted on the elementary volume dx at each point x of the deformed configuration such that the resultant of applied body forces acting on Ω_t is $\int_{\Omega_t} b(x, t)\, dx$. As examples we note the gravitational forces (for which $b(x, t) = -g\rho(x)\varepsilon_3\ \forall x \in \Omega_t, t > 0$, where g is the gravitational constant and the basis vector ε_3 is 'vertical' and oriented 'upward') or the electrostratic forces.

Consider now a part $\mathscr{P}$ of the body which at the moment $t > 0$ occupies the open subset $\omega_t \subset \Omega_t$, and suppose that ω_t has smooth boundary $\partial\omega_t$. The action of the subset $\Omega_t \backslash \omega_t$ of the medium on ω_t is described by *surface forces* acting on $\partial\omega_t$. More precisely, let $x \in \partial\omega_t$ and denote by v the unit outward normal at $\partial\omega_t$. We suppose that there exists a vector $\tau(x, v, t)$ which depends on the point x and also on the direction of the normal v such that an elementary force $\tau(x, v, t)\, da$ is exerted on the elementary area da at the point x. As a consequence, the action of $\Omega_t \backslash \omega_t$ on ω_t gives the resultant surface force $\int_{\partial\omega_t} (x, v, t)\, da$. The vector $\tau(x, v, t)$ is called the *Cauchy stress vector* across an oriented surface element with normal v or the *density of the surface force per unit area in the deformed configuration.* Since $\mathscr{P}$ is an arbitrary part of the body (i.e. ω_t is an arbitrary part of Ω_t) it follows that the stress vector defines a field $\tau(\cdot, \cdot, t)\colon \Omega_t \times \tilde{S} \to \mathbb{R}^3$ where $\tilde{S}$ is the surface of the unit sphere in $\mathbb{R}^3$. In view of Newton's third law we have

$$\tau(x, -v, t) = -\tau(x, v, t) \qquad \forall x \in \Omega_t, \quad v \in \tilde{S}, \quad t > 0. \quad (2.4)$$

The projection of $\tau(x, v, t)$ onto the normal v is called the *normal stress* and the projection of $\tau(x, v, t)$ onto the tangential plane at $\partial\omega_t$ is called the *shear* (or *tangential*) *stress*.

The applied body forces corresponding to the vector field $b(\cdot, t)\colon \Omega_t \to \mathbb{R}^3$ and the surface forces corresponding to the vector field $\tau(\cdot, \cdot, t)\colon \Omega_t \times \tilde{S} \to \mathbb{R}^3$ form a *system of forces*. Two general principles hold for every continuous medium subjected to a system of forces, given by the equalities:

$$\frac{\mathrm{d}}{\mathrm{d}t}\int_{\omega_t} \rho v \, \mathrm{d}x = \int_{\omega_t} \rho b \, \mathrm{d}x + \int_{\partial\omega_t} \tau \, \mathrm{d}a \tag{2.5}$$

$$\frac{\mathrm{d}}{\mathrm{d}t}\int_{\omega_t} x \wedge \rho v \, \mathrm{d}x = \int_{\omega_t} x \wedge \rho b \, \mathrm{d}x + \int_{\partial\omega_t} x \wedge \tau \, \mathrm{d}a \tag{2.6}$$

for every $t > 0$ and $\omega_t \subset \Omega_t$. The equality (2.5) is called the *momentum balance principle* and it represents a version of Newton's second law for continuous media. Indeed, the term $\int_{\omega_t} \rho b \, \mathrm{d}x$ represents the resultant of applied body forces acting on ω_t and the term $\int_{\partial\omega_t} \tau \, \mathrm{d}a$ represents the resultant of the surface forces acting on ω_t; hence the right-hand side of (2.5) represents the *resultant of the system of forces* acting on ω_t. By (2.5), this resultant must be equal to the time derivative of the kinetic resultant on ω_t expressed by the left-hand side $\mathrm{d}(\int_{\omega_t} \rho v \, \mathrm{d}x)/\mathrm{d}t$.

In (2.6) $\wedge$ represents the exterior product in $\mathbb{R}^3$. This equality is called the *balance law of angular momentum* and expresses the fact that the resultant momentum at the origin O of the system of forces acting on ω_t is equal to the time derivative of the kinetic momentum of ω_t at the origin.

2.3 The Cauchy stress tensor

We now derive consequences of paramount importance from the balance laws (2.5) and (2.6). The first consequence, due to Cauchy, is one of the most important results in continuous mechanics. It asserts that the dependence of the Cauchy stress vector $\tau(x, v, t)$ with respect to its second argument $v \in \tilde{S}$ *is linear*, hence it defines a second-order tensor $T(x, t)$. The second consequence asserts that at each point $x \in \Omega_t$ the tensor $T(x, t)$ is *symmetric*; the third consequence, again due to Cauchy, is that the tensor field $T(\cdot, t)$ and the vector field $b(\cdot, t)$ are related by a *partial differential equation* in Ω. More precisely, we have:

Theorem 2.1 (Cauchy's theorem). *Assume that the applied body force density* $b(\cdot, t)$; $\Omega_t \to \mathbb{R}^3$ *is continuous and that the Cauchy stress vector field* $\tau(\cdot, \cdot, t)\colon \Omega_t \times \tilde{S} \to \mathbb{R}^3$ *is continuously differentiable with respect to the variable* $x \in \Omega_t$ *for each* $v \in \tilde{S}$ *and continuous with respect to the variable* $v \in \tilde{S}$ *for each* $x \in \Omega_t$. *Then the momentum balance principle and the balance law of angular momentum imply that there exists a continuously differentiable tensor field*

$T(\cdot, t)\colon \Omega_t \times \tilde{S} \to M_3$ such that

$$\tau(x, v, t) = T(x, t)v \qquad \forall x \in \Omega_t, \quad v \in \tilde{S}, \quad t > 0 \tag{2.7}$$

$$\rho a = \mathrm{Div}_x\, T + \rho b \qquad \forall x \in \Omega_t, \quad t > 0 \tag{2.8}$$

$$T(x, t) = T^{\mathrm{T}}(x, t) \qquad \forall x \in \Omega_t, \quad t > 0, \tag{2.9}$$

where $\mathrm{Div}_x\, T$ *represents the divergence of the tensor T with respect to the spatial coordinates x.*

The symmetric tensor $T(x, t)$ introduced by the previous theorem is called the *Cauchy stress tensor* at the point $x \in \Omega_t$. It is helpful to keep in mind the interpretation of its elements $T_{ij}(x, t)$. Since $\tau(x, e_j, t) = T_{ij}(x, t)e_i$, the elements of the jth row of the tensor $T(x, t)$ represent the components of the Cauchy stress vector $\tau(x, v, t)$ at the points x corresponding to the particular choice $v = e_j$. The knowledge of the three vectors $\tau(x, e_j, t)$ completely determines the Cauchy stress vector $\tau(x, v, t)$ for an arbitrary vector $v = v_i e_i \in \tilde{S}$ (see (2.7)).

Let us notice that if

$$T(x, t) = -p(t)I_3, \tag{2.10}$$

the Cauchy stress tensor is a *pressure* and the real number p is also called a *pressure*. In this case the Cauchy stress vector $\tau(x, v, t) = -p(t)v$ is always normal to the elementary surface elements, its length is constant in space and it is directed inward if $p(t) > 0$ (because of the minus sign) or outward if $p(t) < 0$.

We finish this section by recalling that the vector equation (2.8) satisfied by the Cauchy stress tensor is called the *Cauchy equation of motion in the deformed configuration.*

2.4 The Piola–Kirchhoff stress tensors and the linearized theory

Our final objective is to determine the deformation field and the Cauchy stress tensor field that arise in a body subjected to a given system of applied forces. In this respect, the equations of motion in the deformed configuration are not of much avail since they are expressed in terms of the *Euler variable* x which is precisely one of the unknowns of the problem. To obviate this difficulty, we shall rewrite these equations in terms of the *Lagrange variable* X that is attached to the reference configuration. For this we shall transform the right-hand side $\mathrm{Div}_x\, T$ appearing in the equation of motion on Ω_t into a similar expression over Ω.

We start by introducing the *first Piola–Kirchhoff stress tensor* defined for all $X \in \Omega$ and $t > 0$ by

$$S(X, t) = J(X, t)T(\chi(X, t), t)F^{-\mathrm{T}}(X, t) \qquad (S = JTF^{-\mathrm{T}}), \tag{2.11}$$

i.e.

$$S_{ij} = J T_{ik} F_{jk}^{-1}. \tag{2.12}$$

The main advantage of this transform is to induce a particularly simple relation between the divergences of both tensors:

$$\mathrm{Div}_X S = J \, \mathrm{Div}_x T \tag{2.13}$$

where $\mathrm{Div}_X S$ represents the divergence of the tensor S with respect to the material variables. Now we are able to write the *equation of motion* on the *deformed configuration*: multiplying (2.8) by J and using (2.1) and (2.13), we get

$$\rho_0 a = \mathrm{Div}_X S + \rho_0 b \qquad \forall X \in \Omega, \quad t > 0. \tag{2.14}$$

This equation is analogous to the equation (2.8), replacing ρ, T, Div_x by ρ_0, S, Div_X.

Whereas the Cauchy stress tensor T is symmetric (see (2.9)), the first Piola–Kirchhoff stress tensor S is not symmetric in general; indeed, using (2.9) and (2.11) we get

$$SF^{\mathrm{T}} = FS^{\mathrm{T}}. \tag{2.15}$$

It is nevertheless desirable to define a symmetric stress tensor in the reference configuration, essentially in the study of some constitutive equations. For this we define the *second Piola–Kirchhoff stress tensor* Π by letting

$$\Pi = F^{-1}S = JF^{-1}TF^{-\mathrm{T}}. \tag{2.16}$$

It is easy to see that (2.15) implies

$$\Pi^{\mathrm{T}} = \Pi, \tag{2.17}$$

i.e. Π is a symmetric tensor. Moreover, the equation of motion (2.14) becomes

$$\rho_0 a = \mathrm{Div}_X(F\Pi) + \rho_0 b \qquad \forall X \in \Omega, \quad t > 0. \tag{2.18}$$

We now investigate the linearized theory. We suppose that $\eta = |H| \ll 1$ and we neglect the quantities of order 2 or more on η. In this theory it can be proved that the Cauchy stress tensor and the two Piola–Kirchhoff stress tensors coincide ($T \approx S \approx \Pi$). For this reason, in the following we shall denote by σ the tensor field defined by

$$\sigma(X, t) = S(X, t) = \Pi(X, t) = T(\chi(X, t), t) \qquad \forall X \in \Omega, \quad t > 0$$
$$(\sigma = S = \Pi = T) \tag{2.19}$$

and we shall call σ the *stress tensor*. Using (2.9) we get

$$\sigma = \sigma^{\mathrm{T}} \qquad (\sigma_{ij} = \sigma_{ji}) \tag{2.20}$$

hence σ is a *symmetric tensor*.

Let us now return to the equation of motion (2.14), (2.18) and let us suppose that we study the evolution of a continuous medium in a time interval. Using the previous notation this evolution is described by the equation

$$\rho_0 \ddot{u} = \mathrm{Div}_X \sigma + \rho_0 b \qquad \text{in } \Omega, \quad \text{for } t > 0, \tag{2.21}$$

where the unknowns are the displacement function u and the stress function σ. In (2.21) the dots above represent the second time derivative (see (1.16)) and Div_X represents the divergence operator; the density of mass ρ_0 and the density of applied body forces b are given. Since (2.21) is written on Ω in order to simplify the notations, in the following we shall denote by x the coordinates of a particle in the reference configuration and we shall remove the indices of ρ_0 and Div_X. So, (2.21) becomes

$$\rho \ddot{u} = \mathrm{Div}\, \sigma + \rho b \qquad \text{in } \Omega, \quad \text{for } t > 0. \tag{2.22}$$

In some cases equation (2.22) can be simplified. So, in the case of *static problems* (i.e. none of the quantities depend on time) or in the case of *quasistatic problems* (when the term $\rho\ddot{u}$ is very small and can be neglected) instead of (2.22) we use the equation

$$\mathrm{Div}\, \sigma + \rho b = 0 \qquad \text{in } \Omega, \quad \text{for } t > 0. \tag{2.23}$$

With a change of notation, if we denote by f the density of applied body force per unit volume defined by $f = \rho b$, from (2.22) and (2.23) we obtain

$$\rho \ddot{u} = \mathrm{Div}\, \sigma + f \qquad \text{in } \Omega, \quad \text{for } t > 0, \tag{2.24}$$

$$\mathrm{Div}\, \sigma + f = 0 \qquad \text{in } \Omega, \quad \text{for } t > 0. \tag{2.25}$$

In the following we shall call (2.22) (or (2.24)) the *equation of motion* and (2.23) (or (2.25)) the *equation of equilibrium*.

In conclusion, in the linearized theory of deformation the unknowns of the problem are the functions u and σ which satisfy (2.22) (or (2.24)) in the case of dynamic processes and (2.23) (or (2.25)) in the case of quasistatic processes.

3 Some experiments and models for solids

Balance laws do not permit a complete description of a continuous body. Indeed, the three scalar equations of motion or of equilibrium over the reference configuration, which are valid regardless of the macroscopic continuum (gas, liquid, solid) that they are supposed to model, form an *undetermined system* since there are *nine* unknown functions, namely the three components of the displacement and the six components of the stress tensor (taking into account its symmetry). The six missing equations are provided by assumptions regarding the nature of the constitutive material that is considered. They are called *constitutive laws* or *constitutive equations*. In the following, by a constitutive law we understand a relation between the stress tensor σ, the small strain tensor ε and their temporal derivatives $\dot{\sigma}$ and $\dot{\varepsilon}$. Though the constitutive laws must satisfy some invariance principles, they originate mostly from experiment. For this reason we shall give a general description of several diagnostic experiments which are made with 'standard' universal testing machines and we shall point out what sort of information they may reveal which could be useful for the formulation of the constitutive laws.

3.1 Standard tests and elastic laws

The simplest diagnostic test is the uniaxial tension of a cylindrical specimen. Let us consider a bar of length l_0 and (constant) cross-section S. We take the x_1-axis in the direction of the axis of the bar and subject the bar to the tensile variable force $P(t)$ $(t > 0)$ at both its ends (*loading experiment*). We define the quantity σ by the equality

$$\sigma(t) = \frac{P(t)}{S} \tag{3.1}$$

and, since in the case of one-dimensional bodies the stress tensor reduces to a real number, the quantity σ defined by (3.1) can be interpreted as the stress. For every $t > 0$ we also denote by $l(t)$ the length of the specimen at time t and we define the relative extension ε by the equality

$$\varepsilon(t) = \frac{l(t) - l_0}{l_0}. \tag{3.2}$$

One can see that here the quantity defined by (3.2) plays the role of the small strain tensor which in the one-dimensional case reduces to a single component.

Measuring now the relative extension ε as a function of the stress σ, we obtain a so-called *stress–strain* diagram (see Figure 3.1).

Let us interpret the curve in Figure 3.1. The part of the curve between the points O and A is straight, hence for small stresses the relation between stress and strain is a linear one. The stress σ_A corresponding to A is called the *proportionality limit* and it may be established experimentally (for example, when the angle between the tangent to the stress–strain curve and the stress axis is increased by a certain amount established by convention, we say that the proportionality limit is reached). The part AB of the stress–strain diagram is curved; hence for stresses higher than the proportionality limit the relation between σ and ε is no longer linear.

For the purpose of establishing a mathematical model it is useful to perform *unloading experiments*, i.e. to stop the increase of the force P at a certain point of the curve and to make the force decrease to zero. If this reversal takes place at a sufficiently small stress (which has to be defined)

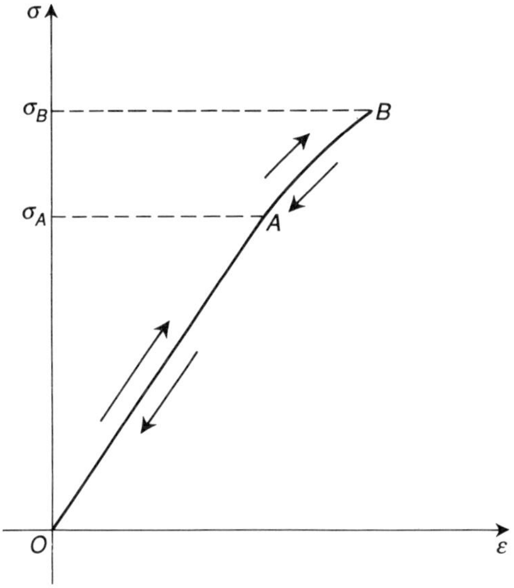

Fig. 3.1 Stress–strain diagram.

the graph is retraced to the origin along the same curve. If this tension test is fulfilled, the behaviour of the material is said to be *elastic*, i.e. the strain vanishes completely if the load is entirely removed.

Elastic deformations are studied in the *theory of elasticity*. In the case of linear theory of elasticity we suppose that the stress σ and the strain ε are mutually proportional, which is expressed by Hooke's law

$$\sigma = E\varepsilon \tag{3.3}$$

where E is the so-called Young's modulus of elasticity. This constitutive law models the behaviour of the material corresponding to the segment OA in Figure 3.1. More generally, we can consider a non-linear relation between the stress σ and the strain ε of the form

$$\sigma = f(\varepsilon), \tag{3.4}$$

where f is a non-linear real-valued function. This constitutive law is able to describe the non-linear behaviour of the material corresponding to the portion AB in Figure 3.1.

The constitutive equations (3.4), (3.5) describe only the 'uniaxial' case. In order to extend Hooke's law to the general case of spatial (triaxial) strain and stress, we suppose that at every point of Ω there exists a linear relation between the stress tensor σ and the small strain tensor ε, i.e.

$$\sigma = \mathscr{A}\varepsilon \qquad (\sigma_{ij} = A_{ijkh}\varepsilon_{kh}) \tag{3.5}$$

where $\mathscr{A} = (A_{ijkh})\colon \mathscr{S}_3 \to \mathscr{S}_3$ is a fourth-order tensor. If this tensor depends on $x \in \Omega$ (i.e. if the coefficients A_{ijkh} depend on $x \in \Omega$) the material of the body is said to be *non-homogeneous*. Otherwise it is said to be *homogeneous*. As usual in elasticity theory we suppose that A is a symmetric tensor, i.e.

$$A_{ijkh} = A_{khij} = A_{ijhk}.$$

Hence there are at most 21 independent constants A_{ijkh} that together characterize the general material at point x. If the constants A_{ijkh} do not depend on the rotations of the reference configuration, the material of the body is said to be *isotropic* at the point x. Otherwise the material is *anisotropic* at the point x. One can show that for homogeneous and isotropic materials we have

$$A_{ijkh} = \lambda\delta_{ij}\delta_{kh} + \mu(\delta_{ik}\delta_{jh} + \delta_{ih}\delta_{jk}) \tag{3.6}$$

where the constants λ and μ are the *Lamé coefficients*. Using now (3.5) and (3.6) we get

$$\sigma_{ij} = \lambda\varepsilon_{kk}\delta_{ij} + 2\mu\varepsilon_{ij} \tag{3.7}$$

which represents the constitutive law of linear homogeneous isotropic elastic materials in the three-dimensional case.

If we invert (3.7) we get

$$\varepsilon_{ij} = \frac{1+\nu}{E}\sigma_{ij} - \frac{\nu}{E}\sigma_{kk}\delta_{ij} \tag{3.8}$$

where the constants K, E, and ν are defined by the equalities

$$K = \frac{3\lambda + 2\mu}{3}, \qquad E = \frac{\mu(3\lambda + 2\mu)}{\lambda + \mu}, \qquad \nu = \frac{\lambda}{2(\lambda + \mu)}. \tag{3.9}$$

These constants are respectively called the *bulk modulus* (K), *Young's modulus* (E), and *Poisson's ratio* (ν), and are used in engineering more often than the Lamé coefficients λ and μ. Some experimental tests lead to restrictions $\lambda > 0$, $\mu > 0$ for the Lamé coefficients and using (3.9) it follows that $K > 0$, $E > 0$, $0 < \nu < \frac{1}{2}$.

We finish this section by remarking that the non-linear elastic constitutive law (3.4) may be extended in the three-dimensional case by taking

$$\sigma = F(\varepsilon) \tag{3.10}$$

where $F\colon \Omega \times \mathscr{S}_3 \to \mathscr{S}_3$ is a given tensorial function.

3.2 Loading and unloading tests. Plastic laws

Let us review the experiment presented in the previous section, referring this time to the stress–strain diagram presented in Figure 3.2, which corresponds to steel.

It may be observed that at point B the strain increases suddenly without any change in the stress. The stress σ_B is called the *yield limit*. To achieve further increases in strain, stress needs to be increased up to point C, after which strain increases with no further increase in stress. The stress σ_C corresponding to this point C is called the *strength limit*. Further stretching would finally cause the bar to break.

Let us imagine now that, after loading as far as point M, which is such that σ_M exceeds the yield limit σ_B, an unloading experiment is started. For most metals the unloading follows the path MN, which is a reasonable approximation to a straight line parallel to OA. We conclude that after the unloading experiment a part of the strain vanishes (the portion NQ in Figure 3.2) although a part of it remains permanently (the portion ON). The portion of the strain that vanishes after unloading is called the *elastic (reversible) strain*, while the portion that remains permanently is called the *plastic (irreversible) strain.*

It is clear that the elastic constitutive laws (3.5), (3.10) cannot model the existence of plastic deformations observed in the unloading tests. For this, *plastic constitutive laws* must be considered. Let us examine some of them in the one-dimensional case.

In the classical theory of plasticity it is assumed that the stress rate is proportional to the strain rate, the proportionality coefficient being a

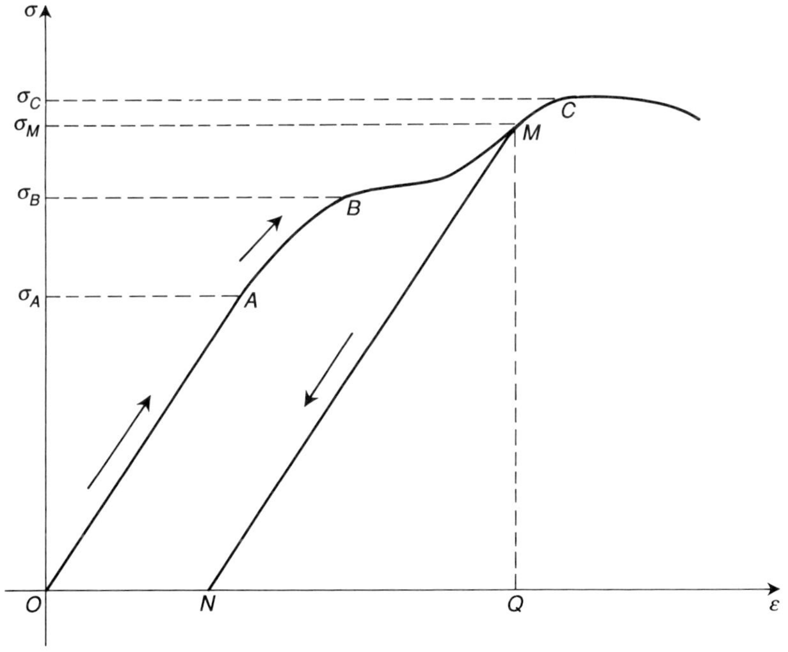

Fig. 3.2 Stress–strain diagram for steel.

function of the stress, the strain, and the strain rate. This means there exists a real-valued function $\mathscr{E}$ such that the stress rate is given by

$$\dot{\sigma} = \mathscr{E}(\sigma, \varepsilon, \dot{\varepsilon})\dot{\varepsilon}. \tag{3.11}$$

One of the most important aspects of classical plasticity theories is the fact that the proportionality coefficient $\mathscr{E}$ is a positively homogeneous function in its last argument, hence the plasticity theory based on a constitutive equation of the form (3.11) is a *time-independent theory* (i.e. a change $\tau = h(t)$ in the time variable does not change the form of the constitutive law).

Another important feature of the constitutive equations of classical plasticity lies in their property of having a domain of (linear) elastic behaviour, that is a domain where the constitutive equation $\dot{\sigma} = E\dot{\varepsilon}$ is satisfied.

In order to present a concrete example of plastic constitutive law we take in (3.11) the function $\mathscr{E}$ defined by

$$\mathscr{E}(\sigma, \varepsilon, \dot{\varepsilon}) = \begin{cases} E & \text{if } |\sigma| < \sigma_Y \text{ or } (|\sigma| = \sigma_Y \text{ and } \sigma \cdot \dot{\varepsilon} < 0) \\ 0 & \text{if } |\sigma| = \sigma_Y \text{ and } \sigma \cdot \dot{\varepsilon} \geq 0 \end{cases} \tag{3.12}$$

where σ_Y is a constant called the *yield limit* (here the negative stresses σ are interpreted as compressions).

It is easy to see that for stresses σ satisfying $|\sigma| < \sigma_Y$ from (3.11), (3.12) we get $\dot{\sigma} = E\dot{\varepsilon}$, i.e. the domain $|\sigma| < \sigma_Y$ is a domain of elastic behaviour. We also remark that in every unloading process the same equality is satisfied; hence the unloading is made following straight lines parallel with the elastic line $\sigma = E\varepsilon$. Moreover, it follows that the constitutive law (3.11), (3.12) may describe the existence of plastic deformations observed in the unloading experiments. One can also see that plastic deformations occur only for $|\sigma| = \sigma_Y$, i.e. only at a *fixed value* for the stress. For this, the model (3.11), (3.12) is called a *perfectly plastic model*. Note also that the stresses σ such that $|\sigma| > \sigma_Y$ have no physical interpretation.

We finish this section by recalling that one can imagine plastic constitutive laws of the form (3.11) for which the yield limit depends also on the deformation (elastic–plastic constitutive equations with *hardening*, see for example Cristescu and Suliciu (1982)).

3.3 Long-range tests and viscoplastic laws

Let us assume now that the experiment described in Section 3.1, which was performed to get the stress–strain curve of Figure 3.2 with a standard testing machine, is stopped at a certain stress and strain state (point P in Figure 3.3, say). With a soft testing machine the stress can further be maintained constant. In this case, if strain is measured in a long following

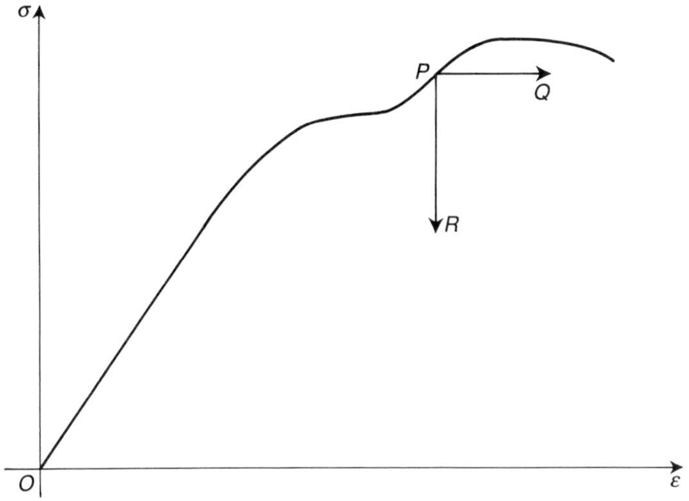

Fig. 3.3 Long-range test (creep and relaxation).

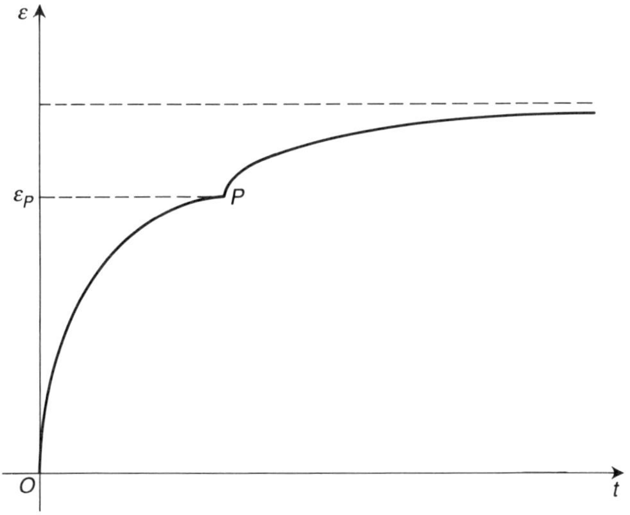

Fig. 3.4 Time evolution of strain during the creep phenomenon.

interval, a slow increase of strain occurs, so that the diagram in Figure 3.3 is continuing from P in the direction PQ. This is the *creep* phenomenon.

A standard creep strain–time curve is shown in Figure 3.4. If the applied stress is small, then after a long time this curve becomes horizontal (has a horizontal asymptote); we say that the *strain is stabilizing*. However, if the applied stress surpasses a certain limit, then the strain–time curve is strictly increasing and sooner or later the fracture of the specimen will occur.

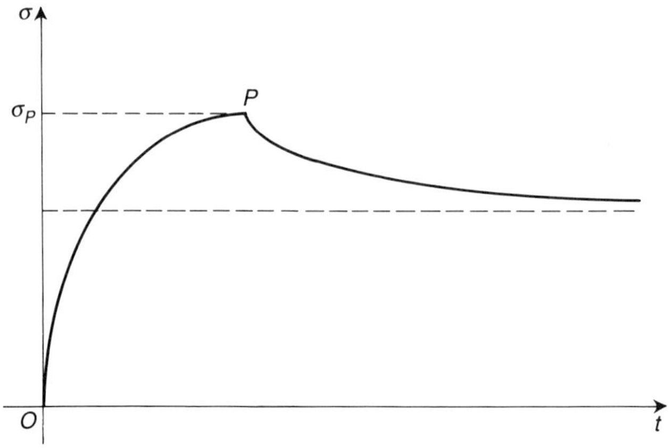

Fig. 3.5 Time evolution of stress during the relaxation phenomenon.

Another diagnostic test can be done with a hard testing machine if after a continuously increasing loading to the state P (Figure 3.3) the strain is kept constant while the stress variation can be measured. The experiments show that the stress decreases in time. This decrease, which occurs in a time interval much longer than those in the loading experiments, is called *relaxation.* The stress variation follows the path PR in Figure 3.3 and its evolution in time is given in Figure 3.5.

The stress may decrease in long time intervals down to nearly zero or (for most materials mainly for relatively high initial stress states), the curve in Figure 3.5 may possess a horizontal asymptote which depends on the initial stress–strain state. In this case, after a long period of time, nearly constant remanent stresses will continue to be present in the specimen.

Let us now remark that the creep phenomenon and the relaxation phenomenon cannot be modelled either by the elastic constitutive laws presented in Section 3.1 or by the plastic constitutive laws presented in Section 3.2. Indeed if we put $\varepsilon(t) = \text{constant}$ for $t \geq t_0$ in (3.10), (3.11) we get $\sigma(t) = \text{constant}$ for $t \geq t_0$ and conversely. Many other phenomena observed in the short-range tests, reloading experiments and so on, can also not be described by the previous elastic constitutive laws or plastic constitutive laws. For this reason, more sophisticated constitutive laws were proposed, for example the *rate-type viscoplastic laws.* The presentation of these models and concrete examples are given in Section 3.1.1. Let us only say that semilinear rate-type viscoplastic laws are of the form

$$\dot{\sigma} = A\dot{\varepsilon} + G(\sigma, \varepsilon)$$

where A is a fourth-order symmetric tensor and G is a given function. Such constitutive laws can describe the previous loading and unloading phenomena as well as the time-dependent phenomena observed during the experimental tests.

Bibliographical notes

The material presented in Sections 1 and 2 is standard and can be found in many books or surveys. More detailed information on the basic concepts of continuous mechanics as well as of classical elasticity theory may be found in the works of Malvern (1969), Truesdell (1974), Germain and Muller (1980), Soós and Teodosiu (1983), Ciarlet (1988).

More details concerning the description of the experiments presented in Section 3 as well as of other experiments in the background theory of viscoplastic constitutive laws can be found in Cristescu and Suliciu (1982), ch. III (see also the references cited there).

For a mathematical treatment of three-dimensional elasticity (the study of elastic constitutive laws) and boundary-value problems, we refer the reader to Hlaváček and Nečas (1981), Ciarlet (1988) and Sofonea (1991).

2

FUNCTIONAL SPACES IN VISCOPLASTICITY

The purpose of this chapter is to provide some fundamental results on functional spaces, as used in viscoplasticity, for later reference. These spaces are various spaces of distributions associated with the deformation operator and the divergence operator, or spaces of vector distributions. To restrict the chapter to a reasonable length most of the basic theorems are stated without proofs. However, for the convenience of the reader some useful properties of the spaces related to the divergence operator are presented here with proofs (Section 2.3). For other results which can be easily found in text books or monographs adequate references are given. The notations stated in this chapter are used all over the book.

1 Functional spaces of scalar-valued functions

In this section we present a brief description of the basic notations and definitions of the Schwartz theory of distributions as well as some basic results concerning the spaces $L^p(\Omega)$. We also introduce the Sobolev spaces of integer order and recall some of their basic properties.

1.1 Test functions, distributions, and L^p spaces

Let Ω be a domain of $\mathbb{R}^N$. We denote by $\mathscr{D}(\Omega)$ the space of all real-valued infinitely differentiable functions defined in Ω and with compact support in Ω. We consider $\mathscr{D}(\Omega)$ to be endowed with the usual topology and we call its elements *test functions*. The dual space of $\mathscr{D}(\Omega)$ is denoted by $\mathscr{D}'(\Omega)$ and it is called the *space of scalar distributions* on Ω. The duality between $\mathscr{D}'(\Omega)$ and $\mathscr{D}(\Omega)$ will be denoted by $\langle\,,\rangle$. $\mathscr{D}'(\Omega)$ is endowed with the weak-star topology, being the dual of $\mathscr{D}(\Omega)$, and is a locally convex topological vector space with this topology.

For every multi-index $\alpha = (\alpha_1, \ldots, \alpha_N) \in \mathbb{N}^N$ we denote by D^α the differential operator $\partial^{|\alpha|}/(\partial_{x_1}^{\alpha_1} \partial_{x_2}^{\alpha_2} \cdots \partial_{x_N}^{\alpha_N})$, where $|\alpha| = \alpha_1 + \cdots + \alpha_N$ is the length of α. If $\theta \in \mathscr{D}'(\Omega)$ and $\alpha \in \mathbb{N}^N$, then the element $D^\alpha\theta$ defined by

$$\langle D^\alpha\theta, \varphi\rangle = (-1)^{|\alpha|}\langle\theta, D^\alpha\varphi\rangle \tag{1.1}$$

is a linear and continuous functional on $\mathscr{D}(\Omega)$. Hence it defines another distribution called the *derivative of order* α *of* θ. So, every distribution in $\mathscr{D}'(\Omega)$ possesses derivatives of arbitrary orders in $\mathscr{D}'(\Omega)$ in the sense of definition (1.1). Furthermore, the mapping D^α from $\mathscr{D}'(\Omega)$ into $\mathscr{D}'(\Omega)$ is continuous.

A function u defined almost everywhere on Ω is said to be *locally integrable* on Ω provided it is measurable and $\int_K |u(x)|\,dx < \infty$ for every compact set $K \subset \Omega$. In this case we write $u \in L^1_{\text{loc}}(\Omega)$. Corresponding to every $u \in L^1_{\text{loc}}(\Omega)$ there is a distribution $\theta_u \in \mathscr{D}'(\Omega)$ defined by

$$\langle \theta_u, \varphi \rangle = \int_\Omega u(x)\varphi(x)\,dx. \tag{1.2}$$

for all $\varphi \in \mathscr{D}(\Omega)$. The distribution θ_u is said to be the *distribution generated by u*. It is possible to show that different locally integrable functions on Ω determine different distributions on Ω, hence the map $u \to \theta_u: L^1_{\text{loc}}(\Omega) \to \mathscr{D}'(\Omega)$ is injective; it is not surjective since not every distribution $\theta \in \mathscr{D}'(\Omega)$ is of the form $\theta = \theta_u$ (defined by (1.2)) for some $u \in L^1_{\text{loc}}(\Omega)$. Consequently, identifying u with θ_u we can consider $L^1_{\text{loc}}(\Omega)$ as a proper subspace of $\mathscr{D}'(\Omega)$.

Let p be a real number, $p \geq 1$. We denote by $L^p(\Omega)$ the classes of all measurable functions u defined on Ω for which

$$\int_\Omega |u(x)|^p\,dx < +\infty. \tag{1.3}$$

We identify in $L^p(\Omega)$ functions that are equal almost everywhere on Ω. The elements of $L^p(\Omega)$ are thus actually equivalence classes of measurable functions satisfying (1.3), two functions being equivalent if they are equal a.e. in Ω. For convenience, however, we ignore this distinction and write $u \in L^p(\Omega)$ if u satisfies (1.3) and $u = 0$ in $L^p(\Omega)$ if $u(x) = 0$ a.e. in Ω. It is clear that $L^p(\Omega)$ is a real vector space and the functional $u \to \|u\|_p$ defined by

$$\|u\|_p = \left(\int_\Omega |u(x)|^p\,dx\right)^{1/p} \tag{1.4}$$

is a norm on $L^p(\Omega)$.

A function u, measurable on Ω is said to be *essentially bounded* on Ω provided there exists a constant k for which $|u(x)| \leq k$ a.e. on Ω. The lower bound of such constants k is called the *essential supremum* of u on Ω and is denoted by $\operatorname{ess\,sup}_{x\in\Omega}|u(x)|$.

We denote by $L^\infty(\Omega)$ the real vector space consisting of all functions u that are essentially bounded on Ω, functions being again identical if they are equal a.e. on Ω. It is easily verified that the functional $u \to \|u\|_\infty$ defined by

$$\|u\|_\infty = \operatorname*{ess\,sup}_{x\in\Omega}|u(x)| \tag{1.5}$$

is a norm on $L^\infty(\Omega)$.

The basic properties of the spaces $L^p(\Omega)$ are the following: $L^p(\Omega)$ is a Banach space if $1 \leq p \leq \infty$; if $1 < p < \infty$ the space $L^p(\Omega)$ is uniformly convex; for every $1 < p < \infty$ the dual of the space $L^p(\Omega)$ can be identified with $L^q(\Omega)$ where q is defined by $(1/p) + (1/q) = 1$; the dual of the space $L^1(\Omega)$ is the space $L^\infty(\Omega)$; $L^p(\Omega)$ is reflexive if and only if $1 < p < \infty$; the space $L^p(\Omega)$ is separable if $1 \le p < \infty$ and $L^\infty(\Omega)$ is not separable; for $p = 2$, $L^p(\Omega)$ is a real Hilbert space with respect to the inner product

$$\langle u, v\rangle_2 = \int_\Omega u(x)v(x)\,\mathrm{d}x. \tag{1.6}$$

Note that $L^p(\Omega) \subset L^1_{\mathrm{loc}}(\Omega)$ for every $1 \le p \le \infty$; hence we can identify $L^p(\Omega)$ with a subspace of $\mathscr{D}'(\Omega)$:

$$L^p(\Omega) \subset \mathscr{D}'(\Omega). \tag{1.7}$$

Moreover, the injection in (1.7) is continuous since from $u_n \to 0$ in $L^p(\Omega)$ we get $u_n \to 0$ in $\mathscr{D}'(\Omega)$.

We finish this section by presenting a well-known result on regularization for L^p spaces; let $\omega \in \mathscr{D}(\mathbb{R}^N)$ be a *mollifier* defined for instance by

$$\omega(x) = \begin{cases} c\exp(|x|^2/(|x|^2 - 1)) & \text{if } |x| < 1 \\ 0 & \text{if } |x| \ge 1, \end{cases} \tag{1.8}$$

where c is a constant such that $\int_{\mathbb{R}^N} \omega(x)\,\mathrm{d}x = 1$. If we put

$$\omega_h(x) = \frac{1}{h^N}\,\omega\left(\frac{x}{h}\right) \tag{1.9}$$

then the convolution $\omega_h * u$ defined by $(\omega_h * u)(x) = \int_{\mathbb{R}^N} \omega_h(x - y)u(y)\,\mathrm{d}y$ for every $x \in \mathbb{R}^N$ and $u \in L^1_{\mathrm{loc}}(\mathbb{R}^N)$ is called a *mollification* or *regularization* of u. If $1 \le p < \infty$ and $u \in L^p(\Omega)$ it follows that $\omega_h * u \in C^\infty(\mathbb{R}^N)$, $\operatorname{supp}(\omega_h * u) \subset \operatorname{supp} u + B(0, h)$ and $\|\omega_h * u - u\|_p \to 0$ when $h \to 0$.

1.2 Sobolev spaces of integer order

Let m be a non-negative integer and $1 \le p \le \infty$; we define the space

$$W^{m,p}(\Omega) = \{u \in L^p(\Omega) \mid D^\alpha u \in L^p(\Omega) \text{ for } 0 \le |\alpha| \le m\} \tag{1.10}$$

where $D^\alpha u$ is the distributional derivative defined by (1.1) for $u \in L^p(\Omega) \subset \mathscr{D}'(\Omega)$. In (1.10) by $D^\alpha u \in L^p(\Omega)$ we mean that there exists a function u_α belonging to $L^p(\Omega)$ such that $D^\alpha u$ is generated by u_α:

$$\langle D^\alpha u, \varphi\rangle = \langle u_\alpha, \varphi\rangle \tag{1.11}$$

for all $\varphi \in \mathscr{D}(\Omega)$. As usual, we identify $D^\alpha u$ with u_α.

The space $W^{m,p}(\Omega)$ is called the *Sobolev space of order m* on $L^p(\Omega)$ and is a real Banach space with the norm

$$\|u\|_{m,p} = \left[\sum_{0 \le |\alpha| \le m} \|D^\alpha u\|_p^p \right]^{1/p} \qquad \text{if } 1 \le p < \infty \tag{1.12}$$

and

$$\|u\|_{m,\infty} = \max_{0 \le |\alpha| \le m} \|D^\alpha u\|_\infty \qquad \text{if } p = \infty. \tag{1.13}$$

Clearly $W^{0,p}(\Omega) = L^p(\Omega)$; moreover, the case $p = 2$ is fundamental. To simplify the notation we put $W^{m,2}(\Omega) = H^m(\Omega)$ and we observe that $H^m(\Omega)$ is a real Hilbert space with the inner product

$$\langle u, v \rangle_{m,2} = \sum_{|\alpha| \le m} \langle D^\alpha u, D^\alpha v \rangle_2. \tag{1.14}$$

The closure of $\mathscr{D}(\Omega)$ in the norm $\|\cdot\|_{m,p}$ will give a closed subspace of $W^{m,p}(\Omega)$ denoted by $W_0^{m,p}(\Omega)$. We shall also use the notation $W_0^{m,2}(\Omega) = H_0^m(\Omega)$. Many properties of Sobolev spaces defined on a domain Ω and in particular the embedding properties of these spaces depend on regularity properties of Ω. Such regularity is normally expressed in terms of geometrical conditions that may or may not be satisfied by any given domain.

We say that Ω has the *segment property* if there exists a locally finite open cover $\{U_j\}$ (i.e. any compact $K \subset \mathbb{R}^N$ can intersect at most a finite number of elements of $\{U_j\}$) of $\Gamma = \partial\Omega$, the boundary of Ω, and a corresponding sequence $\{y_j\} \subset \mathbb{R}^N$ of non-zero vectors such that if $x \in \bar{\Omega} \cap U_j$ for some j, then $x + ty_j \in \Omega$ for $t \in (0, 1)$.

We say that Ω has the *cone property* if there exists a finite cone C of vertex 0 such that each point x of Ω is the vertex of a finite cone C_x contained in Ω and congruent to C (i.e. C_x is obtained from C by a rigid motion). Here by *finite cone of vertex* x we mean a set of the form

$$C_x = B_1 \cap \{x + \lambda(y - x) \mid y \in B_2, \lambda > 0\},$$

where $B_1 \subset \mathbb{R}^N$ is an open ball with centre x and $B_2 \subset \mathbb{R}^N$ is an open ball not containing x.

We say that the bounded domain Ω has the *strong local Lipschitz property* if each point $x \in \Gamma = \partial\Omega$ should have a neighbourhood U_x such that $\Gamma \cap U_x$ is the graph of a Lipschitz function, i.e. there exists a coordinate system $y_1, \ldots, y_N$ and a Lipschitz function defined on an $(N-1)$-dimensional cube B such that

$$\Gamma \cap U_x = \{y = (y'; y_N) \mid y_N = a(y'), y' \in B\}$$

and

$$\Omega \cap U_x \supset \{y = (y', y_N) \mid a(y') - \alpha < y_N < a(y'), y' \in B\},$$

$$(\mathbb{R}^N \backslash \bar{\Omega}) \cap U_x \supset \{y = (y', y_N) \mid a(y') < y_N < a(y') + \alpha, y' \in B\}$$

for some $\alpha > 0$.

Remark 1.1 *If Ω has the strong local Lipschitz property then it has also the cone and the segment property.*

The basic properties of Sobolev spaces $W^{k,p}(\Omega)$ can be summarized as follows:

- *if Ω has the segment property then*

$$C^\infty(\bar{\Omega}) \text{ is dense in } W^{m,p}(\Omega); \tag{1.15}$$

- *if Ω has the cone property then*

$$N > mp \quad \text{and} \quad \frac{1}{q} = \frac{1}{p} - \frac{m}{N} \qquad \text{or} \qquad N = mp \quad \text{and} \quad 1 \le q \le \infty \tag{1.16}$$

implies $W^{m,p}(\Omega) \subset L^q(\Omega)$ with a continuous embedding,

$$N > mp \text{ and } 1 \le q \le \frac{Np}{N - mp} \qquad \text{or} \qquad N \le mp \text{ and } 1 \le q \le \infty \tag{1.17}$$

implies $W^{m,p}(\Omega) \subset L^q(\Omega)$ with a completely continuous embedding;

- *if Ω has the strong local Lipschitz property then*

$$\frac{N}{p} + k < m \le \frac{N}{p} + k + 1 \text{ implies } W^{m,p}(\Omega) \subset C^k(\bar{\Omega}) \tag{1.18}$$

with a continuous embedding; there exists a linear and continuous map

$$\begin{aligned} &\gamma_1 \colon W^{1,p}(\Omega) \to L^p(\Gamma) \text{ such that } \gamma_1 u = u|_\Gamma \\ &\text{for all } u \in C^1(\bar{\Omega}), \text{ surjective in the case } p = 1. \end{aligned} \tag{1.19}$$

Remark 1.2 (1) *In* (1.16)–(1.18) *the embedding is defined as the identical mapping. A* complete continuous mapping *is a continuous mapping that maps any bounded set onto a precompact set;*

(2) *In* (1.19) *$L^p(\Gamma)$ is defined in the same way as $L^p(\Omega)$ using the $(N-1)$-dimensional Lebesgue measure on Γ; the map γ_1 is called* the trace map *on $W^{1,p}(\Omega)$;*

(3) *The surjectivity in* (1.19) *in the special case $p = 1$ was proved by Gagliardo* (1957).

Let us now suppose that Ω has the strong local Lipschitz property. It may be proved that the trace map $\gamma_1 \colon H^1(\Omega) \to L^2(\Gamma)$ defined by (1.19) is not surjective; for this, we denote by $H^{1/2}(\Gamma)$ the Sobolev space of non-integer order $s = \frac{1}{2}$ defined on Γ; it is well known that $H^{1/2}(\Gamma)$ is the image space of $H^1(\Omega)$ under the trace map γ_1 and we endow $H^{1/2}(\Gamma)$ with the Hilbert structure transported by γ_1. We obtain that $\gamma_1 \colon H^1(\Omega) \to H^{1/2}(\Gamma)$ is a linear, continuous and surjective map and moreover there exists a linear and continuous map $z_1: H^{1/2}(\Gamma) \to H^1(\Omega)$ such that

$$\gamma_1(z_1(U)) = U \qquad \text{for all} \quad U \in H^{1/2}(\Gamma). \tag{1.20}$$

The dual space of $H^{1/2}(\Gamma)$ will be denoted by $H^{-1/2}(\Gamma)$. Finally, we recall that if Ω has the strong local Lipschitz property, then the kernel of the trace map on $H^1(\Omega)$ is $H_0^1(\Omega)$.

2 Functional spaces attached to some linear differential operators of first order

In this section we introduce the deformation and divergence operators for distributions, which are the basic operators in viscoplasticity problems. We also define some functional spaces attached to these operators and we present their basic properties which will be useful in Chapters 3–5.

2.1 Linear differential operators of first order

Since the deformation operator is defined for vector-valued functions or distributions and the divergence operator is defined for tensor-valued functions or distributions, for simplicity we introduce the following notation:

$$\boldsymbol{D} = [\mathscr{D}(\Omega)]^N = \{\varphi = (\varphi_i) \mid \varphi_i \in \mathscr{D}(\Omega),\, i = \overline{1, N}\},$$

$$\mathscr{D} = [\mathscr{D}(\Omega)]_s^{N \times N} = \{\phi = \phi_{ij}) \mid \phi_{ij} = \phi_{ji} \in \mathscr{D}(\Omega),\, i, j = \overline{1, N}\}.$$

Similarly, we denote by $\boldsymbol{D}'$ and $\mathscr{D}'$ the following distributional spaces on Ω:

$$\boldsymbol{D}' = [\mathscr{D}'(\Omega)]^N = \{v = (v_i) \mid v_i \in \mathscr{D}'(\Omega),\, i = \overline{1, N}\}.$$

$$\mathscr{D}' = \{\sigma = (\sigma_{ij}) \mid \sigma_{ij} = \sigma_{ji} \in \mathscr{D}'(\Omega),\, i, j = \overline{1, N}\}.$$

Let $\langle , \rangle_{\boldsymbol{D}', \boldsymbol{D}}$, $\langle , \rangle_{\mathscr{D}', \mathscr{D}}$ be the duality mapping on the above-mentioned spaces, defined by

$$\langle v, \varphi \rangle_{\boldsymbol{D}', \boldsymbol{D}} = \sum_{i=1}^{N} \langle v_i, \varphi_i \rangle \tag{2.1}$$

$$\langle \sigma, \phi \rangle_{\mathscr{D}', \mathscr{D}} = \sum_{i,j=1}^{N} \langle \sigma_{ij}, \phi_{ij} \rangle. \tag{2.2}$$

Denoting by $\psi_{,i}$ the partial derivative $\partial\psi/\partial x_i$ of every test function $\psi \in \mathscr{D}(\Omega)$ we introduce the following differential operators of first order:

$$\varepsilon\colon \boldsymbol{D} \to \mathscr{D}, \quad \varepsilon(\varphi) = (\varepsilon_{ij}(\varphi)), \quad \varepsilon_{ij}(\varphi) = \tfrac{1}{2}(\varphi_{i,j} + \varphi_{j,i}) \qquad \forall i, j = \overline{1, N},\, \varphi \in \boldsymbol{D} \tag{2.3}$$

$$\text{Div}\colon \mathscr{D} \to \boldsymbol{D}, \quad \text{Div}\, \phi = \left(\sum_{j=1}^{N} \phi_{ij,j} \right) \qquad \forall i = \overline{1, N}, \quad \phi \in \mathscr{D}. \tag{2.4}$$

We shall call the operator defined by (2.3) the *deformation operator*, and the operator defined by (2.4) the *divergence operator* for test functions. We denote also by $\theta_{,i}$ the derivative of the distribution θ with respect to the coordinate x_i; using (1.1) we have

$$\langle \theta_{,i}, \varphi \rangle = -\langle \theta, \varphi_{,i} \rangle \tag{2.5}$$

for all $\theta \in \mathscr{D}'(\Omega)$, $\varphi \in \mathscr{D}(\Omega)$ and $i = \overline{1, N}$. Equality (2.5) allows us to define differential operators of the form (2.3), (2.4) for distributions. Since no ambiguity can occur, we use the same notation as in (2.3), (2.4):

$$\varepsilon: \boldsymbol{D}' \to \mathscr{D}', \quad \varepsilon(v) = (\varepsilon_{ij}(v)), \varepsilon_{ij}(v) = \tfrac{1}{2}(v_{i,j} + v_{j,i}) \qquad \forall i, j = \overline{1, N}, \quad v \in \boldsymbol{D}' \tag{2.6}$$

$$\text{Div}: \mathscr{D}' \to \boldsymbol{D}', \quad \text{Div}\,\sigma = \left(\sum_{i=1}^{N} \sigma_{ij,j} \right) \qquad \forall \sigma \in \mathscr{D}'. \tag{2.7}$$

The operators defined by (2.6) and (2.7) are called the *deformation operator* and respectively the *divergence operator* for distributions. Using (2.1)–(2.7) we get

$$\langle \varepsilon(v), \phi \rangle_{\mathscr{D}', \mathscr{D}} = -\langle v, \text{Div}\,\phi \rangle_{\boldsymbol{D}', \boldsymbol{D}} \qquad \forall v \in \boldsymbol{D}', \phi \in \mathscr{D}, \tag{2.8}$$

$$\langle \text{Div}\,\sigma, \varphi \rangle_{\boldsymbol{D}', \boldsymbol{D}} = -\langle \sigma, \varepsilon(\varphi) \rangle_{\mathscr{D}', \mathscr{D}} \qquad \forall \sigma \in \mathscr{D}', \varphi \in \boldsymbol{D}. \tag{2.9}$$

Remark 2.1 *In the same way the* gradient operator $\nabla: \boldsymbol{D}' \to [\mathscr{D}'(\Omega)]^{N \times N}$ *can be defined by* $\nabla v = (v_{i,j})$ *for all* $i, j = 1, N$ *and* $v \in \boldsymbol{D}$. *If we denote by* ∇^{T} *the transpose of* ∇ *from* (2.6) *we get that the deformation operator is the symmetric part of the gradient operator*:

$$\varepsilon(v) = \tfrac{1}{2}(\nabla v + \nabla^{\mathrm{T}} v) \qquad \text{for all} \quad v \in \boldsymbol{D}'. \tag{2.10}$$

The kernel of the deformation operator on $\boldsymbol{D}'$ will be denoted by $\mathscr{R}$ and will be called the *set of rigid infinitesimal displacements* (note that such a displacement can be thought of as a rigid behaviour of a body only in the small strain case). This kernel can be described as follows:

$$u \in \boldsymbol{D}', \qquad \varepsilon(u) = 0 \quad \text{iff} \quad u(x) = a + Bx \quad \forall x \in \Omega \tag{2.11}$$

where $a \in \mathbb{R}^N$ and B is the antisymmetric $N \times N$ matrix. In particular if $N = 3$ we have

$$u \in \boldsymbol{D}', \quad \varepsilon(u) = 0 \quad \text{iff} \quad u(x) = a + b \wedge x \quad \forall x \in \Omega, \tag{2.12}$$

where $a, b \in \mathbb{R}^N$ and $\wedge$ denotes the vector product in $\mathbb{R}^3$.

In later chapters the following notations will be used:

$$\boldsymbol{H} = [L^2(\Omega)]^N = \{v = (v_i) | v_i \in L^2(\Omega), \text{i} = \overline{1, N}\} \tag{2.13}$$

$$\mathscr{H} = [L^2(\Omega)]_s^{N \times N} = \{\sigma = (\sigma_{ij}) | \sigma_{ij} = \sigma_{ji} \in L^2(\Omega), i, j = \overline{1, N}\} \tag{2.14}$$

and, for symmetry, we also denote

$$H = L^2(\Omega). \tag{2.15}$$

The spaces H, $\boldsymbol{H}$, and $\mathscr{H}$ are real Hilbert spaces endowed with the following inner products:

$$\langle \theta, \delta \rangle_H = \int_\Omega \theta(x)\delta(x)\,\mathrm{d}x \tag{2.16}$$

$$\langle v, w \rangle_{\boldsymbol{H}} = \int_\Omega v_i(x) \cdot w_i(x)\,\mathrm{d}x \tag{2.17}$$

$$\langle \sigma, \tau \rangle_{\mathscr{H}} = \int_\Omega \sigma_{ij}(x) \cdot \tau_{ij}(x)\,\mathrm{d}x. \tag{2.18}$$

The norms generated by (2.16)–(2.18) will be denoted by $\|\cdot\|_H$, $\|\cdot\|_{\boldsymbol{H}}$ and $\|\cdot\|_{\mathscr{H}}$ respectively.

Moreover, using (1.7) we can identify $\boldsymbol{H}$ and $\mathscr{H}$ with two subspaces of $\boldsymbol{D}'$ and $\mathscr{D}'$ respectively. So, the operators ε and Div defined for distributions can also be defined on the spaces $\boldsymbol{H}$ and $\mathscr{H}$ respectively. Imposing some regularity restrictions on the range of these operators, in Sections 2.2 and 2.3 some useful functional spaces will be defined.

2.2 Functional spaces associated with the deformation operator

Let Ω be a domain in $\mathbb{R}^N$; for all $1 \le p \le \infty$ let us consider the space

$$LD^p(\Omega) = \{u \in [L^p(\Omega)]^N \,|\, \varepsilon(u) \in [L^p(\Omega)]_s^{N \times N}\} \tag{2.19}$$

endowed with the norm $\|\cdot\|_{D,p}$ given by

$$\|u\|_{D,p}^p = \sum_{i=1}^N \|u_i\|_{L^p(\Omega)}^p + \sum_{i,j=1}^N \|\varepsilon_{ij}(u)\|_{L^p(\Omega)}^p. \tag{2.20}$$

It is very easy to verify that $LD^p(\Omega)$ is a Banach space and $[W^{1,p}(\Omega)]^N \subset LD^p(\Omega)$ for all $1 \le p \le \infty$ with a continuous embedding; more precisely, we have

$$\|u\|_{D,p}^p \le \|u\|_{[W^{1,p}(\Omega)]^N}^p \qquad \forall u \in LD^p(\Omega) \tag{2.21}$$

and all $1 \le p \le \infty$.

If for some p we have $LD^p(\Omega) = [W^{1,p}(\Omega)]^N$ then from the closed-graph theorem we get that $LD^p(\Omega)$ is continuously embedded in $[W^{1,p}(\Omega)]^N$; hence the following *deformation coercivity inequality* holds:

$$\|u\|_{[W^{1,p}(\Omega)]^N}^p \le C\|u\|_{D,p}^p \qquad \forall u \in [W^{1,p}(\Omega)]^N, \tag{2.22}$$

where $C > 0$ is a constant which depends only on Ω and p.

As follows from Geymonat and Suquet (1984) we have

$$LD^p(\Omega) = [W^{1,p}(\Omega)]^N \qquad \forall 1 < p < \infty; \tag{2.23}$$

hence (2.22) holds for all $1 < p < \infty$.

The special case $p = 1$ will give us a functional space larger than $[W^{1,1}(\Omega)]^N$. Following Temam (1983, p. 116) one can use a result of Ornstein (1962) in order to prove that (2.22) does not hold for $p = 1$, hence $[W^{1,1}(\Omega)]^N \subsetneqq LD^1(\Omega)$. For simplicity we shall use the notation $LD(\Omega)$ instead of $LD^1(\Omega)$.

In some problems of perfect plasticity a larger space than $LD(\Omega)$ must be considered. Let us denote by $M_1(\Omega)$ the space of bounded measures on Ω which can be identified with the dual of $C_0(\Omega)$. Hence $M_1(\Omega) = (C_0(\Omega))'$ consists of all distributions $u \in \mathscr{D}'(\Omega)$ which are continuous with respect to the topology of $C_0(\Omega)$. One can easily remark that $L^1(\Omega)$ is a closed subspace of $M_1(\Omega)$. Hence if we define the space of bounded deformation functions as follows:

$$BD(\Omega) = \{u \in [L^1(\Omega)]^N | \varepsilon_{ij}(u) \in M_1(\Omega) \ \forall i, j = \overline{1, N}\} \tag{2.24}$$

endowed with the norm $\|\cdot\|_{BD}$ given by

$$\|u\|_{BD} = \sum_{i=1}^{N} \|u_i\|_{L^1(\Omega)} + \sum_{i,j=1}^{N} \|\varepsilon_{ij}(u)\|_{M_1(\Omega)}, \tag{2.25}$$

then $LD(\Omega) \subset BD(\Omega)$ is a closed subspace of $BD(\Omega)$ and

$$\|u\|_{D,1} = \|u\|_{BD} \qquad \forall u \in LD(\Omega).$$

In the following pages we present some basic properties of the spaces $LD^p(\Omega)$ ($p = 1, 2$) and $BD(\Omega)$. We start with the Hilbertian case $p = 2$ and, for simplicity, we use the notation $LD^2(\Omega) = \boldsymbol{H}_1$.

Using (2.19), (2.13), and (2.14) we have

$$\boldsymbol{H}_1 = \{v \in \boldsymbol{H} | \varepsilon(v) \in \mathscr{H}\}. \tag{2.26}$$

$\boldsymbol{H}_1$ is a real Hilbert space with respect to the inner product

$$\langle u, v\rangle_{\boldsymbol{H}_1} = \langle u, v\rangle_{\boldsymbol{H}} + \langle \varepsilon(u), (v)\rangle_{\mathscr{H}} \tag{2.27}$$

which generates the norm $\|\cdot\|_{\boldsymbol{H}_1}$ defined by

$$\|v\|^2_{\boldsymbol{H}_1} = \|v\|^2_{\boldsymbol{H}} + \|\varepsilon(v)\|^2_{\mathscr{H}}. \tag{2.28}$$

Moreover from (2.23) we have $\boldsymbol{H}_1 = [H^1(\Omega)]^N$ and from (2.21), (2.22) we get that (2.28) is equivalent with the product Hilbert norm in $[H^1(\Omega)]^N$.

Let $\boldsymbol{H}_\Gamma = [H^{1/2}(\Gamma)]^N$ be endowed with the canonical product structure. Using the results presented in Section 1.2 we can define the trace map $\gamma_1: \boldsymbol{H}_1 \to \boldsymbol{H}_\Gamma$ such that $\gamma_1 v = v|_\Gamma$ for all $v \in [C^1(\bar{\Omega})]^N$; it follows that γ_1 is a linear, continuous and surjective operator and moreover, there exists a linear and continuous map $z_1 = \boldsymbol{H}_\Gamma \to \boldsymbol{H}_1$ such that

$$\gamma_1(z_1(U)) = U \qquad \forall U \in \boldsymbol{H}_\Gamma \tag{2.29}$$

(see (1.20)). For simplicity, when no ambiguity can occur, we sometimes write v instead of $\gamma_1 v$ for $v \in \boldsymbol{H}_1$. We note that the continuity of z_1 implies that there exists $C > 0$ which depends only on Ω such that

$$\|z_1(U)\|_{\boldsymbol{H}_1} \le C\|U\|_{\boldsymbol{H}_\Gamma} \qquad \forall U \in \boldsymbol{H}_\Gamma. \tag{2.30}$$

Remark 2.2 *In Chapter* 5 *we also consider the* divergence operator *on the space* $\boldsymbol{H}_1$ *defined by* $\operatorname{div} v = \sum_{i=1}^{N} v_{i,i}$ *for all* $v \in \boldsymbol{H}_1$.

Now let Γ_1 be an open subset of Γ and $\Gamma_2 = \Gamma\backslash\bar{\Gamma}_1$; we define the space

$$V_1 = \{v \in \boldsymbol{H}_1 \mid \gamma_1 v = 0 \text{ a.e. on } \Gamma_1\}. \tag{2.31}$$

V_1 is a closed subspace of $\boldsymbol{H}_1$, hence is a Hilbert space with the inner product $\langle , \rangle_{\boldsymbol{H}_1}$, in particular if $\Gamma_1 = \varnothing$ we have $V_1 = \boldsymbol{H}_1$ and if $\Gamma_1 = \Gamma$ we have $V_1 = [H_0^1(\Omega)]^N$.

In solving the problems of linear and non-linear elasticity as well as the problems of viscoplasticity, the basic role is played by Korn's inequality (see for instance Duvaut and Lions (1972, p. 115) or Hlaváček and Nečas (1981, p. 79): if meas $\Gamma_1 > 0$ then there exists $C > 0$ which depends on Ω and Γ_1) such that

$$\|\varepsilon(v)\|_{\mathscr{H}} \ge C\|v\|_{\boldsymbol{H}_1} \qquad \forall v \in V_1. \tag{2.32}$$

In the next chapters we will also use the notation V_{Γ_1} for the space defined by

$$V_{\Gamma_1} = \gamma_1(V_1) = \{\xi \in \boldsymbol{H}_\Gamma \mid \xi = 0 \text{ on } \Gamma_1\}. \tag{2.33}$$

V_{Γ_1} is a closed subspace on $\boldsymbol{H}_\Gamma$, hence it is a Hilbert space with the inner product of $\boldsymbol{H}_\Gamma$. This space and its dual V'_{Γ_1} will be useful in Section 2.3.

The basic properties of the spaces $LD(\Omega)$ and $BD(\Omega)$ can be summarized as follows:

- if Ω has the segment property then

$$[\mathscr{D}(\mathbb{R}^N)]^N \text{ is dense in } LD(\Omega); \tag{2.34}$$

for all $u \in BD(\Omega)$ there exists $(u_n)_{n \ge 0} \subset [\mathscr{D}(\mathbb{R}^N)]^N$ such that $u_n \to u$ in $[L^1(\Omega)]^N$, $\varepsilon_{ij}(u_n) \to \varepsilon_{ij}(u)$ weak* in $M_1(\Omega)$ and $\|\varepsilon_{ij}(u_n)\|_{L^1(\Omega)} \to \|\varepsilon_{ij}(u)\|_{M^1(\Omega)}$ for all $i, j = \overline{1, N}$; (2.35)

- if Ω has the strong local Lipschitz property then there exists γ_1: $BD(\Omega) \to [L^1(\Gamma)]^N$, a linear continuous and surjective trace map, such that

$$\gamma_1(u) = u|_\Gamma \qquad \forall u \in BD(\Omega) \cap C(\bar{\Omega}); \tag{2.36}$$

$BD(\Omega)$ is continuously embedded in $[L^p(\Omega)]^N$ for all $1 \leq p \leq N/(N-1)$ and the embedding is completely continuous for all $1 \leq p < N/(N-1)$. If $N = 1$ and Ω is an interval then $BD(\Omega)$ is continuously embedded in $L^p(\Omega)$ for $1 \leq p \leq \infty$ and the embedding is completely continuous for $1 \leq p < \infty$; (2.37)

if $u \in [\mathscr{D}^1(\Omega)]^N$ such that $\varepsilon_{ij}(u) \in M_1(\Omega)$ $(\varepsilon_{ij}(u) \in L^1(\Omega))$ for all $i, j = \overline{1, N}$, then $u \in BD(\Omega)$ $(u \in LD(\Omega))$; (2.38)

there exists a constant $C > 0$ such that for all $u \in BD(\Omega)$ $(u \in LD(\Omega))$ there exists $r = r(u) \in \mathscr{R}$ such that

$$\|u - r\|_{BD} \leq C \sum_{i,j=1}^{N} \|\varepsilon_{ij}(u)\|_{M_1(\Omega)} \Big(\|u - r\|_{LD} \leq C \sum_{i,j=1}^{N} \|\varepsilon_{ij}(u)\|_{L^1(\Omega)}\Big); \quad (2.39)$$

if $p\colon BD(\Omega) \to R_+$ is a continuous seminorm on $BD(\Omega)$ (on $LD(\Omega)$) which is a norm on $\mathscr{R}$ then

$$u \to p(u) + \sum_{i,j=1}^{N} \|\varepsilon_{ij}(u)\|_{M_1(\Omega)} \qquad \left(u \to p(u) + \sum_{i,j=1}^{N} \|\varepsilon_{ij}(u)\|_{L_1(\Omega)}\right) \quad (2.40)$$

is an equivalent norm on $BD(\Omega)$ (on $LD(\Omega)$).

Remark 2.3 *For simplicity, when no ambiguity can occur, we shall write sometimes v instead of $\gamma_1 v$ for $v \in BD(\Omega)$ or $v \in LD(\Omega)$.*

Remark 2.4 *The space $BD(\Omega)$ is the strong dual of a real normed space (see Temam and Strang* (1980)).

2.3 A Hilbert space associated with the divergence operator

Let us denote by $\mathscr{H}_2$ the space defined by

$$\mathscr{H}_2 = \{\sigma \in \mathscr{H} \mid \operatorname{Div} \sigma \in \boldsymbol{H}\}. \quad (2.41)$$

Then $\mathscr{H}_2$ is a real Hilbert space with respect to the inner product

$$\langle \sigma, \tau \rangle_{\mathscr{H}_2} = \langle \sigma, \tau \rangle_{\mathscr{H}} + \langle \operatorname{Div} \sigma, \operatorname{Div} \tau \rangle_{\boldsymbol{H}}, \quad (2.42)$$

which generates the norm $\|\cdot\|_{\mathscr{H}_2}$ defined by

$$\|\sigma\|^2_{\mathscr{H}_2} = \|\sigma\|^2_{\mathscr{H}} + \|\operatorname{Div} \sigma\|^2_{\boldsymbol{H}}. \quad (2.43)$$

In this norm both the embedding operator of $\mathscr{H}_2$ in $\mathscr{H}$ and the divergence operator Div: $\mathscr{H}_2 \to \boldsymbol{H}$ are continuous. Let $\boldsymbol{H}'_\Gamma$ be the dual of the space $\boldsymbol{H}_\Gamma$; as in the case of the space $\boldsymbol{H}_1$ we can define the trace of any element $\sigma \in \mathscr{H}_2$ using the following result:

Theorem 2.1 *There exists a linear, continuous and surjective operator*

γ_2: $\mathscr{H}_2 \to \boldsymbol{H}'_\Gamma$ such that for all $\sigma \in [C^1(\bar{\Omega})]_S^{N \times N}$ we have

$$\langle \gamma_2 \sigma, w \rangle_{\boldsymbol{H}'_\Gamma, \boldsymbol{H}_\Gamma} = \int_\Gamma \sigma v \cdot w \, d\Gamma \qquad \forall w \in \boldsymbol{H}_\Gamma$$

and

$$\langle \gamma_2 \sigma, \gamma_1 v \rangle_{\boldsymbol{H}'_\Gamma, \boldsymbol{H}_\Gamma} = \langle \sigma, \varepsilon(v) \rangle_{\mathscr{H}} + \langle \operatorname{Div} \sigma, v \rangle_{\boldsymbol{H}} \qquad \forall \sigma \in \mathscr{H}_2, \quad v \in \boldsymbol{H}_1. \tag{2.44}$$

Moreover, there exists a linear and continuous map z_2: $\boldsymbol{H}'_\Gamma \to \mathscr{H}_2$ such that

$$\gamma_2(z_2(\Sigma)) = \Sigma \qquad \forall \Sigma \in \boldsymbol{H}'_\Gamma. \tag{2.45}$$

PROOF. Let $\sigma \in \mathscr{H}_2$; let us remark that for all $\varphi \in \boldsymbol{D}$ by (2.8) we have $\langle \sigma, \varepsilon(\varphi) \rangle_{\mathscr{H}} + \langle \operatorname{Div} \sigma, \varphi \rangle_{\boldsymbol{H}} = 0$; since the embedding $\boldsymbol{H}_1 \subset \boldsymbol{H}$ and the deformation operator ε: $\boldsymbol{H}_1 \to \mathscr{H}$ are continuous and $\boldsymbol{D}$ is dense in $[H_0^1(\Omega)]^N$ we get $\langle \sigma, \varepsilon(\varphi) \rangle_{\mathscr{H}} + \langle \operatorname{Div} \sigma, \varphi \rangle_{\boldsymbol{H}} = 0$ for all $\varphi \in [H_0^1(\Omega)]^N$. Hence, if v and $\tilde{v}$ belong to $\boldsymbol{H}_1$ such that $\gamma_1 v = \gamma_1 \tilde{v}$ we have $v - \tilde{v} \in [H_0^1(\Omega)]^N$ and it follows that

$$\langle \sigma, \varepsilon(v) \rangle_{\mathscr{H}} + \langle \operatorname{Div} \sigma, v \rangle_{\boldsymbol{H}} = \langle \sigma, \varepsilon(\tilde{v}) \rangle_{\mathscr{H}} + \langle \operatorname{Div} \sigma, \tilde{v} \rangle_{\boldsymbol{H}}.$$

So, using the surjectivity of the trace map γ_1: $\boldsymbol{H}_1 \to \boldsymbol{H}_\Gamma$ we can define the linear functional F_σ: $\boldsymbol{H}_\Gamma \to \mathbb{R}$ by

$$F_\sigma(\xi) = \langle \sigma, \mathscr{E}(v) \rangle_{\mathscr{H}} + \langle \operatorname{Div} \sigma, v \rangle_{\mathscr{H}} \qquad \forall \xi = \gamma_1(v) \in \boldsymbol{H}_\Gamma. \tag{2.46}$$

Since from (2.28), (2.29), (2.30), and (2.43) we deduce

$$|F_\sigma(\xi)| \leq C \|\sigma\|_{\mathscr{H}_2} \|\xi\|_{\boldsymbol{H}_\Gamma}, \tag{2.47}$$

we conclude that F_σ is bounded on $\boldsymbol{H}_\Gamma$; so, it defines an element $\Sigma \in \boldsymbol{H}'_\Gamma$ such that

$$F_\sigma(\xi) = \langle \Sigma, \xi \rangle_{\boldsymbol{H}'_\Gamma, \boldsymbol{H}_\Gamma} \qquad \forall \xi = \gamma_1(v) \in \boldsymbol{H}_\Gamma \tag{2.48}$$

$$\|\Sigma\|_{\boldsymbol{H}'_\Gamma} \leq C \|\sigma\|_{\mathscr{H}_2}. \tag{2.49}$$

Denoting by γ_2: $H_2 \to \boldsymbol{H}'_\Gamma$ the correspondence $\sigma \to \Sigma$ we get that γ_2 is a linear operator; from (2.48), (2.46) we obtain (2.44) and from (2.49) we get that γ_2 is a bounded operator.

Now let Σ be an arbitrary element of $\boldsymbol{H}'_\Gamma$; we define the linear functional F_Σ: $\boldsymbol{H}_1 \to \mathbb{R}$ by

$$F_\Sigma(v) = \langle \Sigma, \gamma_1 v \rangle_{\boldsymbol{H}'_\Gamma, \boldsymbol{H}_\Gamma} \qquad \forall v \in \boldsymbol{H}_1. \tag{2.50}$$

Since $|F_\Sigma(v)| \leq \|\Sigma\|_{\boldsymbol{H}'_\Gamma} \|\gamma_1 v\|_{\boldsymbol{H}_\Gamma} \leq C \|\Sigma\|_{\boldsymbol{H}'_\Gamma} \|v\|_{\boldsymbol{H}_1}$ for all $v \in \boldsymbol{H}_1$ we deduce that F_Σ is a bounded functional on $\boldsymbol{H}_1$ and

$$\|F_\Sigma\|_{\boldsymbol{H}'_1} \leq C \|\Sigma\|_{\boldsymbol{H}'_\Gamma}. \tag{2.51}$$

Using Riesz's representation theorem there exists a unique element $w \in \boldsymbol{H}_1$

such that

$$F_\Sigma(v) = \langle w, v\rangle_{\boldsymbol{H}_1} \qquad \forall v \in \boldsymbol{H}_1, \tag{2.52}$$

$$\|F_\Sigma\|_{\boldsymbol{H}_1'} = \|w\|_{\boldsymbol{H}_1}. \tag{2.53}$$

From (2.52), (2.50), and (2.27) we obtain

$$\langle \Sigma, \gamma_1 v\rangle_{\boldsymbol{H}_\Gamma', \boldsymbol{H}_\Gamma} = \langle w, v\rangle_{\boldsymbol{H}} + \langle \varepsilon(w), \varepsilon(v)\rangle_{\mathscr{H}} \qquad \forall v \in \boldsymbol{H}_1. \tag{2.54}$$

Let $\sigma = \varepsilon(w) \in \mathscr{H}$; taking $v = \varphi \in \boldsymbol{D}$ in (2.54) we deduce $\operatorname{Div} \sigma = w \in \boldsymbol{H}$, hence $\sigma \in \mathscr{H}_2$, so, (2.54) becomes

$$\langle \Sigma, \gamma_1 v\rangle_{\boldsymbol{H}_\Gamma', \boldsymbol{H}_\Gamma} = \langle \operatorname{Div} \sigma, v\rangle_{\boldsymbol{H}} + \langle \sigma, \varepsilon(v)\rangle_{\mathscr{H}} \qquad \forall v \in \boldsymbol{H}_1$$

and by (2.44) we get $\gamma_2\sigma = \Sigma$, which means γ_2 is surjective. Moreover from (2.43) and (2.28) we get $\|\sigma\|_{\mathscr{H}_2} = \|w\|_{\boldsymbol{H}_1}$ and from (2.53), (2.51) we obtain

$$\|\sigma\|_{\mathscr{H}_2} \leq C\|\Sigma\|_{\boldsymbol{H}_\Gamma'}. \tag{2.55}$$

Denoting by $z_2\colon \boldsymbol{H}_\Gamma' \to \mathscr{H}_2$ the correspondence $\Sigma \to \sigma$ we get that z_2 is a linear and continuous map (see (2.55)). Since $\gamma_2\sigma = \Sigma$ we get (2.45).

Let now Γ_1 be an open set in Γ and $\Gamma_2 = \Gamma\backslash\bar{\Gamma}_1$; for every $\sigma \in \mathscr{H}_2$ we denote by $\sigma v|_{\Gamma_2}$ the element of $\boldsymbol{V}_{\Gamma_1}'$ which is the restriction of $\gamma_2\sigma$ at $\boldsymbol{V}_{\Gamma_1}$. So we have defined the linear operator $\sigma \to \sigma v|_{\Gamma_2}\colon \mathscr{H}_2 \to \boldsymbol{V}_{\Gamma_1}'$ and hence

$$\|\sigma v|_{\Gamma_2}\|_{\boldsymbol{V}_{\Gamma_1}'} \leq \|\gamma_2\sigma\|_{\boldsymbol{H}_\Gamma'} \leq C\|\sigma\|_{\mathscr{H}_2}$$

(see (2.49)) we get that $\sigma \to \sigma v|_{\Gamma_2}$ is continuous. Using this operator we introduce the following closed subspaces of $\mathscr{H}_2$:

$$\mathscr{V}_1 = \{\sigma \in \mathscr{H}_2 | \sigma v|_{\Gamma_2} = 0\}, \tag{2.56}$$

$$\mathscr{V}_2 = \{\sigma \in \mathscr{H}_2 | \operatorname{Div} \sigma = 0, \sigma v|_{\Gamma_2} = 0\}. \tag{2.57}$$

$\mathscr{V}_1$ and $\mathscr{V}_2$ are Hilbert spaces with respect to (2.42) and, by definition of $\sigma v|_{\Gamma_2}$, the following Green-type formula holds:

$$\langle \sigma, \varepsilon(v)\rangle_{\mathscr{H}} + \langle \operatorname{Div} \sigma, v\rangle_{\boldsymbol{H}} = 0 \qquad \forall \sigma \in \mathscr{V}_1, \quad v \in \boldsymbol{V}_1. \tag{2.58}$$

In the case of $\mathscr{V}_2$ the above equality reduces to the following orthogonality property:

$$\langle \sigma, \varepsilon(v)\rangle_{\mathscr{H}} = 0 \qquad \forall \sigma \in \mathscr{V}_2, \quad v \in \boldsymbol{V}_1 \tag{2.59}$$

Moreover, we have the following result:

Lemma 2.1 *If* meas $\Gamma_1 > 0$ *then $\varepsilon(\boldsymbol{V}_1)$ is a closed subspace $\mathscr{H}$ having $\mathscr{V}_2$ as orthogonal complement in $\mathscr{H}$.*

PROOF. Let $(v_n)_n \subset \boldsymbol{V}_1$ and $\sigma \in \mathscr{H}$ so that $\varepsilon(v_n) \to \sigma$ in $\mathscr{H}$; using (2.32) we get that $(v_n)_n$ is a Cauchy sequence in $\boldsymbol{V}_1$, hence there exists $v \in \boldsymbol{V}_1$ such that $v_n \to v$ in $\boldsymbol{V}_1$; by continuity of the deformation operator we get $\varepsilon(v_n) \to \varepsilon(v)$ in $\mathscr{H}$ and hence $\sigma = \varepsilon(v)$, which means $\sigma \in \varepsilon(\boldsymbol{V}_1)$; so, $\varepsilon(\boldsymbol{V}_1)$ is a closed subspace of $\mathscr{H}$.

Let now $\varepsilon(V_1)^\perp$ be the orthogonal complement of $\varepsilon(V_1)$ in $\mathscr{H}$; from (2.59) we have $\mathscr{V}_2 \subset \varepsilon(V_1)^\perp$ and for the converse inclusion let $\sigma \in \varepsilon(V_1)^\perp$; we have

$$\langle \sigma, \varepsilon(v) \rangle_{\mathscr{H}} = 0 \qquad \forall v \in V_1. \tag{2.60}$$

By (2.44) and (2.60) we get $\langle \gamma_2\sigma, \gamma_1 v\rangle_{H'_\Gamma, H_\Gamma} = \langle \text{Div}\,\sigma, v\rangle_H$ for all $v \in V_1$ and taking $v = \varphi \in D \subset V_1$ we get $\text{Div}\,\sigma = 0$; so, by (2.60) and (2.44) we get $\langle \gamma_2\sigma, \gamma_1 v\rangle_{H'_\Gamma, H_\Gamma} = 0$ for all $v \in V_1$, which means $\sigma v|_{\Gamma_2} = 0$. It follows that $\sigma \in \mathscr{V}_2$ and $\varepsilon(V_1)^\perp \subset \mathscr{V}_2$.

Remark 2.5 *Lemma* 2.1 *holds even if* $\Gamma_1 = \varnothing$ *(see for instance Léné* (1974) *or Suquet* (1982, *p.* 16)).

In solving the problems of linear elasticity as well as the problems of plasticity the following result is very useful.

Lemma 2.2 *There exists a linear and continuous map* $z_{2,\Gamma_1}\colon V'_{\Gamma_1} \to \mathscr{H}_2$ such that

$$z_{2,\Gamma_1}(F)v|_{\Gamma_2} = F \qquad \forall F \in V'_{\Gamma_1}. \tag{2.61}$$

PROOF. Since V_{Γ_1} is a closed subspace of H_Γ for all $v \in H_\Gamma$ there exists a unique couple of functions $v_1 \in V_{\Gamma_1}$, $v_2 \in V_{\Gamma_1}^\perp$ such that $v = v_1 + v_2$. For all $F \in V'_{\Gamma_1}$ we can define $\bar{F} \in H'_\Gamma$ given by $\bar{F}(v) = F(v_1)$ for all $v \in H_\Gamma$, $v = v_1 + v_2$, $v_1 \in V_{\Gamma_1}$, $v_2 \in V_{\Gamma_2}^\perp$. If we keep in mind that the map $F \to \bar{F}$ is continuous from V'_{Γ_1} to H'_Γ, and we put $z_{2,\Gamma_1}(F) = z_2(\bar{F})$ (see Theorem 2.1) we deduce that z_{2,Γ_1} is a linear and continuous map from V'_{Γ_1} to $\mathscr{H}_2$ and $\gamma_2(z_{2,\Gamma_1}(F))(v) = \langle F, v\rangle_{V'_{\Gamma_1}, V_{\Gamma_1}}$ for all $v \in V_{\Gamma_1}$, hence (2.61) holds.

3 Functional spaces of vector-valued functions defined on real intervals

This Section presents the notations, definitions and other necessary background information on vector-valued functions required for the following treatment. Most of the terminology and basic results used here are well known and so will be used without further comment and demonstrations.

Throughout this section $[0, T]$ will be a fixed real interval $(0 < T < +\infty)$ and $(X, \|\cdot\|_X)$ a real Banach space; we denote by $(X', \|\cdot\|_{X'})$ its strong dual and by $\langle\, , \rangle_{X',X}$ the duality pairing map between X' and X.

3.1 Weak and strong measurability and L^p spaces

A function $x\colon [0, T] \to X$ is said to be *strongly measurable* on $[0, T]$ if there exists a sequence $(x_n)_n$ of finite-valued functions which converges strongly almost everywhere on $(0, T)$ to x; a function $x\colon [0, T] \to X$ is *weakly measurable* if the real-valued function $t \to \langle x', x(t) \rangle_{X',X}$ is measurable for each $x' \in X'$; a function $x: [0, T] \to X$ is called *almost separable valued* if there is a set A_0 of measure zero such that the range $x([0, T] \backslash A_0)$ is separable in X. The basic result in this respect is the following theorem:

Theorem 3.1 (Pettis, 1938). *A vector-valued function $x: [0, T] \to X$ is strongly measurable if and only if it is weakly measurable and almost separable valued.*

A function $x: [0, T] \to X$ is said to be a *Bochner integrable* if there exists a sequence $(x_n)_n$ of finitely valued functions on $[0, T]$ to X which converges strongly almost everywhere on $(0, T)$ to x in such a way that

$$\lim_n \int_0^T \|x(t) - x_n(t)\|_X \,\mathrm{d}t = 0.$$

In this case, by definition, the Bochner integral of x over any measurable set $A \subset [0, T]$ is the element of X given by

$$\int_A x(t)\,\mathrm{d}t = \lim_n \int_A x_n(t)\,\mathrm{d}t.$$

The following result can be proved:

Theorem 3.2 (Bochner, 1933). *A necessary and sufficient condition that $x: [0, T] \to X$ is Bochner integrable is that x is strongly measurable and $\int_0^T \|x(t)\|_X \,\mathrm{d}t < +\infty$.*

Let p be a real number, $p \geq 1$. We denote by $L^p(0, T, X)$ the space of all strongly measurable functions x on $[0, T]$ to X such that $\int_0^T \|x(t)\|_X^p \,\mathrm{d}t < +\infty$; $L^p(0, T, X)$ is a real Banach space with the norm

$$\|x\|_{p,X} = \left(\int_0^T \|x(t)\|_X^p \,\mathrm{d}t \right)^{1/p}. \tag{3.1}$$

We also denote by $L^\infty(0, T, X)$ the space of all strongly measurable functions x on $[0, T]$ to X such that $t \to \|x(t)\|_X$ is essentially bounded on $(0, T)$; $L^\infty(0, T, X)$ is a real Banach space under the norm

$$\|x\|_{0,T,X} = \operatorname*{ess\,sup}_{t\in[0,T]} \|x(t)\|_X. \tag{3.2}$$

For every $1 \leq p \leq \infty$ we identify in $L^p(0, T, X)$ functions that are equal almost everywhere on $(0, T)$ and we refer to elements of $L^p(0, T, X)$ as functions.

A function $x': [0, T] \to X'$ is said to be *weakly star measurable* if the real-valued function $t \to \langle x'(t), x \rangle_{X',X}$ is measurable for each $x \in X$. We also denote by $L^p_w(0, T, X')$ the space of all (classes of) weakly star measurable functions x' on $[0, T]$ to X' such that $\int_0^T \|x'(t)\|_{X'}^p \, dt < +\infty$ for $1 \leq p < \infty$; $L_w^p(0, T, X')$ is a real Banach space with the norm $\|\cdot\|_{p,X'}$ which is defined similarly to $\|\cdot\|_{p,X}$ (see (3.1)). In the same way we denote by $L_w^\infty(0, T, X')$ the space of all (classes of) weakly star measurable functions x on $[0, T]$ to X' such that $t \to \|x(t)\|_{X'}$ is essentially bounded on $(0, T)$; $L_w^\infty(0, T, X)$ is a real Banach space under the norm $\|\cdot\|_{0,T,X'}$ which is defined similarly to $\|\cdot\|_{0,T,X}$ (see (3.2)).

Since every strong measurable function x': $[0, T] \to X'$ is weakly star measurable we have $L^p(0, T, X') \subset L_w^p(0, T, X')$ for all $1 \leq p \leq \infty$; if in addition X' is a separable space then $L^p(0, T, X') = L_w^p(0, T, X')$ for all $1 \leq p \leq \infty$. Moreover, we have the following results:

Theorem 3.3 *Let $p \in \mathbb{R}$, $1 < p < \infty$, and let q be defined by $1/p + 1/q = 1$. Then*

(1) *the dual of the space $L^p(0, T, X)$ is the space $L_w^q(0, T, X')$;*
(2) *the dual of the space $L^1(0, T, X)$ is the space $L_w^\infty(0, T, X')$;*

If, moreover, X' is a reflexive or separable space, then

(3) *the dual of the space $L^p(0, T, X)$ is the space $L^q(0, T, X')$;*
(4) *the dual of the space $L^1(0, T, X)$ is the space $L^\infty(0, T, X')$.*

In all these cases the duality pairing mapping between $L^p(0, T, X')$ and its dual is given by

$$\langle x', x \rangle = \int_0^T \langle x'(t), x(t) \rangle_{X',X} \, dt. \tag{3.3}$$

Finally let us notice that if $(X, \langle , \rangle_X)$ is a real Hilbert space then $L^2(0, T, X)$ is also a real Hilbert space with the inner product given by

$$\langle x, y \rangle_{2,X} = \int_0^T \langle x(t), y(t) \rangle_X \, dt. \tag{3.4}$$

3.2 Absolutely continuous vectorial functions and $A^{k,p}$ spaces

Let x: $[0, T] \to X$; the function x is said to be (strongly) *differentiable almost everywhere* (a.e.) if there exists a function denoted by dx/dt defined on $[0, T]$ to X such that

$$\left\| \frac{x(t+h) - x(t)}{h} - \frac{dx}{dt}(t) \right\|_X \to 0 \tag{3.5}$$

when $h \to 0$, a.e. on $(0, T)$. The function dx/dt is called *the* (strong) *derivative* of x. By recurrence the derivative of higher order d^jx/dt^j, $j = 1, 2, \ldots$ can be defined. For simplicity, we denote $dx/dt = \dot{x}$ and we consider $d^jx/dt^j = x$ for $j = 0$.

The function $x\colon [0, T] \to X$ is said to be *absolutely continuous* on $[0, T]$ if for each $\varepsilon > 0$ there exists $\delta(\varepsilon) > 0$ such that $\sum_n \|x(b_n) - x(a_n)\|_X \leq \varepsilon$ whenever $\sum_n (b_n - a_n) \leq \delta(\varepsilon)$ and $(a_n, b_n) \cap (a_m, b_m) = \varnothing$ for $m \neq n$. Here $(a_n, b_n)_n$ is a denumerable family of arbitrary intervals contained in $[0, T]$.

It is well known that every real-valued absolutely continuous function on a real interval $[0, T]$ is almost everywhere differentiable on $(0, T)$ and it is expressed as the indefinite integral of the derivative. Some simple examples (see for instance Barbu (1976, p. 15)) show that this fails when x is absolutely continuous from $[0, T]$ to a general Banach space. However, the following theorem due to Komura (1967) can be proved:

Theorem 3.4 *Let X be a reflexive Banach space. Then every X-valued absolutely continuous function x on $[0, T]$ is a.e. differentiable on $(0, T)$, its derivative belongs to $L^1(0, T, X)$ and*

$$x(s) = x(0) + \int_0^s \dot{x}(t)\,\mathrm{d}t \qquad \forall s \in [0, T]. \tag{3.6}$$

Using the above notations for every $k \in \mathbb{N}$ and $1 \leq p \leq \infty$ we denote by $A^{k,p}(0, T, X)$ the space of all absolutely continuous functions x from $[0, T]$ to X whose derivatives $\mathrm{d}^j x/\mathrm{d}t^j$ (defined almost everywhere) are absolutely continuous for $j = 1, 2, \ldots, k - 1$ and belong to $L^p(0, T, X)$ for $j = 0, 1, \ldots, k$. In particular $A^{1,p}(0, T, X)$ consists of all absolutely continuous functions $x\colon [0, T] \to X$ with the property that the function $t \to \dot{x}(t)$ exists a.e. on $(0, T)$ and belongs to $L^p(0, T, X)$. Theorem 3.4 implies that if X is reflexive, then the function $x\colon [0, T] \to X$ belongs to $A^{1,p}(0, T, X)$ if and only if there exists a function $g \in L^p(0, T, X)$ such that $x(s) = x(0) + \int_0^s g(t)\,\mathrm{d}t$ for every $s \in [0, T]$.

3.3 Vectorial distributions and $W^{k,p}$ spaces

Let $\mathscr{D}(0, T)$ denote the space of all real-valued functions defined on $[0, T]$ which are infinitely differentiable on $[0, T]$ and have compact support in $(0, T)$. The space $\mathscr{D}(0, T)$ is considered with the usual topology (see Section 1.1). If X is a Banach space we denote $\mathscr{D}'(0, T, X)$ the space of all linear continuous operators from $\mathscr{D}(0, T)$ to X. An element f of $\mathscr{D}'(0, T, X)$ is called an *X-valued distribution on* $(0, T)$.

If $f \in \mathscr{D}'(0, T, X)$ and j is a positive integer, then

$$D_j f(\varphi) = (-1)^j f(D_j \varphi) \qquad \forall \varphi \in \mathscr{D}(0, T) \tag{3.7}$$

defines another vectorial distribution $D_j f \in \mathscr{D}'(0, T, X)$ called the *derivative of order j of f*. In (3.7) by $D_j\varphi$ we mean the ordinary derivative of order j of φ. In particular $D_j f = f$ for $j = 0$.

As in the scalar case to every function $x \in L^p(0, T, X)$ $(1 \leq p \leq \infty)$

corresponds a distribution $\tilde{x}$ defined by

$$\tilde{x}(\varphi) = \int_0^T \varphi(t)x(t)\,dt \qquad \forall \varphi \in \mathscr{D}(0, T), \tag{3.8}$$

the integral being in the Bochner sense. Since the embedding $x \to \tilde{x}$ is linear continuous and injective from $L^p(0, T, X)$ to $\mathscr{D}'(0, T, X)$, we can regard $L^p(0, T, X)$ as a linear subspace of $\mathscr{D}'(0, T, X)$. In all that follows, a function $x \in L^p(0, T, X)$ will be identified with the corresponding vectorial distribution. Moreover we say that a distribution $f \in \mathscr{D}'(0, T, X)$ belongs to $L^p(0, T, X)$ if there exists a function $x \in L^p(0, T, X)$ such that $f = \tilde{x}$, where $\tilde{x}$ is the distribution defined by (3.8).

Let k be a positive integer and $1 \leq p \leq \infty$. We denote by $W^{k,p}(0, T, X)$ the space of all vectorial distributions $x \in \mathscr{D}'(0, T, X)$ with the property that $D_j x \in L^p(0, T, X)$ for $j = 0, 1, \dots, k$. If $1 \leq p < \infty$, $W^{k,p}(0, T, X)$ is a real Banach space with the norm

$$\|x\|_{k,p,X} = \left(\sum_{j=0}^{k} \|D_j x\|_{p,X}^p \right)^{1/p} \tag{3.9}$$

and the space $W^{k,\infty}(0, T, X)$ is a real Banach space with the norm

$$\|x\|_{k,T,X} = \sum_{j=0}^{k} \|D_j x\|_{0,T,X}. \tag{3.10}$$

In particular, for $k = 0$ we get $W^{k,p}(0, T, X) = L^p(0, T, X)$ and $\|x\|_{0,p,X} = \|x\|_{p,X}$ for all $x \in L^p(0, T, X)$, $1 \leq p < \infty$.

Let us also notice that if X is a real Hilbert space with the inner product $\langle , \rangle_X$ then for any positive integer k the space $W^{k,2}(0, T, X)$ is a real Hilbert space with respect to the inner product

$$\langle x, y \rangle_{k,2,X} = \sum_{j=0}^{k} \int_0^T \langle D_j x(t), D_j v(t) \rangle_X \,dt. \tag{3.11}$$

The link between the spaces $W^{k,p}(0, T, X)$ and the spaces $A^{k,p}(0, T, X)$ is given by the following result.

Theorem 3.5 *Let X be a real Banach space, k a positive integer, $1 \leq p \leq \infty$ and let $x \in L^p(0, T, X)$. Then the following conditions are equivalent:*

(*a*) $x \in W^{k,p}(0, T, X)$
(*b*) *there exists $x_1 \in A^{k,p}(0, T, X)$ such that $x(t) = x_1(t)$ a.e. on $(0, T)$.*

The above theorem allows us to identify the spaces $W^{k,p}(0, T, X)$ and $A^{k,p}(0, T, X)$ for every positive integer k and $1 \leq p \leq \infty$. Moreover, as follows from the proof of this theorem, for every $x \in W^{k,p}(0, T, X)$ the derivative in the sense of vectorial distributions coincides with the ordinary derivatives: $D_j x = d^j x/dt^j$ a.e. on $(0, T)$ for every $j = 1, \dots, k$.

Using Theorems 3.4 and 3.5 the following result can be easily proved:

Theorem 3.6 *Let X be a real reflexive Banach space and x: $[0, T] \to X$. Then:*

(1) *x belongs to $W^{1,1}(0, T, X)$ iff x is an absolutely continuous function on $[0, T]$ to X;*
(2) *x belongs to $W^{1,\infty}(0, T, X)$ iff x is a Lipschitz function on $[0, T]$ to X.*

Again using Theorem 3.4, the following result can be proved:

Theorem 3.7 *If $(X, \langle , \rangle_X)$ is a real Hilbert space and $x \in W^{1,2}(0, T, X)$, then the real-valued function $t \to \frac{1}{2}\|x(t)\|_X^2$ is absolutely continuous on $[0, T]$,*

$$\frac{\mathrm{d}}{\mathrm{d}t}\left(\tfrac{1}{2}\|x(t)\|_X^2\right) = \langle \dot{x}(t), x(t)\rangle_X \qquad \textit{a.e. on } (0, T)$$

and the following equality holds:

$$\tfrac{1}{2}\|x(s)\|_X^2 = \tfrac{1}{2}\|x(0)\|_X^2 + \int_0^s \langle \dot{x}(t), x(t)\rangle_X \,\mathrm{d}t \tag{3.12}$$

for all $s \in [0, T]$.

Remark 3.1 *Using the definitions of the spaces $W^{k,p}$ it is easy to obtain the following result, which will be frequently used in Chapters 3 and 4: let X_1 and X_2 be real Banach spaces and A: $X_1 \to X_2$ a linear and continuous operator; if for every integer k, $1 \le p \le \infty$ and $x \in W^{k,p}(0, T, X)$ we consider the function $A^{k,p}x$ defined on $[0, T]$ by*

$$(A^{k,p}x)(t) = Ax(t) \qquad \textit{for all } t \in [0, T], \tag{3.13}$$

then $A^{k,p}x$ belongs to $W^{k,p}(0, T, X_2)$ and

$$\left\|\frac{\mathrm{d}^j}{\mathrm{d}t^j} A^{k,p}x(t)\right\|_{X_2} \le \|A\|_{B(X_1,X_2)} \left\|\frac{\mathrm{d}^j}{\mathrm{d}t^j} x(t)\right\|_{X_1} \tag{3.14}$$

for all $t \in [0, T]$ and $j = 0, 1, \ldots, k-1$. Moreover, the operator

$$A^{k,p}\colon W^{k,p}(0, T, X_1) \to W^{k,p}(0, T, X_2)$$

defined by (3.13) *is a linear and continuous operator.*

In Chapter 3 we also use the spaces $C^k([0, T], X)$; for every positive integer k, $C^k([0, T], X)$ consists of all functions x defined on $[0, T]$ to X such that for every $j = 0, \ldots, k$ there exists the (strong) derivatives $\mathrm{d}^jx/\mathrm{d}t^j$ *everywhere* on $[0, T]$ and these derivatives are continuous from $[0, T]$ to X. The space $C^k([0, T], X)$ is a Banach space with the norm

$$\|x\|_{k,T,X} = \sum_{j=0}^{k} \max_{t\in[0,T]} \left\|\frac{\mathrm{d}^j}{\mathrm{d}t^j} x(t)\right\|_X . \tag{3.15}$$

In particular, the norms on the spaces $C^0([0, T], X)$ and $C^1([0, T], X)$ are given by

$$\|x\|_{0,T,X} = \max_{t\in[0,T]} \|x(t)\|_X \tag{3.16}$$

$$\|x\|_{1,T,X} = \|x\|_{0,T,X} + \|\dot{x}\|_{0,T,X}. \tag{3.17}$$

The spaces $C^k([0, T], X)$ and $C^k(R_+, X)$ are defined in a similar way.

Bibliographical notes

For a complete treatment of the spaces introduced in Section 1, as well as for useful generalizations, the reader is referred to Schwartz (1967), Nečas (1967), Lions and Magenes (1968), Yosida (1971), Rudin (1973), and Adams (1975).

The interest on the space $BD(\Omega)$ introduced in Section 2.2 was developed in connection with evolution problems for elastic–plastic solids. This functional space attached to the deformation operator was introduced by Suquet (1978*a*, *b*), Temam and Strang (1978), Matthies *et al.* (1979). The first trace theorem for the space $BD(\Omega)$ was given by Suquet (1978*a*, *b*), then reinforced on the form (2.36) by Temam and Strang (1978, 1980). A different proof of (2.36) was also given by Suquet (1982, Ch. 3). The embedding result (2.37) on $BD(\Omega)$ was proved by Suquet (1978*b*) (the completely continuous case) and Temam and Strang (1980) (the case $p = N/(N-1)$). The regularity result (2.38) was given by Temam and Strang (1980). The proofs of all results on spaces $BD(\Omega)$ and $LD(\Omega)$ can be found in the book of Temam (1983, Ch. 2) (for basic information on the space of bounded measures on Ω the reader is referred for instance to Rudin (1966, Ch. 6)).

The results presented in Section 2.3 on the space $\mathscr{H}_2 = H(\mathrm{Div}, \Omega)$ are well known; Theorem 2.1 and Lemma 2.1 can be found for instance in the paper of Léné (1974); other comments on $\mathscr{H}_2$ are made for instance in the work of Temam (1983, Ch. 1). Some interesting generalizations and results of duality, trace maps, density and orthogonality are given in Geymonat and Suquet (1984).

The material presented in Section 3 is standard and can be found in many books or surveys. For more detailed information on it we refer the reader to Schwarz (1957, 1958) Dunford and Schwarz (1958), Nicolescu (1960), Yosida (1971), Ionescu-Tulcea and Ionescu-Tulcea (1972), Brezis (1973), and Barbu (1976).

3

QUASISTATIC PROCESSES FOR RATE-TYPE VISCOPLASTIC MATERIALS

In this chapter we study quasistatic problems for semilinear rate-type constitutive equations. The existence, uniqueness, and finite-time stability of the solution are given using some arguments of ordinary differential equations in Hilbert spaces. The case of viscoelastic materials is also investigated with respect to asymptotic stability, existence of periodic solutions, approach to non-linear elasticity, and long-term behaviour of the solution. By passing to the limit, the elastic–perfectly plastic case is derived. A Euler method, internal and external approximation techniques are used in order to reduce the continuous problem to a sequence of linear algebraic systems. Existence and uniqueness results in quasistatic processes for rate-type constitutive equations involving a hardening parameter are also given.

1 Discussion of a quasistatic elastic–viscoplastic problem

We start this section by presenting the terminology connected with rate-type constitutive equations used in order to describe the mechanical properties of elastic–viscoplastic materials.

1.1 Rate-type constitutive equations

Rate-type constitutive equations are the simplest constitutive equations that may describe some fundamental phenomena (creep, relaxation) that cannot be described by the constitutive equations of classical plasticity of Chapter 1. In order to present rate-type constitutive equations, let us denote by $\mathscr{S}_N$ ($N = 1, 2, 3$) the space of symmetric second-order tensors on $\mathbb{R}^N$ and let $\mathscr{D} \subset \mathscr{S}_N \times \mathscr{S}_N$ be a domain containing the origin, called the *state domain*; the points $(\sigma, \varepsilon) \in \mathscr{D}$ will be called *states*. A smooth function (say C^1) $t \to \sigma(t)$ (or $t \to \varepsilon(t)$) with $t \in [0, T]$, $T > 0$ is called a *stress history* (or a *strain history*) of a particle of the body. The time derivatives $\dot{\sigma} = \dot{\sigma}(t)$ and $\dot{\varepsilon} = \dot{\varepsilon}(t)$ are called the *stress rate* and *strain rate* respectively at the state $(\sigma = \sigma(t), \varepsilon = \varepsilon(t))$. Throughout this chapter, by a constitutive equation is meant a rule which, at a given state $(\sigma, \varepsilon) \in \mathscr{D}$, determines the stress rate $\dot{\sigma}$ (or the strain rate) when the strain rate $\dot{\varepsilon}$ (or the stress rate) is known.

Let us consider $\mathscr{A}: \mathscr{D} \to B(\mathscr{S}_N)$ and $G: \mathscr{D} \to \mathscr{S}_N$, two functions called *material functions*; the domain $\mathscr{D}$ together with the functions $\mathscr{A}$ and G characterize a given material. So, the material behaviour in $\mathscr{D}$ is said to be described by a *quasilinear rate-type constitutive equation* if, for any state $(\sigma, \varepsilon) \in \mathscr{D}$ and for any strain rate $\dot{\varepsilon}$, the stress rate at the state (σ, ε), is given by

$$\dot{\sigma} = \mathscr{A}(\sigma, \varepsilon)\dot{\varepsilon} + G(\sigma, \varepsilon). \tag{1.1}$$

If in particular $\mathscr{A}$ is constant in $\mathscr{D}$, then (1.1) becomes

$$\dot{\sigma} = \mathscr{A}\dot{\varepsilon} + G(\sigma, \varepsilon). \tag{1.2}$$

Equations of this type are called *rate-type semilinear constitutive equations.* In the study of equations (1.2) we shall assume that $\mathscr{A}$ is an invertible tensor; hence the role of $\dot{\varepsilon}$ can be interchanged with that of $\dot{\sigma}$. It follows that every rate-type semilinear constitutive equation (1.2) may be written in an equivalent form

$$\dot{\varepsilon} = \tilde{\mathscr{A}}\dot{\sigma} + \tilde{G}(\sigma, \varepsilon), \tag{1.3}$$

where $\tilde{\mathscr{A}} = \mathscr{A}^{-1}$ and $\tilde{G}(\sigma, \varepsilon) = -\mathscr{A}^{-1}G(\sigma, \varepsilon)$.

This formula shows that the rate of deformation can be decomposed into two parts: the elastic (reversible) component $\dot{\varepsilon}^{\mathrm{E}}$ defined by

$$\dot{\varepsilon}^{\mathrm{E}} = \tilde{\mathscr{A}}\dot{\sigma} \tag{1.4}$$

and the viscoplastic (irreversible) one $\dot{\varepsilon}^{\mathrm{I}}$ given by

$$\dot{\varepsilon}^{\mathrm{I}} = \tilde{G}(\sigma, \varepsilon). \tag{1.5}$$

In the next paragraphs we shall consider for instance semilinear equations written in the form (1.2). The equivalent form (1.3) will also be used in Section 3 of this chapter and in Sections 3 and 5 of Chapter 4. All points $(\sigma_0, \varepsilon_0) \in \mathscr{D}$ with $G(\sigma_0, \varepsilon_0) = 0$ will be called *equilibrium points* or *points of elastic behaviour*. If the set of equilibrium points may be described by an explicit equation $\sigma = F(\varepsilon)$, we say that (1.1) or (1.2) represents a *viscoelastic constitutive equation*; if the set of equilibrium points has a non-empty interior in $\mathscr{D}$ we say that (1.1) or (1.2) represents a *viscoplastic constitutive equation.*

The fact that rate-type materials may describe time-dependent phenomena (see Section 3.3 of Chapter 1) can easily be proved in the one-dimensional case:

$$\dot{\sigma} = E\dot{\varepsilon} + G(\sigma, \varepsilon), \tag{1.6}$$

where $E > 0$ is Young's modulus and $G: \mathscr{D} \subset \mathbb{R}^2 \to \mathbb{R}$. Taking a stress history $\sigma(t) = \sigma_0 \in \mathscr{D}$ for all $t \in [0, T]$, (1.6) implies $\dot{\varepsilon}(t) = -(1/E)G(\sigma_0, \varepsilon(t))$ and assuming $\varepsilon(0) = \varepsilon_0 \in \mathscr{D}$, $G(\sigma_0, \varepsilon_0) < 0$ we obtain that ε grows in a neighbourhood of $t = 0$, and this implies that at the state $(\sigma_0, \varepsilon_0)$ equation (1.6) has the *creep property*. Taking now a state $(\sigma_0, \varepsilon_0) \in \mathscr{D}$ such that $G(\sigma_0, \varepsilon_0) < 0$ and a strain history $\varepsilon(t) = \varepsilon_0$ for all $t \in [0, T]$, (1.6) gives us $\dot{\sigma}(t) = G(\sigma(t), \varepsilon_0)$, hence σ is decreasing in a neighbourhood of $t = 0$; it means that (1.6) has the *relaxation property* at the state $(\sigma_0, \varepsilon_0)$.

If one gives strain history $\varepsilon(t)$ with $\varepsilon(0) = \tilde{\varepsilon} \neq \bar{\varepsilon}$ such that at the state $(\tilde{\sigma}, \tilde{\varepsilon})$ the strain jumps in time from $\tilde{\varepsilon}$ to $\bar{\varepsilon}$, then the initial condition $\sigma(0+) = \bar{\sigma}$

to be attached to equation (1.6) will also have the property $\bar{\sigma} \neq \tilde{\sigma}$. All such states $(\bar{\sigma}, \bar{\varepsilon}) \in \mathscr{D}$ will be called states of instantaneous response relative to the state $(\tilde{\sigma}, \tilde{\varepsilon}) \in \mathscr{D}$, and they satisfy the equation $\bar{\sigma} = \tilde{\sigma} + E(\bar{\varepsilon} - \tilde{\varepsilon})$. This straight line is called the instantaneous response straight line at the state $(\tilde{\sigma}, \tilde{\varepsilon})$.

The above-mentioned properties show an essential difference between rate-type constitutive equations and the classical models of plasticity (for more details see Cristescu and Suliciu (1982, Ch. II).

The first constitutive equation (1.6) (with $G(\sigma, \varepsilon) = -\sigma$) was proposed by Maxwell in 1867 and this explains why certain authors use the term Maxwellian materials, for all materials that can be described by a constitutive equation (1.1). However, almost a century later it became clear that a constitutive equation (1.1) may also describe the viscoplastic properties of a material.

Subsequently we present some concrete semilinear constitutive equations of the form (1.6):

Example 1.1 (proposed by Sokolovski) $\mathscr{D} = \mathbb{R}^2$,

$$G(\sigma, \varepsilon) = \begin{cases} 0 & \text{if } |\sigma| < \sigma_Y \\ (-\operatorname{sign} \sigma) F(|\sigma| - \sigma_Y) & \text{if } |\sigma| \geq \sigma_Y, \end{cases}$$

where $\sigma_Y > 0$ is the *yield limit* and $F: \mathbb{R}_+ \to \mathbb{R}$ is a smooth function with $F(0) = 0$, $F'(r) > 0$ for all $r \in \mathbb{R}_+$. The simplest example of such a function considered in the literature is $F(r) = kEr$, where $k > 0$ is a *viscosity constant*.

Example 1.2 (Malvern's model) $\mathscr{D} = \{(\sigma, \varepsilon) \in \mathbb{R}^2 | \sigma \geq 0, \varepsilon \geq 0\}$,

$$G(\sigma, \varepsilon) = \begin{cases} 0 & \text{if } 0 \leq \sigma \leq f(\varepsilon) \\ -kF(\sigma - f(\varepsilon)) & \text{if } \sigma > f(\varepsilon) \end{cases}$$

where F has the same properties as the function F in Example 1.1 and $k > 0$ is a *viscosity constant*. The continuous curve $\sigma = f(\varepsilon)$, $\varepsilon \in \mathbb{R}_+$ is called the quasistatic loading curve for the state (0, 0). In most cases on the interval $[0, \varepsilon_Y = \sigma_Y/E]$ (where σ_Y is the yield stress) $f(\varepsilon)$ may be identified with $E\varepsilon$. The role played by the function f is given by the following property: the material behaves elastically for $\sigma < f(\varepsilon)$ and any loading process with relatively small strain rates leads to a $(\sigma(t), \varepsilon(t))$ curve that is close to $(\sigma = f(\varepsilon), \varepsilon)$. Here are some examples of functions F considered in the literature: $F(r) = r$, $F(r) = e^{\lambda r} - 1$, where $\lambda > 0$, $F(r) = (r/a)^n$, where $a > 0$, $n > 0$.

Example 1.3 (Cristescu and Suliciu, 1982, p. 35), $\mathscr{D} = \mathbb{R}^2$,

$$G(\sigma, \varepsilon) = \begin{cases} -k_1 F_1(\sigma - f(\varepsilon)) & \text{if } \sigma \geq f(\varepsilon) \\ 0 & \text{if } g(\varepsilon) < \sigma < f(\varepsilon) \\ k_2 F_2(g(\varepsilon) - \sigma) & \text{if } \sigma \leq g(\varepsilon) \end{cases}$$

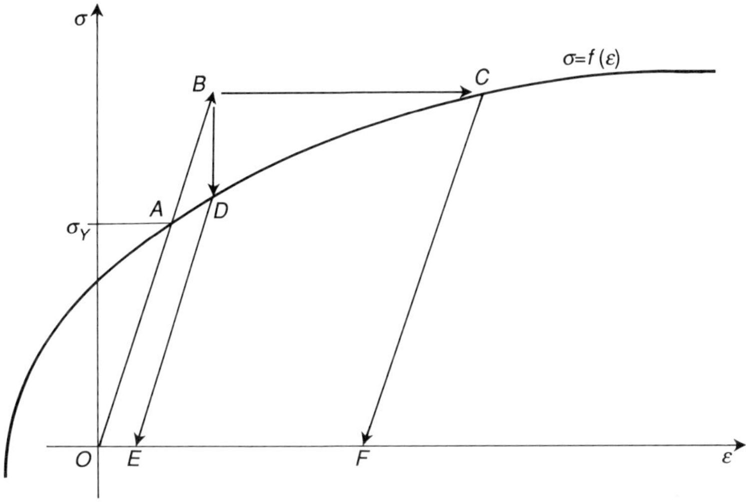

Fig. 1.1 Loading and unloading processes for materials with a constitutive equation given in Example 1.3.

where $k_1, k_2 > 0$ are two viscosity constants, the smooth functions F_1, F_2: $\mathbb{R}_+ \to \mathbb{R}$ have the properties $F_1(r)$, $F_2(r) > 0$, $F'_1(r)$, $F'_2(r) > 0$ for $r > 0$ and the smooth functions f, g: $\mathbb{R} \to \mathbb{R}$ satisfy the conditions $f(\varepsilon) > g(\varepsilon)$, $f(0) > 0 > g(0)$, $E > f'(\varepsilon) \geq 0$, $E > g'(\varepsilon) \geq 0$, $f''(\varepsilon) \leq 0$, $g''(\varepsilon) \geq 0$ for all $\varepsilon \in \mathbb{R}_+$. Note that the present case includes Sokolovski's constitutive equation as well as Maxwell's.

Using Figure 1.1 let us emphasize in the following the basic mechanical phenomena which can be described by the above model. Every loading that starts from 0 contains the elastic line OA. Let us imagine now that we have an instantaneous loading with $\sigma > \sigma_Y$, so that we have an elastic response and we reach point B. Further on, if we maintain constant stress, the strain will increase in time (the creep phenomenon) until point C. If we unload from C, then the behaviour of the material is elastic, and we reach point F. We remark that in this way we obtain an irreversible strain $|OF|$. If we start again from B and we keep a constant strain, then the stress decreases in time (relaxation) until point D. If we unload from D the material behaves in an elastic way and we reach point E. We remark again the irreversible strain $|OE|$.

In the three-dimensional case, a classical example of the elastic–viscoplastic constitutive equation is:

Example 1.4 $\mathscr{D} = \mathbb{R}^3 \times \mathbb{R}^3$ and let $K \subset \mathscr{S}_3$ be a closed convex set such that $0_{\mathscr{S}_3}$ belongs to the interior of K and let P_K: $\mathscr{S}_3 \to K$ be the projector map on K. We define $\tilde{G}$: $\mathscr{S}_3 \to \mathscr{S}_3$ as follows:

$$\tilde{G}(\sigma) = \frac{1}{\mu}(\sigma - P_K\sigma). \tag{1.7}$$

where $\mu > 0$ is a *viscosity constant*. Considering an invertible fourth-order tensor $\mathscr{A}$ we obtain the following equation of the form (1.3):

$$\dot{\varepsilon} = \tilde{\mathscr{A}}\dot{\sigma} + \tilde{G}(\sigma) \tag{1.8}$$

where $\tilde{\mathscr{A}} = \mathscr{A}^{-1}$.

Since $\sigma = P_K\sigma$ iff $\sigma \in K$, from (1.5), (1.7) we obtain that in the case of (1.8), viscoplastic deformations occur only for stress tensors σ which do not belong to K.

The convex set K in the previous example is usually defined by a convex and continuous function $\mathscr{F}: \mathscr{S}_3 \to \mathbb{R}$ which satisfies $\mathscr{F}(0_3) < 0$, as follows:

$$K = \{\sigma \in \mathscr{S}_3 \,|\, F(\sigma) \leq 0\}. \tag{1.9}$$

The function $\mathscr{F}$ is called the *yield function* and equation $\mathscr{F}(\sigma) = 0$ is called the *yield condition*. Concrete examples of yield functions are given by

$$\mathscr{F}(\sigma) = \tfrac{1}{2}\sigma' \cdot \sigma' - k^2 \qquad \text{(von Mises)}, \tag{1.10}$$

$$\mathscr{F}(\sigma) = \max_{i,j=1,2,3} |\sigma_i - \sigma_j| - k^2 \qquad \text{(Tresca)}, \tag{1.11}$$

$$\mathscr{F}(\sigma) = \max(|2\sigma_1 - \sigma_2 - \sigma_3|, |2\sigma_2 - \sigma_3 - \sigma_1|, |2\sigma_3 - \sigma_1 - \sigma_2|) - k^2 \qquad \text{(Yang)} \tag{1.12}$$

The notation in (1.10)–(1.12) has the following meaning:

- σ' is the deviator of σ, defined by $\sigma' = \sigma - (\operatorname{tr}\sigma/3)I_3$, where $\operatorname{tr}\sigma$ is the trace of σ and I_3 is the identity map of $\mathscr{S}_3$;
- σ_i are the eigenvalues of σ, $i = 1, 2, 3$;
- k is a strictly positive constant, called the *yield limit*.

In the particular case of (1.10), the convex set K defined by (1.9) becomes the von Mises convex set given by

$$K = \{\sigma \in \mathscr{S}_3 \,|\, \sqrt{(\sigma_{\mathrm{II}})} \leq k\}, \tag{1.13}$$

where for all tensors $\sigma \in \mathscr{S}_3$ we denote $\sigma_{\mathrm{II}} = \frac{1}{2}\sigma' \cdot \sigma'$.

In this particular case we can give the explicit expression of the projector map on K:

Lemma 1.1 *Let $K \subset \mathscr{S}_3$ be the convex set defined by* (1.13). *Then, for every $\sigma \in \mathscr{S}_3$ we have:*

$$P_K\sigma = \begin{cases} \sigma & \text{if } \sigma \in K \\ \dfrac{\operatorname{tr}\sigma}{3} I_3 + \dfrac{k}{\sqrt{(\sigma_{\mathrm{II}})}}\sigma' & \text{if } \sigma \notin K. \end{cases} \tag{1.14}$$

PROOF. Let $\sigma \in \mathscr{S}_3$ and $\tilde{\sigma} \in \mathscr{S}_3$ be the tensor defined by

$$\tilde{\sigma} = \begin{cases} \sigma & \text{if } \sigma \in K \\ \dfrac{\operatorname{tr}\sigma}{3} I_3 + \dfrac{k}{\sqrt{(\sigma_{\mathrm{II}})}} \sigma' & \text{if } \sigma \notin K. \end{cases} \tag{1.15}$$

Using the variational characterization of the projector map we have

$$\tilde{\sigma} = P_K \sigma \quad \text{iff } \tilde{\sigma} \in K \quad \text{and} \quad (\sigma - \tilde{\sigma})\cdot(\tau - \tilde{\sigma}) \leq 0 \qquad \forall \tau \in K. \tag{1.16}$$

The equality (1.14) follows now from (1.15) and (1.16). More precisely, if $\sigma \notin K$ from (1.15) we have $\sqrt{(\tilde{\sigma}_{\mathrm{II}})} = k$, hence $\tilde{\sigma} \in K$ by (1.13); moreover, for every $\tau \in K$, using the Cauchy–Schwarz inequality, we get

$$\begin{aligned}(\sigma - \tilde{\sigma})\cdot(\tau - \tilde{\sigma}) &= \left(1 - \frac{k}{\sqrt{(\sigma_{\mathrm{II}})}}\right)\sigma' \cdot \left(\tau - \frac{\operatorname{tr}\sigma}{3} I_3 - \frac{k}{\sqrt{(\sigma_{\mathrm{II}})}}\sigma'\right) \\ &= \left(1 - \frac{k}{\sqrt{(\sigma_{\mathrm{II}})}}\right)(\sigma'\cdot\tau' - 2k\sqrt{(\sigma_{\mathrm{II}})}) \\ &\leq 2\left(1 - \frac{k}{\sqrt{(\sigma_{\mathrm{II}})}}\right)\sqrt{(\sigma_{\mathrm{II}})}\cdot(\sqrt{(\tau_{\mathrm{II}})} - k) \leq 0.\end{aligned}$$

Hence $\tilde{\sigma} = P_k \sigma$ by (1.16). This last equality holds obviously in the case $\sigma \in K$. Hence $P_K \sigma = \tilde{\sigma}$ and using (1.15) we get (1.14).

Remark 1.1 *Using the above lemma, the function* $\tilde{G}$ *in* (1.7) *can be computed in the case of the von Mises plasticity convex set*:

$$\tilde{G}(\sigma) = \begin{cases} 0 & \text{if } \sqrt{(\sigma_{\mathrm{II}})} \leq k \\ \dfrac{1}{\mu}\left(1 - \dfrac{k}{\sqrt{(\sigma_{\mathrm{II}})}}\right)\sigma' & \text{if } \sqrt{(\sigma_{\mathrm{II}})} > k. \end{cases} \tag{1.17}$$

Remark 1.2 *In the following sections we shall consider also semilinear rate-type constitutive equations of the form*

$$\dot{\sigma} = \mathscr{A}\dot{\varepsilon} + G(t, \sigma, \varepsilon) \tag{1.18}$$

or, equivalently

$$\dot{\varepsilon} = \tilde{\mathscr{A}}\dot{\sigma} + \tilde{G}(t, \sigma, \varepsilon), \tag{1.19}$$

for which the viscoplastic rate of strain depends also on time by virtue of the temperature field, denoted by θ. *This field can be easily obtained from the heat equation in the case of uncoupled thermomechanical processes.*

1.2 Statement of the problem

Let Ω be a bounded domain in $\mathbb{R}^N$ ($N = 1, 2, 3$) with a smooth boundary $\partial\Omega = \Gamma$, and let Γ_1 be an open subset of Γ. Let $\Gamma_2 = \Gamma \backslash \bar{\Gamma}_1$ and suppose meas $\Gamma_1 > 0$. Let us consider the following mixed problem:

Find the displacement function $u: \mathbb{R}_+ \times \Omega \to \mathbb{R}^N$ and the stress function $\sigma: \mathbb{R}_+ \times \Omega \to \mathcal{S}_N$ such that

$$\operatorname{Div} \sigma(t) + b(t) = 0 \tag{1.20}$$

$$\dot{\sigma}(t) = \mathcal{A}\varepsilon(\dot{u}(t)) + G(t, \sigma(t), \varepsilon(u(t))) \qquad \text{in } \Omega \tag{1.21}$$

$$u(t) = g(t) \text{ on } \Gamma_1, \quad \sigma(t)\nu = f \qquad \text{on } \Gamma \text{ for } t > 0, \tag{1.22}$$

$$u(0) = u_0, \quad \sigma(0) = \sigma_0 \qquad \text{in } \Omega. \tag{1.23}$$

Problem (1.20)–(1.23) represents a quasistatic problem in the study of semilinear rate-type materials having the constitutive equation (1.21), already discussed in Section 1.1 (see also Remark 1.2). Equation (1.20) represents the equilibrium equation in which b is the given body force (see Section 2 in Chapter 1); the functions g and f in (1.22) are the given boundary data and finally the functions u_0, σ_0 in (1.23) are the initial data.

In the study of problem (1.20)–(1.23) the following assumptions are made:

$\mathcal{A}: \Omega \to B(\mathcal{S}_N)$ is a symmetric and positively defined bounded tensor, i.e.

(1.24)
(a) there exists $Q > 0$ such that $|\mathcal{A}(x)\sigma| \leq Q|\sigma|$ for all $\sigma \in \mathcal{S}_N$, a.e. in Ω
(b) $\mathcal{A}(x)\sigma \cdot \tau = \sigma \cdot \mathcal{A}(x)\tau$ for all $\sigma, \tau \in \mathcal{S}_N$, a.e. in Ω
(c) there exists $\alpha > 0$ such that $\mathcal{A}(x)\sigma \cdot \sigma \geq \alpha|\sigma|^2$ for all $\sigma \in \mathcal{S}_N$, a.e. in Ω;

$G: \Omega \times \mathbb{R}_+ \times \mathcal{S}_N \times \mathcal{S}_N \to \mathcal{S}_N$ has the following properties:

(1.25)
(a) there exists $L > 0$ and $\theta \in C(\mathbb{R}_+, H)$ such that

$$|G(x, t_1, \sigma_1, \varepsilon_1) - G(x, t_2, \sigma_2, \varepsilon_2)| \leq L(|\theta(x, t_1) - \theta(x, t_2)| + |\sigma_1 - \sigma_2| + |\varepsilon_1 - \varepsilon_2|)$$
$$\forall t_1, t_2 \in \mathbb{R}_+, \sigma_1, \sigma_2, \varepsilon_1, \varepsilon_2 \in \mathcal{S}_N \text{ a.e.in } \Omega.$$

(b) $x \to G(x, t, \sigma, \varepsilon)$ is a measurable function with respect to the Lebesgue measure on Ω, for all $t \in \mathbb{R}_+$, $\sigma, \varepsilon \in \mathcal{S}_N$
(c) $x \to G(x, t, 0_N, 0_N) \in \mathcal{H} \quad \forall t \in \mathbb{R}_+$

$$b \in C^1(\mathbb{R}_+, \boldsymbol{H}) \tag{1.26}$$

$$g \in C^1(\mathbb{R}_+, \boldsymbol{H}_\Gamma), \qquad f \in C^1(\mathbb{R}_+, \boldsymbol{V}'_{\Gamma_1}) \tag{1.27}$$

$$u_0 \in \boldsymbol{H}_1, \qquad \sigma_0 \in \mathcal{H}_2 \tag{1.28}$$

$$\operatorname{Div} \sigma_0 + b(0) = 0 \text{ in } \Omega, \qquad u_0 = g(0) \text{ on } \Gamma_1, \qquad \sigma_0\nu = f(0) \text{ on } \Gamma_2. \tag{1.29}$$

Concerning the constitutive assumptions (1.24)–(1.25), note the following:

Remark 1.3 (1) (1.24) *shows that the tensor* $\mathscr{A}$ *may depend on spatial coordinate* $x \in \Omega$ *and may be a non-isotropic tensor;*

(2) (1.25) (*a*) *represents a Lipschitz-type assumption on the function G and is useful in applying existence results for abstract differential equations in Hilbert spaces involving Lipschitz operators;* (1.25) (*b, c*) *have a technical purpose. Hence, the assumption of* (1.25) (*a, b, c*) *is sufficient to guarantee the following property:* $\sigma \in \mathscr{H}, \varepsilon \in \mathscr{H} \Rightarrow x \to G(x, t, \sigma(x), \varepsilon(x)) \in \mathscr{H}$ *for all* $t \in \mathbb{R}_+$. *Indeed, from* (1.25, *a, b*) *we find that the above-mentioned function has the Carathéodory property; hence it is a measurable function; moreover, the inequality* $\int_\Omega |G(x, t, \sigma(x), \varepsilon(x))|^2 \, dx < +\infty$ *results from* (1.25*a, c*) *since* $\sigma \in \mathscr{H}$ *and* $\varepsilon \in \mathscr{H}$;

(3) *the constitutive assumption* (1.25) *is satisfied for the function G in Example* 1.3 *if* F_1 *and* F_2 *are Lipschitz functions such that* $F_1(0) = F_2(0) = 0$.

Moreover, this hypothesis holds also for the function $\tilde{G}$ *given by* (1.7); *for this, it is sufficient to have in mind the non-expansivity property of the projector map.*

(4) *Bearing in mind that the constitutive function G is obtained by interpolation of the experimental measurements in a bounded domain of states* $(\sigma, \varepsilon) \in \mathscr{S}_N \times \mathscr{S}_N$, *the constitutive assumption* (1.25) *is reasonable for almost all the elastic–viscoplastic materials. However, this assumption is not appropriate if we intend to model special phenomena for which a local Lipschitz property seems to be more realistic.*

Concerning the regularity hypotheses (1.26)–(1.29) on the data, note the following:

Remark 1.4 (1) (1.26) *and* (1.27) *allow us to consider an auxiliary elastic problem for which the existence and uniqueness of the solution can be proved* (*see Lemmas* 1.2 *and* (1.40)).

(2) (1.28) *gives the regularity of the initial data and finally* (1.29) *ensures the compatibility between the initial data and the boundary data.*

1.3 An existence and uniqueness result

The main result of this section is the following:

Theorem 1.1 *Suppose that* (1.24)–(1.29) *hold. Then, there exists a unique solution of the problem* (1.20)–(1.23) *such that*

$$u \in C^1(\mathbb{R}_+, \boldsymbol{H}_1) \tag{1.30}$$

$$\sigma \in C^1(\mathbb{R}_+, \mathscr{H}_2). \tag{1.31}$$

Remark 1.5 *Note that if the problem* (1.20)–(1.23) *has a solution* (u, σ) *having the regularity* (1.30), (1.31), *then the hypotheses* (1.26)–(1.29) *are fulfilled. Indeed, if we put* $h = \gamma_1 u$, *then from* (1.30) *and the continuity of* γ_1 *we* *get* $h \in C^1(0, T, \boldsymbol{H}_\Gamma)$ *and by* (1.22) *we have* $h = g$ *on* $\Gamma_1 \times (0, T)$; *so the function* g *is extended by* h *and we may consider* $g = h \in C^1(0, T, \boldsymbol{H}_\Gamma)$; *since* $\sigma \in C^1(0, T, \mathscr{H}_2)$ *from* (1.20) *we get* $b \in C^1(0, T, \boldsymbol{H})$ *and since* $\tau \to \tau v|_{\Gamma_2}$ *is a linear and continuous map from* $\mathscr{H}_2$ *to* V'_{Γ_1}, $(1.22)_2$ *implies* $(1.27)_2$. *Finally, taking the limit in* (1.20) *and* (1.22) *when* $t \to 0$ *and using* (1.23) *we get* (1.29).

In order to prove Theorem 1.1 we need some preliminary results:

Lemma 1.2 *Let* (1.24) *hold. Then there exists an operator*

$$D \in B(\boldsymbol{H} \times \boldsymbol{H}_\Gamma \times V'_{\Gamma_1}, \boldsymbol{H}_1 \times \mathscr{H}_2)$$

such that for all $\bar{b} \in \boldsymbol{H}$, $\bar{g} \in \boldsymbol{H}_\Gamma$, $\bar{f} \in V'_{\Gamma_1}$ *the couple* $(v; \tau) = D(\bar{b}, \bar{g}, \bar{f})$ *is the unique solution of the following elastic problem*:

$$\operatorname{Div} \tau + \bar{b} = 0 \tag{1.32}$$

$$\tau = \mathscr{A}\varepsilon(v) \qquad \text{in } \Omega \tag{1.33}$$

$$v = \bar{g} \text{ on } \Gamma_1, \qquad \tau v = \bar{f} \text{ on } \Gamma_2. \tag{1.34}$$

PROOF. Let $\hat{v} = z_1(g)$ and let us consider a: $\boldsymbol{V}_1 \times \boldsymbol{V}_1 \to \mathbb{R}$ the bilinear form given by

$$a(u, w) = \langle \mathscr{A}\varepsilon(u), \varepsilon(w)\rangle_{\mathscr{H}} \tag{1.35}$$

and $l \in V'_{\Gamma_1}$ given by

$$l(w) = -\langle \mathscr{A}\varepsilon(\hat{v}), \varepsilon(w)\rangle_{\mathscr{H}} + \langle \bar{b}, w\rangle_{\boldsymbol{H}} + \langle \bar{f}, \gamma_1(w)\rangle_{V'_{\Gamma_1}, V_{\Gamma_1}} \tag{1.36}$$

for all $w \in \boldsymbol{V}_1$. From (1.24) and (2.32) of Chapter 2, we deduce that a is a symmetric and positive definite, hence there exists (see Corollary A2.6) a unique $\tilde{v} \in \boldsymbol{V}_1$ such that

$$a(\tilde{v}, w) = l(w) \qquad \forall w \in \boldsymbol{V}_1. \tag{1.37}$$

Let $v = \hat{v} + \tilde{v}$ and $\tau = \mathscr{A}\varepsilon(v)$; then from (1.37) we get

$$\langle \tau, \varepsilon(w)\rangle_{\mathscr{H}} = \langle \bar{b}, w\rangle_{\boldsymbol{H}} + \langle \bar{f}, \gamma_1(w)\rangle_{V'_{\Gamma_1}, V_{\Gamma_1}} \tag{1.38}$$

for all $w \in V_{\Gamma_1}$. If we take $w \in [D(\Omega)]^N$ in (1.38), then we get (1.32). From (1.32), (1.38), and (2.44) of Chapter 2 we can deduce $\tau v|_{\Gamma_2} = f$. If we take $w = \tilde{v}$ in (1.37), then we obtain

$$\|\tilde{v}\|_{\boldsymbol{H}_1} \le C(\|\hat{v}\|_{\boldsymbol{H}_1} + \|\bar{b}\|_{\boldsymbol{H}} + \|\bar{f}\|_{V'_{\Gamma_1}}),$$

where C is a generic constant which depends only on Ω, Γ_1 and $\mathscr{A}$. If we have in mind that $\|\hat{v}\|_{\boldsymbol{H}_1} \le C\|\bar{g}\|_{\boldsymbol{H}_\Gamma}$, then we get

$$\|v\|_{\boldsymbol{H}_1} \le C(\|\bar{g}\|_{\boldsymbol{H}_\Gamma} + \|\bar{b}\|_{\boldsymbol{H}} + \|\bar{f}\|_{V'_{\Gamma_1}})$$

and

$$\|\tau\|_{\mathscr{H}_2} = (\|\tau\|^2_{\mathscr{H}} + \|\operatorname{Div} \tau\|^2_{\boldsymbol{H}})^{1/2} \le C(\|v\|_{\boldsymbol{H}_1} + \|\bar{b}\|_{\boldsymbol{H}}).$$

We have just obtained that

$$\|v\|_{\boldsymbol{H}_1} + \|\tau\|_{\mathscr{H}_2} \le C(\|\bar{g}\|_{\boldsymbol{H}} + \|\bar{b}\|_{\boldsymbol{H}} + \|\bar{f}\|_{V'_{\Gamma_1}});$$

hence $D \in B(\boldsymbol{H} \times \boldsymbol{H}_\Gamma \times V'_{\Gamma_1}, \boldsymbol{H}_1 \times \mathscr{H}_2)$.

Let $X = V_1 \times \mathscr{V}_2$, endowed with the following scalar product:

$$\langle x, y\rangle_X = \langle \mathscr{A}\varepsilon(u), \varepsilon(v)\rangle_{\mathscr{H}} + \langle \mathscr{A}^{-1}\sigma, \tau\rangle_{\mathscr{H}} \tag{1.39}$$

for all $x = (u; \sigma)$, $y = (v; \tau) \in X$. The norm $\| \ \|_X$ generated by (1.39) is equivalent with the natural norm of X (see (1.24) and Chapter 2, 2.32)). Let us also consider $\tilde{u}$: $\mathbb{R}_+ \to H_1$, and $\tilde{\sigma}$: $\mathbb{R}_+ \to \mathscr{H}_2$ given by

$$(\tilde{u}(t); \tilde{\sigma}(t)) = D(b(t), g(t), f(t)) \qquad \forall t \in \mathbb{R}_+, \tag{1.40}$$

where D is given by Lemma 1.2. From (1.26), (1.27) we can deduce that

$$\tilde{u} \in C^1(\mathbb{R}_+, H_1), \qquad \tilde{\sigma} \in C^1(\mathbb{R}_+, \mathscr{H}_2) \tag{1.41}$$

and from (1.32) to (1.34) we have

$$\operatorname{Div} \tilde{\sigma}(t) + b(t) = 0 \tag{1.42}$$

$$\tilde{\sigma}(t) = \mathscr{A}\varepsilon(\tilde{u}(t)) \qquad \text{in } \Omega \tag{1.43}$$

$$\tilde{u}(t) = g(t) \quad \text{on } \Gamma_1, \qquad \tilde{\sigma}(t)\nu = f(t) \quad \text{on } \Gamma_2 \tag{1.44}$$

for all $t \in \mathbb{R}_+$.

The next step in the proof of Theorem 1.1 is to homogenize the boundary conditions; denoting by $u^* = u - \tilde{u}$, $\sigma^* = \sigma - \tilde{\sigma}$, $u_0^* = u_0 - \tilde{u}(0)$, $\sigma_0^* = \sigma_0 - \sigma(0)$, one can easily deduce the following lemma:

Lemma 1.3 *The pair* $(u; \sigma)$ *is a solution of* (1.20)–(1.23) *such that* (1.30), (1.31) *hold iff* $u^* \in C^1(\mathbb{R}_+, V_1)$, $\sigma^* \in C^1(\mathbb{R}_+, V_2)$ *and*

$$\dot{\sigma}^*(t) = \mathscr{A}\varepsilon(\dot{u}^*(t)) + G(t, \sigma^*(t) + \tilde{\sigma}(t), \varepsilon(u^*(t)) + \varepsilon(\tilde{u}(t))) \qquad \text{in } \Omega, \quad \text{for } t > 0 \tag{1.45}$$

$$u^*(0) = u_0^*, \qquad \sigma^*(0) = \sigma_0^* \qquad \text{in } \Omega. \tag{1.46}$$

Let us consider A: $\mathbb{R}_+ \times X \to X$, the following operator defined by Riesz's representation theorem by

$$\begin{aligned}\langle A(t, x), y\rangle_X = & -\langle G(t, \sigma + \tilde{\sigma}(t), \varepsilon(u) + \tilde{\varepsilon}(u(t))), \varepsilon(v)\rangle_{\mathscr{H}} \\ & + \langle \mathscr{A}^{-1}G(t, \sigma + \tilde{\sigma}(t), \varepsilon(v) + \varepsilon(\tilde{u}(t))), \tau\rangle_{\mathscr{H}}\end{aligned} \tag{1.47}$$

for all $x = (u; \sigma) \in X$, $y = (v; \tau) \in X$ and $t \in \mathbb{R}_+$.

We use the notation

$$x_0 = (u_0^*; \sigma_0^*) \tag{1.48}$$

and we note the following result:

Lemma 1.4 *Let* $x = (u^*; \sigma^*) \in C^1(\mathbb{R}_+, X)$; *hence* x *is a solution for* (1.45), (1.46) *iff* x *is a solution for the following Cauchy problem on* X:

$$\dot{x}(t) = A(t, x(t)) \qquad \forall t > 0 \tag{1.49}$$

$$x(0) = x_0. \tag{1.50}$$

PROOF. Let x be a solution for (1.45), (1.46) and let $y = (v; \tau) \in X$. Using (2.59) of Chapter 2, from (1.45) we get

$$\langle \mathscr{A}\varepsilon(\dot{u}^*(t)), \varepsilon(v)\rangle_{\mathscr{H}} = -\langle G(t, \sigma^*(t) + \tilde{\sigma}(t), \varepsilon(u^*(t)) + \varepsilon(\tilde{u}(t))), \varepsilon(v)\rangle_{\mathscr{H}}$$

$$\langle \mathscr{A}^{-1}\dot{\sigma}^*(t), \tau\rangle_{\mathscr{H}} = \langle \mathscr{A}^{-1}G(t, \sigma^*(t) + \tilde{\sigma}(t), \varepsilon(u^*(t)) + \tilde{\varepsilon}(u(t))), \tau\rangle_{\mathscr{H}}$$

for all $t > 0$. By adding these equalities and using (1.39), (1.47) we get

$$\langle \dot{x}(t), y\rangle_X = \langle A(t, x(t)), y\rangle_X \tag{1.51}$$

for all $t > 0$. Since y is an arbitrary element of X, (1.51) implies (1.49) and (1.50) results from (1.46) and (1.48).

Conversely, let us suppose that x is a solution for the Cauchy problem (1.49)–(1.50); for every $t > 0$ let $z(t) \in \mathscr{H}$ be given by

$$z(t) = \dot{\sigma}^*(t) - \mathscr{A}\varepsilon(\dot{u}^*(t)) - G(t, \sigma^*(t) + \tilde{\sigma}(t), \varepsilon(u^*(t)) + \varepsilon(\tilde{u}(t))). \tag{1.52}$$

Taking $y = (v; 0) \in X$ in (1.51) and using (2.59) of Chapter 2, we get

$$\langle z(t), \varepsilon(v)\rangle_{\mathscr{H}} = 0 \qquad \forall v \in V_1 \quad \text{and} \quad t > 0. \tag{1.53}$$

If we put $y = (0; \tau) \in X$ in (1.51) and we again use (2.59) of Chapter 2, we obtain

$$\langle \mathscr{A}^{-1}z(t), \tau\rangle_{\mathscr{H}} = 0 \qquad \forall \tau \in \mathscr{V}_2 \quad \text{and} \quad t > 0. \tag{1.54}$$

Using now Lemma 2.1 of Chapter 2, from (1.53) we get $z(t) \in \mathscr{V}_2$ for all $t > 0$ and taking $\tau = z(t)$ in (1.54), by (1.24) we deduce $z(t) = 0$ for all $t > 0$. Hence, (1.45) holds and (1.46) results from (1.50) and (1.48).

Using Lemma 1.4, the problem (1.45), (1.46) was replaced by the Cauchy problem (1.49), (1.50) in the Hilbert space X. In order to prove the existence and the uniqueness of the solution for this problem, let us study the properties of the operator A given by (1.47):

Lemma 1.5 *The operator A given by* (1.47) *is continuous and there exists $L > 0$ such that*

$$\|A(t, x_1) - A(t, x_2)\|_X \leq L\|x_1 - x_2\|_X \qquad \forall x_1, x_2 \in X \quad \textit{and} \quad t \in \mathbb{R}_+. \tag{1.55}$$

PROOF. Let $t_1, t_2 \in \mathbb{R}_+$ and $x_1 = (u_1; \sigma_1)$, $x_2 = (u_2; \sigma_2) \in X$. For all $y = (v; \tau) \in X$ we have

$$|\langle A(t_1, x_1) - A(t_2, x_2), v\rangle_X| \leq$$
$$|\langle G(t_1, \sigma_1 + \tilde{\sigma}(t_1), \varepsilon(u_1) + \varepsilon(\tilde{u}(t_1))) - G(t_2, \sigma_2 + \tilde{\sigma}(t_2), \varepsilon(u_2) + \varepsilon(\tilde{u}(t_2))), \varepsilon(v)\rangle_{\mathscr{H}}|$$
$$+ |\langle \mathscr{A}^{-1}G(t_1, \sigma_1 + \tilde{\sigma}(t_1), \varepsilon(u_1) + \varepsilon(\tilde{u}(t_1))) - \mathscr{A}^{-1}G(t_2, \sigma_2 + \tilde{\sigma}(t_2), \varepsilon(u_2) + \varepsilon(\tilde{u}(t_2))), \tau\rangle_{\mathscr{H}}|$$

Using (1.24) and (1.25) in the above inequality, after some estimations we get

$$|\langle A(t_1, x_1) - A(t_2, x_2), y\rangle_X| \leq C(\|\theta(t_1) - \theta(t_2)\|_H + \|x_1 - x_2\|_X + \|\tilde{\sigma}(t_1) - \tilde{\sigma}(t_2)\|_{\mathscr{H}} + \|\tilde{u}(t_1) - \tilde{u}(t_2)\|_{H_1})\|y\|_X;$$

hence

$$\|A(t_1, x_1) - A(t_2, x_2)\|_X \leq C(\|\theta(t_1) - \theta(t_2)_H + \|x_1 - x_2\|_X + \|\tilde{\sigma}(t_1) - \tilde{\sigma}(t_2)\|_{\mathscr{H}} + \|\tilde{u}(t_1) - \tilde{u}(t_2)\|_{H_1}) \quad (1.56)$$

The continuity of A follows from (1.56), (1.41) and the regularity $\theta \in C^0(\mathbb{R}_+, H)$ (see (1.25a)). Moreover, taking $t_1 = t_2 = t \in \mathbb{R}_+$ from (1.56) we get (1.55).

PROOF OF THEOREM 1.1. Let us notice that from (1.48), (1.46), (1.28), and (1.29) we deduce $x_0 \in X$. Since A has the properties given in Lemma 1.5 we can apply Theorem A.3.2. Hence follow the existence and uniqueness of the solution for the problem (1.49), (1.50) with the regularity $x \in C^1(\mathbb{R}_+, X)$. Hence, Theorem 1.1 follows from Lemmas 1.3, 1.4.

Remark 1.6 *Theorem* 1.1 *can be stated even under weaker assumptions on the function G; namely,* (1.25(*a*)) *can be replaced by*

(i) $G(x, \cdot, \cdot, \cdot)\colon \mathbb{R}_+ \times \mathscr{S}_N \times \mathscr{S}_N \to \mathscr{S}_N$ *is a continuous function a.e.* $x \in \Omega$.
(ii) *there exist two non-decreasing functions* $L_1, L_2\colon \mathbb{R}_+ \to \mathbb{R}_+$ *such that*

$$|G(x, t, \sigma, \varepsilon)|^2 \leq L_1(t) + L_2(t)(|\sigma|^2 + |\varepsilon|^2) \qquad \forall t \in \mathbb{R}_+, \sigma, \varepsilon \in \mathscr{S}_N, \text{ a.e. } x \in \Omega.$$

(iii) *There exists* $L_3 > 0$ *such that*

$$-G(x, t, \sigma_1, \varepsilon_1) - G(x, t, \sigma_2, \varepsilon_2)\cdot(\varepsilon_1 - \varepsilon_2) + \mathscr{A}^{-1}(x)\cdot(G(x, t, \sigma_1, \varepsilon_1) - G(x, t, \sigma_2, \varepsilon_2))\cdot(\sigma_1 - \sigma_2) \leq L_3(|\varepsilon_1 - \varepsilon_2|^2 + |\sigma_1 - \sigma_2|^2)$$
$$\forall t \in \mathbb{R}_+, \sigma_1, \sigma_2, \varepsilon_2, \varepsilon_2 \in \mathscr{S}_N, \text{ a.e. in } \Omega.$$

Indeed, (1.25b,c), and (i), (ii) is a sufficient condition in order to have $x \to G(x, t, \sigma(x), \varepsilon(x)) \in \mathscr{H}$ for all $t \in \mathbb{R}_+$, σ, $\varepsilon \in \mathscr{H}$; (i), (ii) assumes the continuity of A and from (iii) we get (A.3.5). Hence, Theorem A.3.3 can be applied instead of Theorem A.3.2.

Remark 1.7 (*i*) *The hypothesis* (1.26), (1.27) *may be replaced by the following ones*: $b \in W^{1,p}(0, T, \boldsymbol{H})$, $\boldsymbol{g} \in W^{1,p}(0, T, \boldsymbol{H}_\Gamma)$, $f \in W^{1,p}(0, T, V'_{\Gamma_1})$ *where* $T > 0$ and $1 \leq p \leq \infty$; *in this case problem* (1.20)–(1.23) *may be considered on* $\Omega \times (0, T)$ *with boundary conditions on* $\Gamma_1 \times (0, T)$ *and* $\Gamma_2 \times (0, T)$. *Theorem* 1.1 *also holds except* (1.30) *and* (1.31) *which must be replaced by the regularity* $u \in W^{1,p}(0, T, \boldsymbol{H}_1)$, $\sigma \in W^{1,p}(0, T, \mathscr{H}_2)$.

1.4 The dependence of the solution upon the input data

In this section two solutions of the problem (1.20)–(1.23) for two different input data are considered. An estimation of the difference of these solutions is given for finite time intervals which give the continuous dependence of the solution upon all input data (Theorem 1.2). In this way, the finite time stability of the solution is obtained (Corollary 1.1).

Theorem 1.2 *Let* (1.24)–(1.25) *hold and let* $(u_i; \sigma_i)$ *be the solution of* (1.20)–(1.23) *for the data* b_i, g_i, f_i, u_{0i}, σ_{0i}, $i = 1, 2$, *which satisfies* (1.26)–(1.29). *Then, for all* $T > 0$ *there exists* $C > 0$ *which depends only on* Ω, Γ_1, T, $\mathscr{A}$ *and* G *such that*

$$\|u_1 - u_2\|_{j,T,\boldsymbol{H}_1} + \|\sigma_1 - \sigma_2\|_{j,T,\mathscr{H}_2} \leq C(\|u_{01} - u_{02}\|_{\boldsymbol{H}_1} + \|\sigma_{01} - \sigma_{02}\|_{\mathscr{H}_2} + \|b_1 - b_2\|_{j,T,\boldsymbol{H}} + \|g_1 - g_2\|_{j,T,\boldsymbol{H}_\Gamma} + \|f_1 - f_2\|_{j,T,\boldsymbol{V}'_{\Gamma 1}}) \quad \text{for } j = 0, 1. \tag{1.57}$$

In order to prove Theorem 1.2, the following result is useful:

Lemma 1.6 *Let* (1.24) *hold and let* $(\tilde{u}_i, \tilde{\sigma}_i)$ *be given by*

$$(\tilde{u}_i(t); \tilde{\sigma}_i(t)) = D(b_i(t); g_i(t); f_i(t)) \tag{1.58}$$

for the data b_i, g_i, f_i, $i = 1, 2$ *which satisfies* (1.26)–(1.27), *where* D *is given in Lemma* 1.2. *Then, there exists* $C > 0$ *which depends only on* Ω, Γ_1, T *and* $\mathscr{A}$ *such that for all* $t \in \mathbb{R}_+$ *we have*

$$\|\tilde{u}_1(t) - \tilde{u}_2(t)\|_{\boldsymbol{H}_1} + \|\tilde{\sigma}_1(t) - \tilde{\sigma}_2(t)\|_{\mathscr{H}_2} \leq C(\|b_1(t) - b_2(t)\|_{\boldsymbol{H}} + \|g_1(t) - g_2(t)\|_{\boldsymbol{H}_\Gamma} + \|f_1(t) - f_2(t)\|_{\boldsymbol{V}'_{\Gamma 1}}) \tag{1.59}$$

$$\|\dot{\tilde{u}}_1(t) - \dot{\tilde{u}}_2(t)\|_{\boldsymbol{H}_1} + \|\dot{\tilde{\sigma}}_1(t) - \dot{\tilde{\sigma}}_2(t)\|_{\mathscr{H}_2} \leq C(\|\dot{b}_1(t) - \dot{b}_2(t)\|_{\boldsymbol{H}} + \|\dot{g}_1(t) - \dot{g}_2(t)\|_{\boldsymbol{H}_\Gamma} + \|\dot{f}_1(t) - \dot{f}_2(t)\|_{\boldsymbol{V}'_{\Gamma 1}}) \quad \forall t \in [0, T] \tag{1.60}$$

PROOF OF THEOREM 1.2. If we put

$$u_i = u_i^* + \tilde{u}_i, \qquad \sigma_i = \sigma_i^* + \tilde{\sigma}_i \tag{1.61}$$

where $(\tilde{u}_i; \tilde{\sigma}_i)$ are given in Lemma 1.2, then (u_i^*, σ_i^*) satisfies the following equalities:

$$(u_i^*; \sigma_i^*) = x_i \in C^1(\mathbb{R}_+, \boldsymbol{X}) \tag{1.62}$$

$$\dot{x}_i(t) = A_i(t, x_i(t)) \qquad \forall t > 0 \tag{1.63}$$

$$x_i(0) = x_{0i} = (u_{0i} - \tilde{u}_i(0); \sigma_{0i} - \tilde{\sigma}_i(0)). \tag{1.64}$$

The operators A_i in (1.63) are given by (1.47), replacing $\tilde{u}$, $\tilde{\sigma}$ by $\tilde{u}_i$, $\tilde{\sigma}_i$, $i = 1, 2$. For every $y_1, y_2 \in X$ and $t \in \mathbb{R}_+$, from (1.47), (1.24), and (1.25) we have

$$\|A_1(t, y_1) - A_2(t, y_2)\|_X \le C(\|y_1 - y_2\|_X + \|\varepsilon(\tilde{u}_1(t)) - \varepsilon(\tilde{u}_2(t))\|_{\mathscr{H}} + \|\tilde{\sigma}_1(t) - \tilde{\sigma}_2(t)\|_{\mathscr{H}}). \tag{1.65}$$

Hence, from (1.63) and (1.65) we deduce

$$\begin{aligned}\langle \dot{x}_1(t) - \dot{x}_2(t), x_1(t) - x_2(t)\rangle_X &= \langle A_1(t, x_1(t)) - A_2(t, x_2(t)), x_1(t) - x_2(t)\rangle_X \\ &\le \|A_1(t, x_1(t)) - A_2(t, x_2(t))\|_X \|x_1(t) - x_2(t)\|_X \\ &\le C(\|x_1(t) - x_2(t)\|_X + \|\varepsilon(\tilde{u}_1(t)) - \varepsilon(\tilde{u}_2(t))\|_{\mathscr{H}} + \|\tilde{\sigma}_1(t) - \tilde{\sigma}_2(t)\|_{\mathscr{H}}) \\ &\quad \cdot \|x_1(t) - x_2(t)\|_X\end{aligned}$$

for all $t \in \mathbb{R}_+$. By integration and using (1.64) we have

$$\begin{aligned}\tfrac{1}{2}\|x_1(s) - x_2(s)\|_X^2 &\le \tfrac{1}{2}\|x_{01} - x_{02}\|_X^2 \\ &\quad + C\int_0^s (\|\varepsilon(\tilde{u}_1(t)) - \varepsilon(\tilde{u}_2(t))\|_{\mathscr{H}} + \|\tilde{\sigma}_1(t) - \tilde{\sigma}_2(t)\|_{\mathscr{H}}) \\ &\quad\quad \cdot \|x_1(t) - x_2(t)\|_X \,\mathrm{d}t \\ &\quad + C\int_0^s \|x_1(t) - x_2(t)\|_X^2 \,\mathrm{d}t \qquad \forall s \in [0, T].\end{aligned}$$

Using Lemma A.4.13 we now obtain

$$\begin{aligned}\|x_1(s) - x_2(s)\|_X \le C(\|x_{01} - x_{02}\|_X + \int_0^s (\|\varepsilon(\tilde{u}_1(t)) - \varepsilon(\tilde{u}_2(t))\|_{\mathscr{H}} \\ + \|\tilde{\sigma}_1(t) - \tilde{\sigma}_2(t)\|_{\mathscr{H}}) \,\mathrm{d}t \qquad \forall s \in [0, T].\end{aligned}$$

Hence

$$\begin{aligned}&\|x_1 - x_2\|_{0,T,X} \\ &\quad \le C(\|x_{01} - x_{02}\|_X + \|\varepsilon(\tilde{u}_1) - \varepsilon(\tilde{u}_2)\|_{0,T,\mathscr{H}} + \|\tilde{\sigma}_1 - \tilde{\sigma}_2\|_{0,T,\mathscr{H}}).\end{aligned} \tag{1.66}$$

From (2.32) of Chapter 2, (1.39), (1.24), (1.62), and (1.64) we obtain

$$\begin{aligned}&\|u_1^* - u_2^*\|_{0,T,\boldsymbol{H}_1} + \|\sigma_1^* - \sigma_2^*\|_{0,T,\mathscr{H}_2} \\ &\le C(\|u_{01} - u_{02}\|_{\boldsymbol{H}_1} + \|\sigma_{01} - \sigma_{02}\|_{\mathscr{H}_2} + \|\tilde{u}_1 - \tilde{u}_2\|_{0,T,\boldsymbol{H}_1} + \|\tilde{\sigma}_1 - \tilde{\sigma}_2\|_{0,T,\mathscr{H}_2}).\end{aligned} \tag{1.67}$$

Using (1.59) and (1.67) we obtain (1.57) for $j = 0$. Moreover, from (1.63) and

(1.65) we deduce

$$\|\dot{x}_1(t) - \dot{x}_2(t)\|_X = \|A_1(t, x_1(t)) - A_2(t, x_2(t)\|_X$$
$$\leq C(\|x_1(t) - x_2(t)\|_X + \|\varepsilon(\tilde{u}_1(t)) - \varepsilon(\tilde{u}_2(t))\|_{\mathscr{H}} + \|\tilde{\sigma}_1(t) - \tilde{\sigma}_2(t)\|_{\mathscr{H}})$$

for all $t \in \mathbb{R}_+$. Using (1.66) we have

$$\|\dot{x}_1 - \dot{x}_2\|_{0,T,X} \leq C(\|x_{01} - x_{02}\|_X + \|\varepsilon(\tilde{u}_1) - \varepsilon(\tilde{u}_2)\|_{0,T,\mathscr{H}}$$
$$+ \|\tilde{\sigma}_1 - \tilde{\sigma}_2\|_{0,T,\mathscr{H}}). \tag{1.68}$$

Hence, from (1.68), (1.64), (1.39), (1.62), (1.24), and (2.32) from Chapter 2, we obtain

$$\|\dot{u}_1^* - \dot{u}_2^*\|_{0,T,H_1} + \|\dot{\sigma}_1^* - \dot{\sigma}_2^*\|_{0,T,\mathscr{H}_2} \leq C(\|u_{01} - u_{02}\|_{H_1} + \|\sigma_{01} - \sigma_{02}\|_{\mathscr{H}_2}$$
$$+ \|\tilde{u}_1 - \tilde{u}_2\|_{0,T,H_1}$$
$$+ \|\tilde{\sigma}_1 - \tilde{\sigma}_2\|_{0,T,\mathscr{H}}). \tag{1.69}$$

Using (1.61), (1.69), (1.59), and (1.60) we obtain

$$\|\dot{u}_1 - \dot{u}_2\|_{0,T,H_1} + \|\dot{\sigma}_1 - \dot{\sigma}_2\|_{0,T,\mathscr{H}_2} \leq C(\|u_{01} - u_{02}\|_{H_1} + \|\sigma_{01} - \sigma_{02}\|_{\mathscr{H}_2}$$
$$+ \|b_1 - b_2\|_{1,T,H} + \|g_1 - g_2\|_{1,T,H_\Gamma}$$
$$+ \|f_1 - f_2\|_{1,T,V_{\Gamma 1}}). \tag{1.70}$$

From (1.57) for $j = 0$ and (1.70) we obtain (1.57) for $j = 1$.

From Theorem 1.2 it is easy to deduce the following result:

Corollary 1.1 *Let the hypotheses of Theorem* 1.2 *hold. If* $b_1 = b_2$, $g_1 = g_2$, $f_1 = f_2$, *then*

$$\|u_1 - u_2\|_{1,T,H_1} + \|\sigma_1 - \sigma_2\|_{1,T,\mathscr{H}_2} \leq C(\|u_{01} - u_{02}\|_{H_1} + \|\sigma_{01} - \sigma_{02}\|_{\mathscr{H}_2}). \tag{1.71}$$

Further on, we give the definition of *stability*, *finite time stability* and *asymptotic stability* for the problem (1.20)–(1.23) (following for instance Hahn, 1967, Ch. V). A solution $(u; \sigma)$ of the problem (1.20)–(1.23) will be called:

(i) *stable* if there exists m: $\mathbb{R}_+ \to \mathbb{R}_+$ a continuous, increasing function with $m(0) = 0$ such that

$$\|u(t) - u_1(t)\|_{H_1} + \|\sigma(t) - \sigma_1(t)\|_{\mathscr{H}_2} \leq m(\|u_0 - u_{01}\|_{H_1} + \|\sigma_0 - \sigma_{01}\|_{\mathscr{H}_2}) \tag{1.72}$$

for all $t \in \mathbb{R}_+$ and (u_{01}, σ_{01}) satisfying (1.28), (1.29) where $(u_1; \sigma_1)$ is the solution of (1.20)–(1.23) for the initial data $(u_{01}; \sigma_{01})$;

(ii) *finite-time stable* if (1.72) holds for all finite time intervals.

(iii) *asymptotically stable* if there exist m as in (i) and $n\colon \mathbb{R}_+ \to \mathbb{R}_+$ a continuous decreasing, function with $\lim_{t\to\infty} n(t) = 0$ such that

$$\|u(t) - u_1(t)\|_{H_1} + \|\sigma(t) - \sigma_1(t)\|_{\mathscr{H}_2} \le m(\|u_0 - u_{01}\|_{H_1} + \|\sigma_0 - \sigma_{01}\|_{\mathscr{H}_2})n(t)$$

for all $t \in \mathbb{R}_+$ and $(u_{01}; \sigma_{01})$ satisfying (1.28), (1.29), where $(u_1; \sigma_1)$ is the solution of (1.20)–(1.23) for the initial data $(u_{01}; \sigma_{01})$.

Remark 1.8 *From* (1.71) *we deduce the finite-time stability of every solution of* (1.20)–(1.23). *Generally speaking stability does not hold* (*see Remark* 2.2).

2 Behaviour of the solution in the viscoelastic case

As has been pointed out in Section 1 we say that (1.1) or (1.2) represents a viscoelastic constitutive equation if the set of equilibrium states (i.e. the set of all $(\sigma, \varepsilon) \in \mathscr{D}$ such that $G(\sigma, \varepsilon) = 0$) may be described by an explicit equation $\sigma = F(\varepsilon)$. For the sake of simplicity we suppose in this section that

$$G(\sigma, \varepsilon) = -k(\sigma - F(\varepsilon)) \tag{2.1}$$

for all $\sigma, \varepsilon \in \mathscr{S}_N$, where $k > 0$ is a *viscosity constant* and $F\colon \Omega \times \mathscr{S}_N \to \mathscr{S}_N$. In order to satisfy (1.25) we suppose that F has the following properties:

$$\exists L_0 > 0 \text{ such that } |F(x, \varepsilon_1) - F(x, \varepsilon_2)| \le L_0|\varepsilon_1 - \varepsilon_2| \qquad \forall \varepsilon_1, \varepsilon_2 \in \mathscr{S}_N, \textit{ a.e. in } \Omega; \tag{2.2}$$

$$x \to F(x, \varepsilon) \text{ is a measurable function with respect to the Lebesgue measure on } \Omega, \qquad \forall \varepsilon \in \mathscr{S}_N; \tag{2.3}$$

$$x \to F(x, 0) \in \mathscr{H}; \tag{2.4}$$

$$\exists a > 0 \text{ such that } (F(x, \varepsilon_1) - F(x, \varepsilon_2))\cdot(\varepsilon_1 - \varepsilon_2) \ge a|\varepsilon_1 - \varepsilon_2|^2 \qquad \forall \varepsilon_1, \varepsilon_2 \in \mathscr{S}_N, \text{ a.e. in } \Omega. \tag{2.5}$$

Regarding (2.2)–(2.5) similar comments as in Remark 1.3(2) can be given; (2.2) shows that F is a Lipschitz function while (2.5) shows that F is a strongly monotone function. Assumption (2.5) is physically reasonable since it models the fact that for a larger strain we need a larger stress in order to obtain the equilibrium.

In this section we point out some properties of the solution in problem (1.20)–(1.23) in the case when the constitutive function G is given by (2.1).

2.1 Asymptotic stability

In Section 1.4 the finite-time stability of the solution has been proved (Theorem 1.2, Corollary 1.1). In this section we shall prove that in the viscoelastic case asymptotic stability also holds.

Theorem 2.1 *Let* (1.24) *and* (2.2)–(2.5) *hold and let* $(u_i; \sigma_i)$ *be two solutions of* (1.20)–(1.23), (2.1) *for the data* f_i, b_i, g_i, u_{0i}, σ_{0i}, $i = 1, 2$ *for which* (1.26)–(1.29) *hold. Then there exist two constants* $C, \bar{C} > 0$ *(depending only on* Ω, Γ_1, $\mathscr{A}$, L_0, a*) such that for all* $T > 0$ *we have*

$$\begin{aligned}\|u_1(t) - u_2(t)\|_{H_1} + \|\sigma_1(t) - \sigma_2(t)\|_{\mathscr{H}_2} \\ \le \bar{C}[(\|u_{01} - u_{02}\|_{H_1} + \|\sigma_{01} - \sigma_{02}\|_{\mathscr{H}_2}) \exp(-Ckt) + \|b_1 - b_2\|_{0,T,H} \\ + \|g_1 - g_2\|_{0,T,H_\Gamma} + \|f_1 - f_2\|_{0,T,V_{\Gamma 1}}] \qquad \forall t \in [0, T]. \qquad (2.6)\end{aligned}$$

Corollary 2.1 *Let the assumptions of Theorem* 1.1 *hold. If* $b_1 = b_2$, $f_1 = f_2$, $g_1 = g_2$, *then we have*

$$\begin{aligned}\|u_1(t) - u_2(t)\|_{H_1} + \|\sigma_1(t) - \sigma_2(t)\|_{\mathscr{H}_2} \\ \le \bar{C}(\|u_{01} - u_{02}\|_{H_1} + \|\sigma_{01} - \sigma_{02}\|_{\mathscr{H}_2}) \exp(-Ckt) \qquad \forall t \in \mathbb{R}_+. \qquad (2.7)\end{aligned}$$

Remark 2.1 *From* (2.7) *one can deduce that in the viscoelastic case all the solutions are asymptotically stable (see the definition given in Section* 1.4).

Remark 2.2 *If F is not a monotone function* (*usually at the phase transition phenomena*) stability generally cannot hold. *This can be seen in the following one-dimensional example. Let us consider a non-monotone function F given for instance by*

$$F(\varepsilon) = \begin{cases} E\varepsilon & \text{for } \varepsilon \le \alpha \\ -E_1\varepsilon + \gamma & \text{for } \alpha < \varepsilon < \beta \\ E\varepsilon - \Delta & \text{for } \varepsilon \ge \beta \end{cases} \qquad (2.8)$$

where $E > 0$, $\Delta > (\beta - \alpha)E$, $E_1 = \Delta/(\beta - \alpha) - E > 0$, $\gamma = \alpha\Delta/(\beta - \alpha)$. *The graph of F given by* (2.8) *is plotted in Figure* 2.1.

Let us imagine a creep experiment. Consider $\Omega = (0, 1)$, $\Gamma_1 = \{0\}$, $\Gamma_2 = \{1\}$, $b \equiv 0$, $g \equiv 0$, $f(t) = \sigma_0 \in (\sigma_2, \sigma_1)$, $\sigma_0(x) = \sigma_0$, where $\sigma_1 = F(\alpha)$, $\sigma_2 = F(\beta)$. One can easily see that $\sigma(t, x) = \sigma_0$ and $\varepsilon(t, x) = \partial u(t, x)/\partial x$ constitute the solution of the following Cauchy problem

$$\dot{\varepsilon}(t, x) = k\mathscr{A}^{-1}(\sigma_0 - F(\varepsilon(t, x))), \qquad (t, x) \in \mathbb{R}_+ \times (0, 1) \qquad (2.9)$$

$$\varepsilon(0, x) = \varepsilon_0(x) = \frac{du_0(x)}{dx}, \qquad x \in (0, 1). \qquad (2.10)$$

Note that in (2.9)–(2.10) the spatial variable x acts as a parameter. Hence if ε_0 does not depend on x, then (2.9)–(2.10) reduce to an ordinary

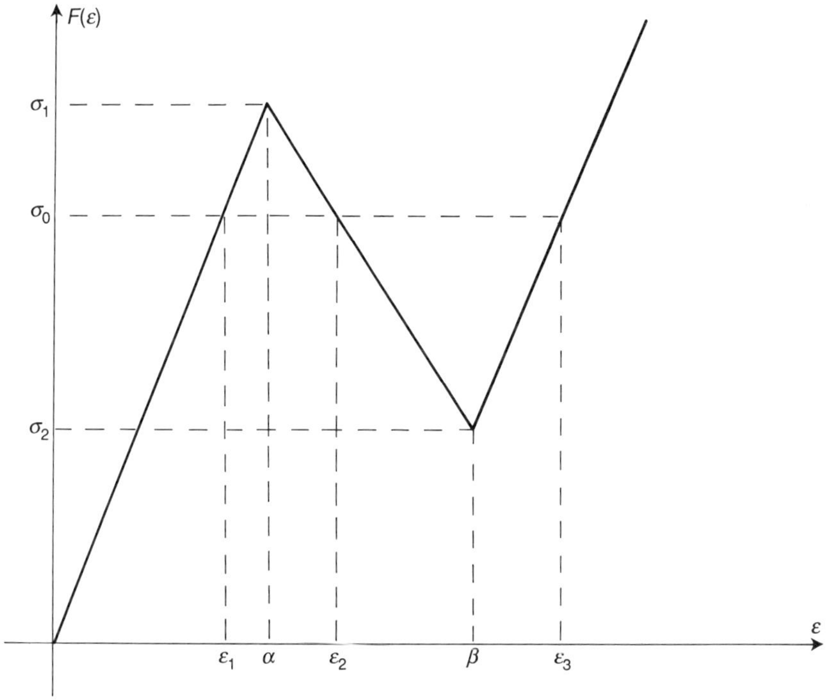

Fig. 2.1 The graphic representation of function F given by (2.8).

differential equation

$$\dot{\varepsilon}(t) = k\mathscr{A}^{-1}(\sigma_0 - F(\varepsilon(t))) \qquad \forall t > 0 \tag{2.9'}$$

$$\varepsilon(0) = \varepsilon_0. \tag{2.10'}$$

Let us denote by ε_1, ε_2, ε_3 the solutions of the equation $F(\varepsilon) = \sigma_0$ given by $\varepsilon_1 = \sigma_0/E$, $\varepsilon_2 = (\gamma - \sigma_0)/E_1$, $\varepsilon_3 = (\sigma_0 + \Delta)/E$. We notice that $\varepsilon_i(t) = \varepsilon_i$ are solutions of (2.9′), (2.10′) when $\varepsilon_0 = \varepsilon_i$; hence they are steady solutions. One can easily prove that ε_1 and ε_3 are asymptotically stable, having their domains of attractivity in the set of initial homogeneous strains $A_1 = (-\infty, \varepsilon_2)$ and $A_3 = (\varepsilon_2, +\infty)$ respectively. More precisely if $\varepsilon_0 \in (-\infty, \varepsilon_2)$, then $\lim_{t\to\infty} \varepsilon(t) = \varepsilon_1$, and if $\varepsilon_0 \in (\varepsilon_2, +\infty)$, and then $\lim_{t\to\infty} \varepsilon(t) = \varepsilon_3$, hence the solution ε_2 is not stable. If we choose $u_{01}(x) = (\varepsilon_2 - \lambda)x$, $u_{02}(x) = \varepsilon_2 x$, $\sigma_{01} = \sigma_{02} = \sigma_0$ with a very small positive λ and if we denote by $(u_i; \sigma_i)$ the solutions of these particular one-dimensional problems, then we have

$$\lim_{t\to\infty} \|u_1(t) - u_2(t)\|_{\boldsymbol{H}_1} = \|\bar{u}_1 - \bar{u}_2\|_{\boldsymbol{H}_1} \qquad \forall \lambda > 0,$$

where $\bar{u}_1(x) = \varepsilon_1 x$, $\bar{u}_2(x) = \varepsilon_2 x$. Hence the solution $u_2(t, x) = \varepsilon_2 x$, $\sigma_2(t, x) = \sigma_0$ is not stable (see (1.72)).

PROOF OF THEOREM 2.1. The same notations as in the proof of Theorem 1.2 are used. From (1.61), (2.1), and (1.21) we get

$$\dot{\sigma}_i^*(t) = \mathscr{A}\varepsilon(\dot{u}_i^*(t)) - k[\sigma_i^*(t) + \tilde{\sigma}_i(t) - F(\varepsilon(u_i(t)))] \qquad \forall t \in \mathbb{R}_+, \quad i = 1, 2. \tag{2.11}$$

If we take the difference in (2.11) for $i = 1$ and $i = 2$ and we put $u^* = u_1^* - u_2^*$, $\sigma^* = \sigma_1^* - \sigma_2^*$, $\tilde{u} = \tilde{u}_1 - \tilde{u}_2$, $\tilde{\sigma} = \tilde{\sigma}_1 - \tilde{\sigma}_2$, we get

$$\dot{\sigma}^*(t) = \mathscr{A}\varepsilon(\dot{u}^*(t)) - k[\sigma^*(t) + \tilde{\sigma}(t) - F(\varepsilon(u_1(t))) + F(\varepsilon(u_2(t)))] \qquad \forall t \in \mathbb{R}_+. \tag{2.12}$$

After multiplying (2.12) by $\varepsilon(u^*(t))$ and integrating the result over Ω we obtain

$$\langle \mathscr{A}\varepsilon(\dot{u}^*(t)\rangle, \varepsilon(u^*(t))\rangle_{\mathscr{H}} = -k(\langle\tilde{\sigma}(t), \varepsilon(u^*(t))\rangle_{\mathscr{H}} + \langle F(\varepsilon(u_1(t))) - F(\varepsilon(u_2(t))), \varepsilon(u^*(t))\rangle_{\mathscr{H}}). \tag{2.13}$$

If we let $\theta(t) = \langle \mathscr{A}\varepsilon(u^*(t)), \varepsilon(u^*(t)\rangle_{\mathscr{H}}$ and

$$\beta(T) = \|b_1 - b_2\|_{0,T,\boldsymbol{H}} + \|g_1 - g_2\|_{0,T,\boldsymbol{H}_\Gamma} + \|f_1 - f_2\|_{0,T,V'_{\Gamma 1}}$$

then after some algebra we can deduce from (2.13), (2.2), (2.5), and (1.59) that

$$\dot{\theta}(t) \le k(-2C_1\theta(t) + 2C_2\beta(T)\sqrt{(\theta(t))} + 2C_2\beta^2(T) \qquad \forall t \in [0, T]$$

with $C_1, C_2 > 0$, $C_1 < 1$. Using Lemma A.4.15, we have

$$\sqrt{(\theta(t))} \le \sqrt{(\theta(0))}\exp(-kC_1 t) + \beta(T)(C_2 + \sqrt{(C_1^2 + 4C_1C_2)})/(2C_1) \tag{2.14}$$

for all $t \in [0, T]$. Using (1.24), (1.59) and Korn's inequality ((2.32) from Chapter 2), from (2.14) we obtain

$$\|u_1(t) - u_2(t)\|_{\boldsymbol{H}_1} \le C_3(\|u_{01} - u_{02}\|_{\boldsymbol{H}_1}\exp(-kC_1 t) + \beta(T)) \qquad \forall t \in [0, T]. \tag{2.15}$$

In a similar way, if we multiply (2.12) by $\mathscr{A}^{-1}$ on the left and by $\sigma^*(t)$ on the right and integrate the result on Ω, we obtain

$$\langle \mathscr{A}^{-1}\dot{\sigma}^*(t), \sigma^*(t)\rangle_{\mathscr{H}} = -k\langle \mathscr{A}^{-1}\sigma^*(t), \sigma^*(t)\rangle_{\mathscr{H}} - k\langle \mathscr{A}^{-1}\tilde{\sigma}(t), \sigma^*(t)\rangle_{\mathscr{H}} + k\langle \mathscr{A}^{-1}F(\varepsilon(u_1(t))) - \mathscr{A}^{-1}F(\varepsilon(u_2(t))), \sigma^*(t)\rangle_{\mathscr{H}} \qquad \forall t \in \mathbb{R}_+. \tag{2.16}$$

If we let $\theta(t) = \langle \mathscr{A}^{-1}\sigma^*(t), \sigma^*(t)\rangle_{\mathscr{H}}$, then from (2.16) and (2.2) we obtain

$$\dot{\theta}(t) \le -2k\theta(t) + 2kC_4(\beta(T) + \|u_1(t) - u_2(t)\|_{\boldsymbol{H}_1})\sqrt{(\theta(t))} \qquad \forall t \in [0, T].$$

Using (2.15), from Lemma A4.14 we deduce

$$\sqrt{(\theta(t))} \leq \sqrt{(\theta(0))} \exp(-kt) + C_5\beta(T) + C_5/(1 - C_1) \exp(-kC_1t) \cdot \|u_{01} - u_{02}\|_{\boldsymbol{H}_1} \qquad \forall t \in [0, T]. \tag{2.17}$$

From (2.17), (1.24), and (1.59) we obtain

$$\|\sigma_1(t) - \sigma_2(t)\|_{\mathscr{H}_2} \leq C_6(\|u_{01} - u_{02}\|_{\boldsymbol{H}_1} + \|\sigma_{01} - \sigma_{02}\|_{\mathscr{H}_2}) \exp(-kC_1t) + C_6\beta(T) \qquad \forall t \in [0, T]. \tag{2.18}$$

The inequality (2.6) easily follows from (2.15) and (2.18).

2.2 Periodic solutions

In this section we are interested in periodic solutions of the problem (1.20)–(1.23), (2.1). The main result of this section is the following:

Theorem 2.2 *Let* (1.24), (2.2)–(2.5) *hold and let the data b, f, g which satisfy* (1.26), (1.27) *be periodic functions with the same period ω. Then there exists a unique initial datum $(u_0; \sigma_0)$ which satisfies* (1.28), (1.29) *such that the solution of* (1.20)–(1.23), (2.1) *is a periodic function with the same period ω.*

Remark 2.3 *With the hypotheses of Theorem* 2.2, *using Corollary* 2.1 *we deduce that for all initial data, the solution of the problem* (1.20)–(1.23), (2.1) *approaches a unique periodic function when $t \to +\infty$. In other words if the external data are oscillating then the body too will 'begin' to oscillate after a while.*

PROOF OF THEOREM 2.2. The result stated in Theorem 2.2 is a direct consequence of asymptotic stability (Corollary 2.1). Indeed, let $\mathscr{C} \subset \boldsymbol{H}_1 \times \mathscr{H}_2$ be given by

$$\mathscr{C} = \{(u; \sigma) \in \boldsymbol{H}_1 \times \mathscr{H}_2 \,|\, u = g(0) \text{ on } \Gamma_1, \sigma\nu = f(0) \text{ on } \Gamma_2, \operatorname{Div} \sigma + b(0) = 0 \text{ in } \Omega\} \tag{2.19}$$

and let $\mathscr{L}: \mathscr{C} \to \boldsymbol{H}_1 \times \mathscr{H}_2$ be the non-linear operator which associates with every $(u_0; \sigma_0)$ the solution $(u(t); \sigma(t))$ of the problem (1.20)–(1.23), (2.1) at the time $t = \omega$. Since g, b, f are periodic functions with the same period ω we find that $\mathscr{L}: \mathscr{C} \to \mathscr{C}$ and $\mathscr{L}^n(u_0, \sigma_0) = (u(n\omega); \sigma(n\omega))$ for all $n \in \mathbb{N}$.

From Corollary 2.1 we get

$$\|\mathscr{L}^n(u_{01}; \sigma_{01}) - \mathscr{L}^n(u_{02}; \sigma_{02})\|_{\boldsymbol{H}_1 \times \mathscr{H}_2} \leq \bar{C}\|(u_{01}; \sigma_{01}) - (u_{02}; \sigma_{02})\|_{\boldsymbol{H}_1 \times \mathscr{H}_2} \exp(-nk\omega C). \tag{2.20}$$

For a large $n \in \mathbb{N}$ we have $\bar{C}\exp(-nk\omega C) < 1$ and using Banach's fixed-point theorem we deduce that there exists a unique couple $(u_0; \sigma_0) \in \mathscr{C}$ such that $\mathscr{L}(u_0; \sigma_0) = (u_0; \sigma_0)$, i.e. for this initial data the solution of (1.20)–(1.23), (2.1) is a periodic function with the period ω.

2.3 An approach to elasticity

In this section we prove the convergence when $k \to \infty$ of the solution $(u_k; \sigma_k)$ of (1.20)–(1.23), (2.1) to the solution of the following boundary-value problem for an elastic body.

Find the displacement function $\hat{u}$: $\mathbb{R}_+ \times \Omega \to \mathbb{R}^N$ and the stress function $\hat{\sigma}$: $\mathbb{R}_+ \times \Omega \to \mathscr{S}_N$ such that

$$\operatorname{Div} \hat{\sigma}(t) + b(t) = 0 \tag{2.21}$$

$$\hat{\sigma}(t) = F(\varepsilon(\hat{u}(t))) \qquad \text{in } \Omega \tag{2.22}$$

$$\hat{u}(t) = g(t) \quad \text{on } \Gamma_1, \qquad \hat{\sigma}(t)\nu = f(t) \quad \text{on } \Gamma_2 \qquad \forall t \in \mathbb{R}_+. \tag{2.23}$$

Lemma 2.1 *Let us suppose that* (1.26), (1.27), (2.2)–(2.5) *hold. Then the problem* (2.21)–(2.33) *has a unique solution* $\hat{u} \in C^0(\mathbb{R}_+, \boldsymbol{H}_1)$, $\hat{\sigma} \in C^0(\mathbb{R}_+, \mathscr{H}_2)$. *Moreover, for all* $T > 0$, $\hat{u}$, $\hat{\sigma}$ *are absolutely continuous functions on* $[0, T]$ *and there exists* $C > 0$ *(which depends only on* Ω, Γ_1, L_0 *and* a*) such that*

$$\|\dot{\hat{u}}(t)\|_{\boldsymbol{H}_1} = \|\dot{\hat{\sigma}}(t)\|_{\mathscr{H}_2} \le C(\|\dot{b}(t)\|_{\boldsymbol{H}} + \|\dot{g}(t)\|_{\boldsymbol{H}_\Gamma} + \|\dot{f}(t)\|_{V'_{\Gamma 1}}). \tag{2.24}$$

In order to prove Lemma 2.1 the following lemma will be useful.

Lemma 2.2 *If* (2.2)–(2.5) *hold, then there exists a Lipschitz continuous operator* E: $\boldsymbol{H} \times \boldsymbol{H}_\Gamma \times V'_{\Gamma_1} \to \boldsymbol{H}_1 \times \mathscr{H}_2$ *such that for all* $\bar{b} \in \boldsymbol{H}$, $\bar{g} \in \boldsymbol{H}_\Gamma$, $\bar{f} \in V'_{\Gamma_1}$ *the couple* $(v; \tau) = E(\bar{b}, \bar{g}, \bar{f})$ *is the unique solution of the following non-linear boundary-value problem*:

$$\operatorname{Div} \tau + \bar{b} = 0 \tag{2.25}$$

$$\tau = F(\varepsilon(v)) \tag{2.26}$$

$$v = \bar{g} \quad \text{on } \Gamma_1, \quad \tau\nu = \bar{f} \quad \text{on } \Gamma_2. \tag{2.27}$$

PROOF. Let $\tilde{v} = z_1(g)$ and let us consider B: $V_1 \to V'_1$ given by

$$\langle B(v), w\rangle_{V'_1, V_1} = \langle F(\varepsilon(\tilde{v}) + \varepsilon(v)), \varepsilon(w)\rangle_{\mathscr{H}}$$

and $l \in V'_1$ given by

$$\langle l, w\rangle_{V'_1, V_1} = \langle \bar{b}, w\rangle_H + \langle f, \gamma_1(w)\rangle_{V'_{\Gamma 1}, V_{\Gamma 1}}$$

for all $v, w \in V_1$. From (2.2) we deduce that B is Lipschitz continuous and from (2.32) of Chapter 2 and (2.5), we obtain that B is a strongly monotone operator (i.e. there exists $m > 0$ such that

$$\langle B(u_1) - B(u_2), u_1 - u_2 \rangle_{V_1', V_1} \geq m \|u_1 - u_2\|_{H_1}^2$$

for all $u_1, u_2 \in V_1$). Hence we can use Theorem A.2.4 to obtain that there exists a unique $\bar{v} \in V_1$ such that $B(\bar{v}) = l$. If we let $v = \bar{v} + \tilde{v}$ and $\tau = F(\varepsilon(v))$, then we have

$$\langle \tau, \varepsilon(w) \rangle_{\mathscr{H}} = \langle \bar{b}, w \rangle_{H} + \langle \bar{f}, \gamma_1(w) \rangle_{V_{\Gamma 1}', V_{\Gamma 1}} \qquad \forall w \in V_1. \tag{2.28}$$

If we put $w \in [\mathscr{D}(\Omega)]^N$ in (2.28) then we get (2.25). Since the boundary conditions (2.27) follow from (2.28) and from (2.44) of Chapter 2, we can put $E(\bar{b}, \bar{g}, \bar{f}) = (v; \tau)$. In order to prove that E is Lipschitz continuous let $(b_i; g_i; f_i) \in H \times H_\Gamma \times V_{\Gamma_1}'$, $(v_i; \tau_i) = E(b_i, g_i, f_i)$, $i = 1, 2$ and let us denote by $\tilde{v}_i = z_1(g_i)$, $\bar{v}_i = v_i - \tilde{v}_i$, $i = 1, 2$.

If we multiply $\operatorname{Div}(\tau_1 - \tau_2) + b_1 - b_2 = 0$ by $\bar{v}_1 - \bar{v}_2 \in V_1$ and integrate the result over Ω, we obtain

$$\begin{aligned}\langle F(\varepsilon(v_1)) - F(\varepsilon(v_2)), \varepsilon(\bar{v}_1 - \bar{v}_2) \rangle_{\mathscr{H}} &= \langle b_1 - b_2, \bar{v}_1 - \bar{v}_2 \rangle_{H} \\ &\quad + \langle f_1 - f_2, \gamma_1(\bar{v}_1 - \bar{v}_2) \rangle_{V_{\Gamma 1}', V_{\Gamma 1}}.\end{aligned}$$

Using (2.2), (2.5) we get

$$\begin{aligned}a\|\varepsilon(v_1) - \varepsilon(v_2)\|_{\mathscr{H}}^2 &\leq L_0 \|\varepsilon(v_1) - \varepsilon(v_2)\|_{\mathscr{H}} \|\varepsilon(\tilde{v}_1) - \varepsilon(\tilde{v}_2)\|_{\mathscr{H}} \\ &\quad + C_1(\|b_1 - b_2\|_{H} + \|f_1 - f_2\|_{V_{\Gamma 1}'}) \|\bar{v}_1 - \bar{v}_2\|_{H_1}.\end{aligned}$$

If we have in mind that

$$\begin{aligned}&\|\varepsilon(\bar{v}_1) - \varepsilon(\bar{v}_2)\|_{\mathscr{H}} - \|\varepsilon(\tilde{v}_1) - \varepsilon(\tilde{v}_2)\|_{\mathscr{H}} \\ &\quad \leq \|\varepsilon(v_1) - \varepsilon(v_2)\|_{\mathscr{H}} \leq \|\varepsilon(\tilde{v}_1) - \varepsilon(\tilde{v}_2)\|_{\mathscr{H}} + \|\varepsilon(\bar{v}_1) - \varepsilon(\bar{v}_2)\|_{\mathscr{H}},\end{aligned}$$

then from (2.32) of Chapter 2 we get

$$\begin{aligned}\|\bar{v}_1 - \bar{v}_2\|_{H_1}^2 &\leq C_2(\|\tilde{v}_1 - \tilde{v}_2\|_{H_1} + \|b_1 - b_2\|_{H} + \|f_1 - f_2\|_{V_{\Gamma 1}'}) \|\bar{v}_1 - \bar{v}_2\|_{H_1} \\ &\quad + C_3 \|\tilde{v}_1 - \tilde{v}_2\|_{H_1}^2.\end{aligned}$$

Since $\|\tilde{v}_1 - \tilde{v}_2\|_{H_1} \leq C_4 \|g_1 - g_2\|_{H_\Gamma}$ we have

$$\|\bar{v}_1 - \bar{v}_2\|_{H_1} \leq C_4(\|g_1 - g_2\|_{H_\Gamma} + \|f_1 - f_2\|_{V_{\Gamma 1}'} + \|b_1 - b_2\|_{H});$$

hence

$$\|v_1 - v_2\|_{H_1} \leq C_5(\|g_1 - g_2\|_{H_\Gamma} + \|f_1 - f_2\|_{V_{\Gamma 1}'} + \|b_1 - b_2\|_{H}). \tag{2.29}$$

From

$$\|\tau_1 - \tau_2\|_{\mathscr{H}} \leq L_0 \|\varepsilon(v_1) - \varepsilon(v_2)\|_{\mathscr{H}} \leq C_6 \|v_1 - v_2\|_{H_1},$$

and

$$\|\mathrm{Div}\,\tau_1 - \mathrm{Div}\,\tau_2\|_{\boldsymbol{H}} = \|b_1 - b_2\|_{\boldsymbol{H}}$$

we deduce

$$\|\tau_1 - \tau_2\|_{\mathscr{H}_2} \leq C_7(\|g_1 - g_2\|_{\boldsymbol{H}_\Gamma} + \|b_1 - b_2\|_{\boldsymbol{H}} + \|f_1 - f_2\|_{V'_{\Gamma 1}}). \quad (2.30)$$

If we use (2.29), (2.30) we deduce that there exists $C > 0$ (which depends only on Ω, Γ_1, L_0 and a) such that

$$\|E(b_1, g_1, f_1) - E(b_2, g_2, f_2)\|_{\boldsymbol{H}_1 \times \mathscr{H}_2}$$
$$\leq C(\|b_1 - b_2\|_{\boldsymbol{H}} + \|g_1 - g_2\|_{\boldsymbol{H}_\Gamma} + \|f_1 - f_2\|_{V'_{\Gamma 1}}) \quad (2.31)$$

for all $b_i \in \boldsymbol{H}$, $g_i \in \boldsymbol{H}_\Gamma$, $f_i \in \boldsymbol{V}'_{\Gamma_1}$, $i = 1, 2$.

PROOF OF LEMMA 2.1. If we put $(\hat{u}(t); \hat{\sigma}(t)) = E(b(t), g(t), f(t))$ for all $t \in \mathbb{R}_+$ with E given by Lemma 2.2 then (2.21)–(2.23) are satisfied for all $t \in \mathbb{R}_+$. Since $t \to (b(t); g(t); f(t))$ is an absolutely continuous function from $[0, T]$ to $\boldsymbol{H} \times \boldsymbol{H}_\Gamma \times \boldsymbol{V}'_{\Gamma_1}$ and E is Lipschitz continuous we deduce that $t \to (\hat{u}(t), \hat{\sigma}(t))$ is an absolutely continuous function from $[0, T]$ to $\boldsymbol{H}_1 \times \mathscr{H}_2$, hence a.e. derivable. From (2.31) we get

$$\|(\hat{u}(t+h) - \hat{u}(t))/h\|_{\boldsymbol{H}_1} + \|(\hat{\sigma}(t+h) - \hat{\sigma}(t))/h\|_{\mathscr{H}_2}$$
$$\leq C(\|(b(t+h) - b(t))/h\|_{\boldsymbol{H}} + \|(g(t+h) - g(t))/h\|_{\boldsymbol{H}_\Gamma}$$
$$+ \|(f(t+h) - f(t))/h\|_{V'_{\Gamma 1}})$$

and if $(\hat{u}; \hat{\sigma})$ are derivable in t then we can pass to the limit with $h \to 0$ to obtain (2.24).

The following lemma evaluates the difference between the solutions of (1.20)–(1.23), (2.1) and of (2.21)–(2.23) for the same data b, f, g:

Lemma 2.3 *Let* (1.24), (2.2)–(2.5), (1.26)–(1.28) *hold and let* $(u; \sigma)$ *be the solution of* (1.20)–(1.23), (2.1). *If* $(\hat{u}; \hat{\sigma})$ *is the solution of* (2.21)–(2.23), *then for all* $t \in \mathbb{R}_+$ *we have*

$$\|u(t) - \hat{u}(t)\|_{\boldsymbol{H}_1} \leq \bar{C}\Big[\|u_0 - \hat{u}(0)\|_{\boldsymbol{H}_1} \exp(-Ckt)$$
$$+ \int_0^t (\|\dot{b}(\dot{s})\|_{\boldsymbol{H}} + \|\dot{g}(s)\|_{\boldsymbol{H}_\Gamma} + \|f(s)\|_{V'_{\Gamma 1}}) \exp(-Ck(t-s))\,\mathrm{ds}\Big] \quad (2.32)$$

$$\|\sigma(t)-\hat{\sigma}(t)\|_{\mathscr{H}_2} \le \bar{C}\Bigg[\|\sigma_0-\hat{\sigma}(0)\|_{\mathscr{H}_2}\exp(-kt) + \int_0^t (k\|u(s)-\hat{u}(s)\|_{\boldsymbol{H}_1} + \|\dot{b}(s)\|_{\boldsymbol{H}} + \|\dot{g}(s)\|_{\boldsymbol{H}_\Gamma} + \|\dot{f}(s)\|_{V'_{\Gamma 1}})\exp(-k(t-s))\,ds\Bigg] \quad (2.33)$$

where the strictly positive constants C, $\bar{C} > 0$, $C < 1$ *depend only on* Ω, Γ_1, $\mathscr{A}$, L_0, and a.

PROOF. If we denote by $\bar{u} = u - \hat{u}$, $\bar{\sigma} = \sigma - \hat{\sigma}$, then from (1.21), (2.1), (2.22) and Lemma 2.1 we get

$$\dot{\hat{\sigma}}(t) + \dot{\bar{\sigma}}(t) = \mathscr{A}\varepsilon(\dot{\bar{u}}(t)) + \mathscr{A}\varepsilon(\dot{\hat{u}}(t)) - k[\bar{\sigma}(t) + F(\varepsilon(\hat{u}(t)) - F(\varepsilon(u(t)))] \quad \text{a.e. } t \in \mathbb{R}_+. \quad (2.34)$$

Let us multiply (2.34) by $\varepsilon(\bar{u}(t))$ and integrate the result on Ω. We obtain

$$\langle\dot{\hat{\sigma}}(t) - \mathscr{A}\varepsilon(\dot{\hat{u}}(t)), \varepsilon(\bar{u}(t))\rangle_{\mathscr{H}} = \langle\mathscr{A}\varepsilon(\dot{\bar{u}}(t)), \varepsilon(\bar{u}(t))\rangle_{\mathscr{H}} + k\langle F(\varepsilon(u(t))) - F(\varepsilon(\hat{u}(t))), \varepsilon(\bar{u}(t))\rangle_{\mathscr{H}} \quad \text{a.e. } t \in \mathbb{R}_+.$$

Hence, from (2.24), (1.24), and (2.5) we deduce

$$\langle\mathscr{A}\varepsilon(\dot{\bar{u}}(t)), \varepsilon(\bar{u}(t))\rangle_{\mathscr{H}} \le -Ck\langle\mathscr{A}\varepsilon(\bar{u}(t)), \varepsilon(\bar{u}(t))\rangle_{\mathscr{H}} + C_1(\|\dot{b}(t)\|_{\boldsymbol{H}} + \|\dot{g}(t)\|_{\boldsymbol{H}_\Gamma} + \|\dot{f}(t)\|_{V'_{\Gamma 1}})(\langle\mathscr{A}\varepsilon(\bar{u}(t)), \varepsilon(\bar{u}(t))\rangle_{\mathscr{H}}^{1/2}) \quad \text{a.e. } t \in \mathbb{R}_+.$$

If we denote by $\theta(t) = \langle\mathscr{A}\varepsilon(\bar{u}(t)), \varepsilon(\bar{u}(t))\rangle_{\mathscr{H}}$ then in the last inequality we can use Lemma A.4.14 and the Korn inequality (2.32) of Chapter 2 to deduce (2.32).

If we multiply (2.34) by $\mathscr{A}^{-1}$ on the right and by $\bar{\sigma}(t)$ on the left and integrate the result over Ω, we obtain

$$\langle\mathscr{A}^{-1}\dot{\bar{\sigma}}(t), \bar{\sigma}(t)\rangle_{\mathscr{H}} = -k\langle\mathscr{A}^{-1}\bar{\sigma}(t)\rangle_{\mathscr{H}} - \langle\mathscr{A}^{-1}\dot{\hat{\sigma}}(t) - \varepsilon(\hat{u}(t)), \bar{\sigma}(t)\rangle_{\mathscr{H}} + k\langle\mathscr{A}^{-1}F(\varepsilon(u(t))) - \mathscr{A}^{-1}F(\varepsilon(\hat{u}))), \bar{\sigma}(t)\rangle_{\mathscr{H}} \quad \text{a.e. } t \in \mathbb{R}_+.$$

We can use now (2.24) to obtain

$$\langle\mathscr{A}^{-1}\dot{\bar{\sigma}}(t), \bar{\sigma}(t)\rangle_{\mathscr{H}} \le -k\langle\mathscr{A}^{-1}\bar{\sigma}(t), \bar{\sigma}(t)\rangle_{\mathscr{H}} + C_3[k\|\bar{u}(t)\|_{\boldsymbol{H}_1} + \|\dot{b}(t)\|_{\boldsymbol{H}} + \|\dot{g}(t)\|_{\boldsymbol{H}_\Gamma} + \|\dot{f}(t)\|_{V'_{\Gamma 1}}] \times \langle\mathscr{A}^{-1}\bar{\sigma}(t), \bar{\sigma}(t)\rangle_{\mathscr{H}}^{1/2} \quad \text{a.e. } t \in \mathbb{R}_+.$$

Using again Lemma A.4.14 for $\theta(t) = \langle\mathscr{A}^{-1}\bar{\sigma}(t), \bar{\sigma}(t)\rangle_{\mathscr{H}}$ we deduce (2.33).

The main result of this section is the following:

Theorem 2.3 *Suppose* (1.24), (1.26)–(1.28), (2.2)–(2.5) *hold. Let* $(u_k; \sigma_k)$ *be the solution of* (1.20)–(1.23), (2.1) *for all* $k > 0$ *and let* $(\hat{u}; \hat{\sigma})$ *be the solution of* (2.21)–(2.23). *Then for all* $t > 0$ *we have*

$$\|u_k(t), - \hat{u}(t)\|_{\boldsymbol{H}_1} \to 0, \quad \|\sigma_k(t) - \hat{\sigma}(t)\|_{\mathscr{H}_2} \to 0 \quad \textit{when } k \to +\infty.$$

PROOF. From (2.32) for $t > 0$ we get

$$\|u_k(t) - \hat{u}(t)\|_{\boldsymbol{H}_1} \leq \bar{C}[\|u_0 - \hat{u}(0)\|_{\boldsymbol{H}_1} \exp(-Ckt) + (Ck)^{-1}(\|\dot{b}\|_{0,t,\boldsymbol{H}} + \|\dot{g}\|_{0,t,\boldsymbol{H}} + \|\dot{f}\|_{0,t,V'_{\Gamma 1}})] \tag{2.35}$$

hence $\|u_k(t) - \hat{u}(t)\|_{\boldsymbol{H}_1} \to 0$ when $k \to +\infty$.

If we use (2.35) in (2.33) we obtain

$$\begin{aligned}\|\sigma_k(t) - \hat{\sigma}(t)\|_{\mathscr{H}_2} \leq \bar{C}[\|\sigma_0 - \hat{\sigma}(0)\|_{\mathscr{H}_2} \exp(-kt) &+ C/(1 - C)\exp(-Ckt)\|u_0 - \hat{u}(0)\|_{\boldsymbol{H}_1} \\ &+ (C + 1)(Ck)^{-1}(\|\dot{b}\|_{0,t,\boldsymbol{H}} + \|\dot{g}\|_{0,t,\boldsymbol{H}_\Gamma} + \|\dot{f}\|_{0,t,V'_{\Gamma 1}})]\end{aligned} \tag{2.36}$$

hence $\|\sigma_k(t) - \hat{\sigma}(t)\|_{\mathscr{H}_2} \to 0$ when $k \to +\infty$.

2.4 Long-term behaviour of the solution

In this section we consider the problem (1.20)–(1.23), (2.1) for a fixed $k > 0$ and we study the behaviour of the solution when $t \to +\infty$. The main result is the following:

Theorem 2.4 *Let* (1.24), (1.26)–(1.28), (2.2)–(2.5) *hold; we denote by* $(u; \sigma)$ *the solution of* (1.20)–(1.23), (2.1) *and by* $(\hat{u}; \hat{\sigma})$ *the solution of* (2.21)–(2.23). *If*

$$\lim_{t \to \infty} (\|\dot{b}(t)\|_{\boldsymbol{H}} + \|\dot{g}(t)\|_{\boldsymbol{H}_\Gamma} + \|\dot{f}(t)\|_{V'_{\Gamma 1}}) = 0$$

then

$$\lim_{t \to \infty} (\|u(t) - \hat{u}(t)\|_{\boldsymbol{H}_1} + \|\sigma(t) - \hat{\sigma}(t)\|_{\mathscr{H}_2}) = 0. \tag{2.37}$$

Remark 2.4 *Note that for all* $t \geq 0$ *the functions* $\hat{u}(t), \hat{\sigma}(t)$ *are uniquely determined by the data* $b(t)$, $f(t)$, $g(t)$. *From Theorem* 2.4 *we find that if* $\|\dot{b}(t)\|_{\boldsymbol{H}} + \|\dot{g}(t)\|_{\boldsymbol{H}_\Gamma} + \|\dot{f}(t)\|_{V'_{\Gamma 1}} \to 0$ *when* $t \to \infty$, *then after a long time the solution* $(u; \sigma)$ *of the viscoelastic problem will be 'determined' only by the present values of* b, g, f. *Hence, in this case, the initial data and the history of external data have 'no influence' upon the long-term behaviour of the solution.*

Remark 2.5 *If* $\lim_{t\to\infty}(\|\dot{b}(t)\|_{\boldsymbol{H}} + \|\dot{g}(t)\|_{\boldsymbol{H}_\Gamma} + \|\dot{f}(t)\|_{V'_{\Gamma_1}}) \neq 0$ *the statement of Theorem 2.4 cannot generally hold. For example, let b, g, f be periodic functions with the same period* ω. *Then* $\hat{u}, \hat{\sigma}$ *are periodic functions and from Theorem 2.2 we find that there exists an initial datum* $(u_0; \sigma_0)$ *such that the solution* $(u; \sigma)$ *of* (1.20)–(1.23), (2.1) *is periodic. If we suppose that* (2.37) *holds then we get* $u \equiv \hat{u}$, $\sigma \equiv \hat{\sigma}$ *and from* (1.21), (2.22) *we get* $\dot{\hat{\sigma}}(t) = \mathscr{A}\varepsilon(\dot{\hat{u}}(t))$ *for all* $t \in \mathbb{R}_+$; *this equality is generally false if* $\dot{\hat{\sigma}} \neq 0$, $\dot{\hat{u}} \neq 0$ *and* $F(\varepsilon) \neq \mathscr{A}\varepsilon$. *Hence if the external data are periodic there exists a phase shift between the periodic solutions of the elastic and the viscoelastic problem.*

Corollary 2.2 *Let the assumption of Theorem* 2.4 *hold. If in addition we suppose that there exist* $\bar{b} \in \boldsymbol{H}$, $\bar{f} \in V'_{\Gamma_1}$, *and* $\bar{g} \in \boldsymbol{H}_\Gamma$ *such that*

$$\lim_{t\to\infty}(\|b(t) - \bar{b}\|_{\boldsymbol{H}} + \|g(t) - \bar{g}\|_{\boldsymbol{H}_\Gamma} + \|f(t) - \bar{f}\|_{V'_{\Gamma_1}}) = 0 \tag{2.38}$$

and if we denote by $(\hat{\bar{u}}; \hat{\bar{\sigma}})$ *the solution of* (2.21)–(2.23) *for the data* $\bar{b}$, $\bar{f}$, $\bar{g}$, *then*

$$\lim_{t\to\infty}(\|u(t) - \hat{\bar{u}}\|_{\boldsymbol{H}_1} + \|\sigma(t) - \hat{\bar{\sigma}}\|_{\mathscr{H}_2}) = 0. \tag{2.39}$$

PROOF. From the continuous dependence of the elastic solution of (2.21)–(2.23) upon the data f, b, g and from (2.38) we get

$$\lim_{t\to\infty}(\|\hat{u}(t) - \hat{\bar{u}}\|_{\boldsymbol{H}_1} + \|\hat{\sigma}(t) - \hat{\bar{\sigma}}\|_{\mathscr{H}_2} = 0.$$

We can use now (2.37) to obtain (2.39).

Remark 2.6 *Let the data b, g, f be constant in time and let* $(u; \sigma)$ *be the solution of* (1.30)–(1.23), (2.1) *and* $(\hat{u}; \hat{\sigma})$ *be the solution of* (2.21)–(2.23). *In this case the differential equation* (1.49) *is an autonomous one and* $(\hat{u} - \tilde{u}; \hat{\sigma} - \tilde{\sigma})$ *is a stationary point of A. From Corollary* 2.1 *we can obtain*

$$\|u(t) - \hat{u}\|_{\boldsymbol{H}_1} + \|\sigma(t) - \hat{\sigma}\|_{\mathscr{H}_2}$$
$$\leq \bar{C}(\|u_0 - \hat{u}\|_{\boldsymbol{H}_1} + \|\sigma_0 - \hat{\sigma}_0\|_{\mathscr{H}_2})\exp(-Ckt) \qquad \forall t \in \mathbb{R}_+. \tag{2.40}$$

PROOF OF THEOREM 2.4 easily follows from (2.32)–(2.33) and Lemma A.4.16.

3 An approach to perfect plasticity

In this section we use the same notations as in Section 1 and we consider the problem (1.20)–(1.23) in which (1.21) is replaced by the constitutive equation (1.8), (1.7). We study the behaviour of the solution $(u_\mu; \sigma_\mu)$ of this problem when the viscosity constant μ tends to zero. More precisely, denoting $\dot{u}_\mu = v_\mu$, we prove that $(v_\mu; \sigma_\mu)$ converges when μ tends to zero to a pair of functions $(v;\sigma)$ satisfying the mixed problem (3.9)–(3.12). We prove that this last problem can be considered as a weak formulation of a quasistatic problem for perfectly plastic materials and we also prove the uniqueness in σ for this problem. Finally, some 'pathological' examples are presented in order to illustrate the 'bad' behaviour of the velocity field in perfect-plasticity problems.

3.1 A convergence result

Let K be a closed convex subset of $\mathscr{S}_N$ such that 0 belongs to the interior of K; denote by $P_K\colon \mathscr{S}_N \to K$ the projector map on K and let $\mathscr{A}$ be a fourth-order tensor. For every viscosity constant $\mu > 0$ we consider the elastic–viscoplastic constitutive law given by

$$\dot{\varepsilon}(u) = \tilde{\mathscr{A}}\dot{\sigma} + \tilde{G}_\mu(\sigma) \tag{3.1}$$

where $\tilde{\mathscr{A}} = \mathscr{A}^{-1}$ and $\tilde{G}_\mu\colon \mathscr{S}_N \to \mathscr{S}_N$ is defined by

$$\tilde{G}_\mu(\sigma) = \frac{1}{\mu}(\sigma - P_K\sigma) \qquad \forall \sigma \in \mathscr{S}_N. \tag{3.2}$$

As it was pointed out in Remark 1.3(3) the function (3.2) satisfies (1.25). Hence, if (1.24), (1.26)–(1.29) are satisfied, by Theorem 1.1 we get that for every $\mu > 0$ there exists a unique pair of functions $(u_\mu;\sigma_\mu)$ satisfying (1.20), (1.22), (1.23), (3.1) and having the regularity $u_\mu \in C^1(\mathbb{R}_+, \boldsymbol{H}_1)$, $\sigma_\mu \in C^1(\mathbb{R}_+, \mathscr{H}_2)$. We use the notation $\dot{u}_\mu = v_\mu$.

In order to study the convergence of $(v_\mu;\sigma_\mu)$ when $\mu \to 0$ we introduce the following notations:

$$\mathscr{K} = \{\sigma \in \mathscr{H} \mid \sigma(x) \in K \text{ a.e. in } \Omega\} \tag{3.3}$$

$$\mathscr{H}_{ad}(t) = \{\tau \in \mathscr{H}_2 \mid \tau v = f(t) \text{ on } \Gamma_2\}, \tag{3.4}$$

for all $t \in \mathbb{R}_+$. Clearly $\mathscr{K}$ is a closed convex subset of $\mathscr{H}$; moreover, using the notation $\mathscr{H}_\infty = [L^\infty(\Omega)]_s^{N\times N}$, where $\mathscr{H}_\infty$ is endowed with the canonical norm $\|\cdot\|_{\mathscr{H}_\infty}$, we suppose that the following 'safety assumption' is also fulfilled:

for every $T > 0$ there exists a function $\chi \in W^{1,\infty}(0, T, \mathscr{H}_\infty)$ having the properties

$$\left.\begin{array}{ll}
\text{(a)} & \operatorname{Div}\chi(t) + b(t) = 0 \text{ in } \Omega\\
\text{(b)} & \chi(t)v = f(t) \text{ on } \Gamma_2\ \forall t \in [0,T]\\
\text{(c)} & \chi(0) = \sigma_0 \text{ in } \Omega\\
\text{(d)} & \text{there exists } \delta > 0 \text{ such that } \chi(t) + \tau \in \mathscr{K}\ \forall t \in [0,T] \text{ and}\\
 & \tau \in \mathscr{H}_\infty,\ \|\tau\|_{\mathscr{H}_\infty} \le \delta.
\end{array}\right\} \tag{3.5}$$

We have the following result:

Theorem 3.1 *Under the hypotheses* (1.24), (1.26)–(1.29), (3.5), *for every* $T > 0$ *there exists a couple of functions* $v \in L^2_w(0, T, BD(\Omega))$, $\sigma \in W^{1,2}(0, T, \mathscr{H})$ *such that (extracting a subsequence denoted again by* (σ_μ), (v_μ)*) we have*

$$\sigma_\mu \to \sigma \text{ weak* in } L^\infty(0, T, \mathscr{H}) \tag{3.6}$$

$$\dot{\sigma}_\mu \to \dot{\sigma} \text{ weak in } L^2(0, T, \mathscr{H}) \tag{3.7}$$

$$v_\mu \to v \text{ weak* in } L^2_w(0, T, BD(\Omega)), \tag{3.8}$$

when $\mu \to 0$. *Moreover,* v *and* σ *satisfying the following problem:*

$$\sigma(t) \in \mathscr{K} \cap \mathscr{H}_{ad}(t) \text{ for all } t \in [0, T] \tag{3.9}$$

$$\langle \tilde{\mathscr{A}}\dot{\sigma}(t), \tau - \sigma(t)\rangle_{\mathscr{H}} + \langle v(t), \operatorname{Div} \tau - \operatorname{Div} \sigma(t)\rangle_{\boldsymbol{H}}$$
$$\geq \langle \gamma_2\tau - \gamma_2\sigma(t), \dot{g}(t)\rangle_{\boldsymbol{H}_\Gamma', \boldsymbol{H}_\Gamma} \qquad \forall \tau \in \mathscr{K} \cap \mathscr{H}_{ad}(t) \text{ a.e. on } (0, T) \tag{3.10}$$

$$\operatorname{Div} \sigma(t) + b(t) = 0 \qquad \text{in } \Omega, \quad \forall t \in [0, T] \tag{3.11}$$

$$\sigma(0) = \sigma_0 \qquad \text{in } \Omega. \tag{3.12}$$

PROOF. Let $T > 0$; for every $\mu > 0$ we have

$$\left.\begin{aligned} &\operatorname{Div} \sigma_\mu + b = 0 &(3.13)\\ &\tilde{\mathscr{A}}\dot{\sigma}_\mu + \tilde{G}_\mu(\sigma_\mu) = \varepsilon(v_\mu) &(3.14)\end{aligned}\right] \quad \text{in } (0, T) \times \Omega$$

$$v_\mu = \dot{g} \quad \text{on} \quad (0, T) \times \Gamma_1, \qquad \sigma_\mu \nu = f \quad \text{on} \quad (0, T) \times \Gamma_2 \tag{3.15}$$

$$\sigma_\mu(0) = \sigma_0. \tag{3.16}$$

We consider the function $\mathscr{G}_\mu : \mathscr{H} \to \mathbb{R}_+$ defined by

$$\mathscr{G}_\mu(\sigma) = \frac{1}{2\mu} \|\sigma - P_K\sigma\|^2_{\mathscr{H}} \tag{3.17}$$

and since $\tilde{G}_\mu$ is the differential of the convex function $\sigma \to (1/2\mu)|\sigma - P_K\sigma|^2$ on $\mathscr{S}_N$, from Lemma A.2.4, we have the inequality

$$\mathscr{G}_\mu(\tau) - \mathscr{G}_\mu(\sigma) \geq \langle \tilde{G}_\mu(\sigma), \tau - \sigma\rangle_{\mathscr{H}} \tag{3.18}$$

for all $\sigma, \tau \in \mathscr{H}$.

A priori estimates I. Let $\bar{\sigma}_\mu = \sigma_\mu - \chi$; taking as test function $\bar{\sigma}_\mu$, from (3.14) we get

$$\langle \tilde{\mathscr{A}}\dot{\bar{\sigma}}_\mu, \bar{\sigma}_\mu\rangle_{\mathscr{H}} + \langle \tilde{G}_\mu(\sigma_\mu), \bar{\sigma}_\mu\rangle_{\mathscr{H}} = \langle \varepsilon(v_\mu), \bar{\sigma}_\mu\rangle_{\mathscr{H}} - \langle \tilde{\mathscr{A}}\dot{\chi}, \bar{\sigma}_\mu\rangle_{\mathscr{H}} \tag{3.19}$$

a.e. in $(0, T)$. But from (3.5) we have

$$\operatorname{Div} \bar{\sigma}_\mu = 0 \qquad \text{in } (0, T) \times \Omega \tag{3.20}$$

$$\bar{\sigma}_\mu \nu = 0 \qquad \text{in } (0, T) \times \Gamma_2 \tag{3.21}$$

and taking $v_1 = z_1\dot{g} \in C(0, T, \boldsymbol{H}_1)$ (see (2.30) in Chapter 2), by (2.29) in

Chapter 2 we have

$$v_\mu - v_1 = 0 \qquad \text{on } \Gamma_1 \times (0, T) \tag{3.22}$$

From (3.20)–(3.22) and (2.59) of Chapter 2 we deduce $\langle \varepsilon(v_\mu), \bar{\sigma}_\mu \rangle_{\mathscr{H}} = \langle \varepsilon(v_1), \bar{\sigma}_\mu \rangle_{\mathscr{H}}$ $\forall t \in [0, T]$; hence (3.19) becomes

$$\langle \tilde{\mathscr{A}} \dot{\bar{\sigma}}_\mu, \bar{\sigma}_\mu \rangle_{\mathscr{H}} + \langle \tilde{G}_\mu(\sigma_\mu), \bar{\sigma}_\mu \rangle = \langle \varepsilon(v_1), \bar{\sigma}_\mu \rangle_{\mathscr{H}} - \langle \tilde{\mathscr{A}} \dot{\chi}, \bar{\sigma}_\mu \rangle_{\mathscr{H}} \text{ a.e. in } (0, T). \tag{3.23}$$

Using (3.5d) and (3.17) we have $\mathscr{G}_\mu(\chi(t)) = 0$ for all $t \in [0, T]$; hence by (3.18) we get $\langle \tilde{G}_\mu \sigma_\mu, \bar{\sigma}_\mu \rangle_{\mathscr{H}} \geq \mathscr{G}_\mu(\sigma_\mu)$ for all $t \in [0, T]$; from (3.23) we deduce

$$\langle \tilde{\mathscr{A}} \dot{\bar{\sigma}}_\mu, \bar{\sigma}_\mu \rangle_{\mathscr{H}} + \mathscr{G}_\mu(\sigma_\mu) \leq \langle \varepsilon(v_1), \bar{\sigma}_\mu \rangle_{\mathscr{H}} - \langle \tilde{\mathscr{A}} \dot{\chi}, \bar{\sigma}_\mu \rangle_{\mathscr{H}} \tag{3.24}$$

a.e. on $(0, T)$ and since $\mathscr{G}_\mu(\sigma_\mu) \geq 0$ we obtain

$$\langle \tilde{\mathscr{A}} \dot{\bar{\sigma}}_\mu, \bar{\sigma}_\mu \rangle_{\mathscr{H}} \leq \|\varepsilon(v_1)\|_{\mathscr{H}} \|\bar{\sigma}_\mu\|_{\mathscr{H}} + \|\tilde{\mathscr{A}} \dot{\chi}\|_{\mathscr{H}} \|\bar{\sigma}_\mu\|_{\mathscr{H}} \tag{3.25}$$

a.e. on $(0, T)$. From (3.5c) and (3.16) we have $\bar{\sigma}_\mu(0) = 0$; hence by integration and using (1.24), (3.25) becomes

$$\|\bar{\sigma}_\mu(t)\|_{\mathscr{H}}^2 \leq C \int_0^t (\|\varepsilon(v_1)\|_{\mathscr{H}} + \|\tilde{\mathscr{A}} \dot{\chi}\|_{\mathscr{H}}) \|\sigma_\mu\|_{\mathscr{H}}$$

for all $t \in [0, T]$. Using Lemma A.4.13 we get

$$(\bar{\sigma}_\mu) \text{ (and } (\sigma_\mu)) \text{ is a bounded sequence in } L^\infty(0, T, \mathscr{H}). \tag{3.26}$$

From (3.24) and (3.26) it results

$$\int_0^T \mathscr{G}_\mu(\sigma_\mu)\, \mathrm{d}t \leq C \tag{3.27}$$

and using (3.23) we get

$$\int_0^T \langle \tilde{G}_\mu(\sigma_\mu), \bar{\sigma}_\mu \rangle_{\mathscr{H}}\, \mathrm{d}t \leq C. \tag{3.28}$$

Let $\tau \in \mathscr{H}_\infty$ such that $\|\tau\|_{\mathscr{H}_\infty} \leq \delta$; by (3.5d) we have $\chi + \tau \in \mathscr{K}$; hence $\mathscr{G}_\mu(\chi + \tau) = 0$ for all $t \in [0, T]$; using (3.18) we get $\langle \tilde{G}_\mu(\sigma_\mu), \chi + \tau - \sigma_\mu \rangle \leq \mathscr{G}_\mu(\chi + \tau) - \mathscr{G}_\mu(\sigma_\mu) \leq 0$ hence

$$\langle \tilde{G}_\mu(\sigma_\mu), \tau \rangle_{\mathscr{H}} \leq \langle \tilde{G}_\mu(\sigma_\mu), \bar{\sigma}_\mu \rangle \qquad \forall t \in [0, T] \tag{3.29}$$

Moreover, we have

$$\|\tilde{G}_\mu(\sigma_\mu)\|_{[L^1(\Omega)]_s^{N \times N}} = \frac{1}{\delta} \sup_{\substack{\tau \in \mathscr{H}_\infty \\ \|\tau\|_{\mathscr{H}_\infty} \leq \delta}} \langle \tilde{G}_\mu(\sigma_\mu), \tau \rangle_{\mathscr{H}} \leq \frac{1}{\delta} \langle \tilde{G}_\mu(\sigma_\mu), \bar{\sigma}_\mu \rangle_{\mathscr{H}} \tag{3.30}$$

for all $t \in [0, T]$ and by (3.28) we deduce

$$(\tilde{G}_\mu(\sigma_\mu)) \text{ is a bounded sequence in } L^1(0, T, [L^1(\Omega)]_s^{N\times N}). \tag{3.31}$$

A priori estimates II. We now take $\dot{\bar{\sigma}}_\mu$ as test function; using (3.14), since by (3.20)–(3.22) we have $\langle \varepsilon(v_\mu), \dot{\bar{\sigma}}_\mu \rangle_{\mathscr{H}} = \langle \varepsilon(v_1), \dot{\bar{\sigma}}_\mu \rangle_{\mathscr{H}}$ it follows that

$$\langle \tilde{\mathscr{A}} \dot{\bar{\sigma}}_\mu, \dot{\bar{\sigma}}_\mu \rangle_{\mathscr{H}} + \langle \tilde{G}_\mu(\sigma_\mu), \dot{\bar{\sigma}}_\mu \rangle_{\mathscr{H}} = \langle \varepsilon(v_1), \dot{\bar{\sigma}}_\mu \rangle_{\mathscr{H}} - \langle \tilde{\mathscr{A}} \chi, \dot{\bar{\sigma}}_\mu \rangle_{\mathscr{H}}$$

a.e. on $(0, T)$. By integration we get

$$\begin{aligned}\int_0^T \langle \tilde{\mathscr{A}} \dot{\bar{\sigma}}_\mu, \dot{\bar{\sigma}}_\mu \rangle_{\mathscr{H}} \,\mathrm{d}t + \int_0^T \langle \tilde{G}_\mu(\sigma_\mu), \dot{\sigma}_\mu \rangle_{\mathscr{H}} \,\mathrm{d}t = \int_0^T \langle \tilde{G}_\mu(\sigma_\mu), \dot{\chi} \rangle_{\mathscr{H}} \,\mathrm{d}t \\ + \int_0^T \langle \varepsilon(v_1), \dot{\bar{\sigma}}_\mu \rangle_{\mathscr{H}} \,\mathrm{d}t \\ - \int_0^T \langle \tilde{\mathscr{A}} \dot{\chi}, \dot{\bar{\sigma}}_\mu \rangle_{\mathscr{H}} \,\mathrm{d}t\end{aligned}$$

and, using (1.24), (3.31) we obtain

$$\alpha |||\dot{\bar{\sigma}}_\mu|||^2 + \int_0^T \langle \tilde{G}_\mu(\sigma_\mu), \dot{\sigma}_\mu \rangle_{\mathscr{H}} \,\mathrm{d}t \le C_1 + C_2 \tag{3.32}$$

where by $|||\cdot|||$ we denote the norm on the space $L^2(0, T, \mathscr{H})$.

But

$$\int_0^T \langle \tilde{G}_\mu(\sigma_\mu), \dot{\sigma}_\mu \rangle_{\mathscr{H}} \,\mathrm{d}t = \mathscr{G}_\mu(\sigma_\mu(T)) - \mathscr{G}_\mu(\sigma_\mu(0)) \ge -\mathscr{G}_\mu(\sigma_0) = 0$$

(see (3.5c,d)). Hence, from (3.32) we deduce

$$(\dot{\bar{\sigma}}_\mu) \text{ and } (\dot{\sigma}_\mu) \text{ are bounded sequences in } L^2(0, T, \mathscr{H}). \tag{3.33}$$

From (3.23) we deduce

$$\langle \tilde{G}_\mu(\sigma_\mu), \bar{\sigma}_\mu \rangle_{\mathscr{H}} \le \|\varepsilon(v_1)\|_{\mathscr{H}} \|\bar{\sigma}_\mu\|_{\mathscr{H}} + \|\tilde{\mathscr{A}} \dot{\chi}\|_{\mathscr{H}} \|\sigma_\mu\|_{\mathscr{H}} + \|\tilde{\mathscr{A}} \dot{\bar{\sigma}}_\mu\|_{\mathscr{H}} \|\bar{\sigma}_\mu\|_{\mathscr{H}}$$

a.e. on $(0, T)$; hence by (3.26) and (3.33) we find that $\langle \tilde{G}_\mu(\sigma_\mu), \bar{\sigma}_\mu \rangle_{\mathscr{H}}$ is a bounded sequence in $L^2(0, T, \mathbb{R})$. Using now (3.30) we get that $(\tilde{G}_\mu(\sigma_\mu))$ is a bounded sequence in $L^2(0, T, [L^1(\Omega)]_s^{N\times N})$. By (3.14) and (3.33) we deduce

$$(\varepsilon(v_\mu)) \text{ is a bounded sequence in } L^2(0, T, [L^1(\Omega)]_s^{N\times N}). \tag{3.34}$$

It follows from (2.40) in Chapter 2 that due to the boundary condition (3.22):

$$(v_\mu) \text{ is a bounded sequence in } L^2(0, T, BD(\Omega)). \tag{3.35}$$

The behaviour of v_μ, σ_μ when $\mu \to 0$. From (3.26) and (3.33) we deduce that there exists $\sigma \in W^{1,2}(0, T, \mathscr{H})$ such that extracting a subsequence, again denoted by (σ_μ), (3.6) and (3.7) hold. Moreover, since $BD(\Omega)$ is the dual of a Banach space C, then $L^2_w(0, T, BD(\Omega))$ is the dual of the space $L^2(0, T, C)$ (see Remark 2.4 and Theorem 3.3(1) in Chapter 2); so, using (3.35) we deduce that there exists $v \in L^2_w(0, T, BD(\Omega))$ such that extracting a subsequence again denoted by (v_μ), (3.8) is fulfilled.

Let us prove that (3.9)–(3.12) are satisfied. Indeed, (3.6) and (3.7) imply that $\sigma_\mu(t) \to \sigma(t)$ weakly in $\mathscr{H}$ for all $t \in [0, T]$. Hence (3.12) follows from (3.16). Moreover, for all $t \in [0, T]$, $\mu > 0$ and $\varphi \in \boldsymbol{D}$, from (2.3) of Chapter 2 and (3.13) we have

$$\langle b(t), \varphi \rangle_{\boldsymbol{H}} = -\langle \operatorname{Div} \sigma_\mu(t), \varphi \rangle_{\boldsymbol{H}} = \langle \sigma_\mu(t), \varepsilon(\varphi) \rangle_{\mathscr{H}};$$

hence

$$\langle b(t), \varphi \rangle_{\boldsymbol{H}} = \langle \sigma(t), \varepsilon(\varphi) \rangle_{\mathscr{H}} = -\langle \operatorname{Div} \sigma(t), \varphi \rangle_{\boldsymbol{D}', \boldsymbol{D}},$$

which implies (3.11). In the same way from (2.44) of Chapter 2, (3.11), (3.6), and (3.7), we deduce $\gamma_2 \sigma_\mu(t) \to \gamma_2 \sigma(t)$ weak $*$ in H'_Γ for all $t \in [0, T]$. Hence from $(3.15)_2$ we get

$$\sigma v|_{\Gamma_2} = f \qquad \text{for all } t \in [0, T]. \tag{3.36}$$

From (3.18) we get that for all $\mu > 0$ we have

$$\mathscr{G}_\mu(\sigma) \leq \mathscr{G}_\mu(\sigma_\mu) - \langle \tilde{G}_\mu(\sigma), \sigma_\mu - \sigma \rangle \qquad \forall t \in [0, T]$$

hence by (3.2) we get $\mu \mathscr{G}_\mu(\sigma) \leq \mu \mathscr{G}_\mu(\sigma_\mu) - \langle \sigma - P_{\boldsymbol{K}} \sigma, \sigma_\mu - \sigma \rangle_{\mathscr{H}}$ for all $t \in [0, T]$. It follows that

$$\int_0^T \mu \mathscr{G}_\mu(\sigma)\, \mathrm{d}t \leq \mu \int_0^T \mathscr{G}_\mu(\sigma_\mu)\, \mathrm{d}t - \int_0^T \langle \sigma - P_{\boldsymbol{K}} \sigma, \sigma_\mu - \sigma \rangle_{\mathscr{H}}\, \mathrm{d}t$$

and using (3.27), (3.6), and (3.17) we get

$$\frac{1}{2} \int_0^T \|\sigma - P_{\boldsymbol{K}} \sigma\|^2_{\mathscr{H}}\, \mathrm{d}t = 0; \qquad \text{therefore } \sigma = P_{\boldsymbol{K}} \sigma \text{ a.e. in } \forall t \in [0, T].$$

Using (3.3) we get

$$\sigma(t) \in \mathscr{K} \qquad \forall t \in [0, T], \tag{3.37}$$

and from (3.37), (3.36), and (3.4) we obtain (3.9). In order to prove (3.10) we multiply (3.14) by $\tau - \sigma_\mu$, where $\tau \in L^2(0, T, \mathscr{H}_2)$, $\tau(t) \in \mathscr{K} \cap \mathscr{H}_{ad}(t)$ a.e. on $(0, T)$, and we use (2.44) of Chapter 2 and (3.22). We deduce

$$\langle \tilde{\mathscr{A}} \dot{\sigma}_\mu, \tau - \sigma_\mu \rangle_{\mathscr{H}} + \langle \tilde{G}_\mu(\sigma_\mu), \tau - \sigma_\mu \rangle_{\mathscr{H}} + \langle v_\mu, \operatorname{Div} \tau - \operatorname{Div} \sigma_\mu \rangle_{\boldsymbol{H}}$$
$$= \langle \gamma_2 \tau - \gamma_2 \sigma_\mu, \dot{g} \rangle_{\boldsymbol{H}'_\Gamma, \boldsymbol{H}_\Gamma} \qquad \text{a.e. on } (0, T).$$

Since from (3.18) it follows that $\langle \tilde{G}_\mu(\sigma_\mu), \tau - \sigma_\mu \rangle \leq 0$ a.e. on $(0, T)$ and from (3.13), (3.11) we have $\operatorname{Div} \sigma_\mu = \operatorname{Div} \sigma$, we deduce

$$\langle \tilde{\mathscr{A}} \dot{\sigma}_\mu, \tau - \sigma_\mu \rangle_{\mathscr{H}} + \langle v_\mu, \operatorname{Div} \tau - \operatorname{Div} \sigma \rangle_{\boldsymbol{H}} \geq \langle \gamma_2 \tau - \gamma_2 \sigma_\mu, \dot{g} \rangle_{\boldsymbol{H}'_\Gamma, \boldsymbol{H}_\Gamma}$$

a.e. on $(0, T)$.

By integration we have

$$\int_0^s \langle \tilde{\mathscr{A}}\dot{\sigma}_\mu, \tau\rangle_{\mathscr{H}}\,dt + \int_0^s \langle v_\mu, \operatorname{Div}\tau - \operatorname{Div}\sigma\rangle_{\boldsymbol{H}}\,dt$$
$$\geq \int_0^s \langle \tilde{\mathscr{A}}\dot{\sigma}_\mu, \sigma_\mu\rangle_{\mathscr{H}}\,ds + \int_0^s \langle \gamma_2\tau - \gamma_2\sigma_\mu, \dot{g}\rangle_{\boldsymbol{H}'_\Gamma,\boldsymbol{H}_\Gamma}\,dt \qquad \forall s \in [0, T]. \quad (3.38)$$

By a lower semicontinuity argument from (3.6) and (3.7) we deduce

$$\varliminf_{\mu} \int_0^s \langle \tilde{\mathscr{A}}\dot{\sigma}_\mu, \sigma_\mu\rangle_{\mathscr{H}}\,dt \geq \int_0^s \langle \tilde{\mathscr{A}}\dot{\sigma}, \sigma\rangle_{\mathscr{H}}\,dt \qquad (3.39)$$

for all $s \in [0, T]$. Let us also remark that from (3.6), (3.13) and (3.11) we have $\gamma_2\sigma_\mu \to \gamma_2\sigma$ weakly* in $L^\infty(0, T, \boldsymbol{H}'_\Gamma)$. Hence, passing to the limit in (3.38) and using (3.6), (3.7), (3.8), (3.39) we get

$$\int_0^s \langle \tilde{\mathscr{A}}\dot{\sigma}, \tau\rangle_{\mathscr{H}}\,dt + \int_0^s \langle v, \operatorname{Div}\tau - \operatorname{Div}\sigma\rangle_{\boldsymbol{H}}\,dt$$
$$\geq \int_0^s \langle \tilde{\mathscr{A}}\dot{\sigma}, \sigma\rangle_{\mathscr{H}}\,dt + \int_0^s \langle \gamma_2\tau - \gamma_2\sigma, \dot{g}\rangle_{\boldsymbol{H}'_\Gamma,\boldsymbol{H}_\Gamma}\,dt$$

for all $s \in [0, T]$. After a classical use of Lebesgue points for an L^1 function we get (3.10). The proof of Theorem 3.1 is complete.

Remark 3.1 *Since* $\sigma \in W^{1,2}(0, T, \mathscr{H})$ *satisfies* (3.11), *from* (1.26) *we get* $\operatorname{Div}\sigma \in W^{1,2}(0, T, \boldsymbol{H})$ *then actually we obtain the regularity* $\sigma \in W^{1,2}(0, T, \mathscr{H}_2)$.

3.2 Quasistatic processes in perfect plasticity

In this section we give the physical motivation of problem (3.9)–(3.12). For this, let us start by considering the constitutive equations of a perfect-plastic material:

$$\dot{\varepsilon} = \dot{\varepsilon}^E + \dot{\varepsilon}^I \qquad (3.40)$$

$$\dot{\varepsilon}^E = \tilde{\mathscr{A}}\dot{\sigma} \qquad (3.41)$$

$$\sigma \in K, \qquad \dot{\varepsilon}^I(\tau - \sigma) \leq 0 \qquad \forall \tau \in K \qquad (3.42)$$

where K is a closed convex of $\mathscr{S}_N$.

Equation (3.40) shows that the total strain rate is the sum of the elastic part $\dot{\varepsilon}^E$ and the non-elastic part $\dot{\varepsilon}^I$; the elastic part $\dot{\varepsilon}^E$ is given by equation (3.41) in which $\tilde{\mathscr{A}}$ denotes the elastic compliance tensor; (3.42) describes the rate of the irreversible part of the strain tensor and is equivalent to the multivalued equation $\dot{\varepsilon}^I \in \partial\psi_K(\sigma)$ where $\partial\psi_K$ represents the subdifferential mapping of the indicator function ψ_K (see Section A.2.1).

The strong formulation of a quasistatic process on a finite time interval $[0, T]$ involving the perfect plastic considered model consists of finding the displacement function $u: [0, T] \times \Omega \to \mathbb{R}^N$ and the stress function $\sigma: [0, T] \times \Omega \to \mathscr{S}_N$ such that (3.40)–(3.42) are satisfied in $\Omega \times (0, T)$ and moreover

$$\varepsilon = \tfrac{1}{2}(\nabla u + \nabla^{\mathrm{T}} u). \tag{3.43}$$

$$\operatorname{Div} \sigma + b = 0 \qquad \text{in } (0, T) \times \Omega \tag{3.44}$$

$$u = g \quad \text{on } (0, T) \times \Gamma_1, \qquad \sigma\nu = f \quad \text{on } (0, T) \times \Gamma_2 \tag{3.45}$$

$$u(0) = u_0, \quad \sigma(0) = \sigma_0 \qquad \text{in } \Omega. \tag{3.46}$$

In (3.44)–(3.46) b represents the body forces, g and f are the boundary data and u_0, σ_0 are the initial data. Supposing that (3.40)–(3.46) has a smooth solution $(u; \sigma)$, let $\dot{u} = v$ and from (3.40)–(3.43) we deduce

$$(\tilde{\mathscr{A}}\dot{\sigma})\cdot(\tau - \sigma) \geq (\varepsilon(v))\cdot(\tau - \sigma) \qquad \forall \tau \in K \quad \text{a.e. on } (0, T) \times \Omega. \tag{3.47}$$

From (3.45) we have $v = \dot{g}$ on $\Gamma_1 \times (0, T)$; defining $\mathscr{K}$ and $\mathscr{H}_{ad}(t)$ by (3.3), (3.4), if we take in (3.47) $\tau \in \mathscr{K} \cap \mathscr{H}_{ad}(t)$ after integrating on Ω and using (2.44) in Chapter 2, we obtain (3.10). From (3.42) and $(3.45)_2$ we have (3.9) and (3.44), $(3.46)_2$ imply (3.11), (3.12). Hence, we arrive at the weak formulation of the problem (3.40)–(3.46):

> Find the velocity field $v: [0, T] \times \Omega \to \mathbb{R}^N$ and the stress tensor $\sigma: [0, T] \times \Omega \to \mathscr{S}_N$ such that (3.9)–(3.12) are satisfied.

The above considerations allow us to consider problem (3.9)–(3.12) as describing *quasistatic processes for elastic–perfectly plastic materials.*

From this point of view Theorem 3.1 also represents an existence result in perfect plasticity and at the same time shows that the solution $(v; \sigma)$ of this 'irregular' problem can be approximate in the sense given by (3.6)–(3.8) by the solution of the 'regular' problem (3.13)–(3.16).

Remark 3.2 *In the study of problems* (3.9)–(3.12) *a uniqueness result in σ can be easily proved. Indeed, if $(v_1; \sigma_1)$, $(v_2; \sigma_2)$ are two solutions of* (3.9)–(3.12) *satisfying the regularity of Theorem* 3.1, *we have*

$$\langle \tilde{\mathscr{A}}\dot{\sigma}_1, \tau - \sigma_1 \rangle_{\mathscr{H}} \geq \langle \gamma_2\tau - \gamma_2\sigma_1, \dot{g} \rangle_{\boldsymbol{H}'_\Gamma, \boldsymbol{H}_\Gamma}$$

$$\langle \tilde{\mathscr{A}}\dot{\sigma}_2, \tau - \sigma_2 \rangle_{\mathscr{H}} \geq \langle \gamma_2\tau - \gamma_2\sigma_2, \dot{g} \rangle_{\boldsymbol{H}'_\Gamma, \boldsymbol{H}_\Gamma}$$

a.e. on $(0, T)$ *for every* $\tau \in \mathscr{K} \cap \mathscr{H}_{ad}(t)$ *with* $\operatorname{Div} \tau + b = 0$ *in* $(0, T) \times \Omega$. *Taking* $\tau = \sigma_2$ *and* $\tau = \sigma_1$ *in the above inequalities and adding the results we get* $\langle \mathscr{A}\dot{\sigma}_1 - \mathscr{A}\dot{\sigma}_2, \sigma_1 - \sigma_2 \rangle_{\mathscr{H}} \leq 0$ *a.e. on* $(0, T)$. *Using* (3.12) *and* (1.24) *we get* $\|\sigma_1(t) - \sigma_2(t)\|^2_{\mathscr{H}} \leq 0$ *for all* $t \in [0, T]$. *It follows that* $\sigma_1 = \sigma_2$.

3.3 Some 'pathological' examples

In this section we present two one-dimensional examples in problem (3.9)–(3.12) inspired from the papers of Suquet (1981b, 1982). These examples show that we have non-uniqueness in the velocity field for problem (3.9)–(3.12) (Example 3.1) and also that the velocity field in problem (3.40)–(3.46) can be a discontinuous function (Example 3.2).

Example 3.1 Let us consider

$$\Omega = (0, 1), \quad \Gamma_1 = \{0, 1\}, \quad \Gamma_2 = \varnothing, \quad \mathscr{A} = 1, \quad K = [-1, 1], \quad T > 0,$$

$$b = 0, \quad \dot{g}(0, t) = 0, \quad \dot{g}(1, t) = \gamma > \frac{1}{T} \quad \text{and} \quad \sigma_0 = 0.$$

The assumptions of Theorem 3.1 are satisfied (for (3.5) take $\chi = 0$ and $\delta = 1$). The elastic–viscoplastic problem (3.13)–(3.16) becomes

$$\left.\begin{array}{ll} \left.\begin{array}{l} \dfrac{d\sigma_\mu}{dx} = 0 \\ \dot{\sigma}_\mu = \varepsilon(v_\mu) \quad \text{if } |\sigma| \le 1 \\ \dot{\sigma}_\mu + \tilde{G}_\mu(\sigma_\mu) = \varepsilon(v_\mu) \quad \text{if } |\sigma| > 1 \end{array}\right\} & \text{for } (t, x) \in (0, T) \times (0, 1) \\ \left.\begin{array}{l} v_\mu(0, t) = 0 \\ v_\mu(1, t) = \gamma \end{array}\right\} \quad \forall t \in [0, T] & \\ \sigma_\mu(x, 0) = 0 \qquad \forall x \in (0, 1) & \end{array}\right\} \tag{3.48}$$

where

$$\tilde{G}_\mu(\sigma) = \frac{1}{\mu}(\sigma - P_K\sigma) = \begin{cases} \dfrac{1}{\mu}(\sigma + 1) & \text{if } \sigma < -1 \\ 0 & \text{if } -1 \le \sigma \le 1 \\ \dfrac{1}{\mu}(\sigma - 1) & \text{if } \sigma > 1. \end{cases}$$

It is easy to see that the solution of (3.48) is given by

$$\sigma_\mu(t, x) = \begin{cases} \gamma t & \text{for } t \in \left[0, \dfrac{1}{\gamma}\right] \\ -\mu\gamma\, e^{(1 - t\gamma)/\mu\gamma} + \mu\gamma + 1 & \text{for } t \in \left[\dfrac{1}{\gamma}, T\right] \end{cases} \quad \forall x \in (0, 1) \tag{3.49}$$

$$v_\mu(t, x) = \gamma x \qquad \text{for all } x \in (0, 1) \quad \text{and} \quad t \in [0, T]. \tag{3.50}$$

Since in this particular case $\mathscr{H}_2 = H^1(\Omega) \subset C(\bar{\Omega})$ (see (1.18) in Chapter 2), $\mathscr{H}_{ad}(t) = H^1(\Omega)$ for all $t \geq 0$, we have $\mathscr{K} \cap \mathscr{H}_{ad}(t) = \{\tau \in H^1(\Omega) \mid |\tau(x)| \leq 1$ for all $x \in \Omega\}$. Problem (3.9)–(3.12) becomes

$$\left.\begin{aligned}
&\left.\begin{aligned}
&\sigma \in H^1(\Omega), \qquad |\sigma(x,t)| \leq 1 \quad \forall x \in (0,1)\\
&\int_0^1 \dot{\sigma}(\tau - \sigma)\,\mathrm{d}x + \int_0^1 v \frac{\mathrm{d}\tau}{\mathrm{d}x}\,\mathrm{d}x \geq \gamma(\tau(1) - \sigma(1))\\
&\qquad \forall \tau \in H^1(\Omega), \quad |\tau(x)| \leq 1 \text{ in } (0,1)
\end{aligned}\right\} \text{ a.e. on } (0,T)\\
&\frac{\mathrm{d}\sigma}{\mathrm{d}x} = 0\\
&\sigma(x,0) = 0 \qquad \forall x \in (0,1).
\end{aligned}\right\} \tag{3.51}$$

Using Theorem 3.1 we have the existence of a solution $(v; \sigma)$ for (3.51) and moreover from (3.6), (3.8), (3.49), and (3.50) we have

$$\sigma(t,x) = \begin{cases} \gamma t & \text{for } t \in \left[0, \dfrac{1}{\gamma}\right] \\ 1 & \text{for } t \in \left[\dfrac{1}{\gamma}, T\right] \end{cases} \qquad \forall x \in (0,1) \tag{3.52}$$

$$v(t,x) = \gamma x \qquad \forall x \in (0,1), \quad t \in [0,T]. \tag{3.53}$$

From (3.52) it follows that σ has 'a blocking property' at the value $\sigma = 1$ for $t \geq 1/\gamma$; moreover, since problem (3.51) has the uniqueness property in σ (see Remark 3.2), (3.51), and (3.52) imply

$$\int_0^1 v \frac{\mathrm{d}\tau}{\mathrm{d}x}\,\mathrm{d}x \geq \gamma(\tau(1) - 1) \quad \forall \tau \in H^1(\Omega), \qquad |\tau(x)| \leq 1 \text{ in } (0,1) \quad \forall t \in \left[\frac{1}{\gamma}, T\right].$$

It is easy to see that this inequality has more than one solution: for instance, $v = 0$, $v = \gamma$, or $v = 0$ if $x \in (0, x_0]$, $v = \gamma$ if $x \in (x_0, 1)$). Hence, problem (3.51) has also the solutions $(v_i; \sigma)$, $i = 1, 2, 3$, where

$$v_1(t,x) = \begin{cases} \gamma x & \text{if } t \in \left[0, \dfrac{1}{\gamma}\right] \\ 0 & \text{if } t \in \left(\dfrac{1}{\gamma}, T\right] \end{cases} \qquad x \in (0,1)$$

$$v_2(t,x) = \begin{cases} \gamma x & \text{if } t \in \left[0, \dfrac{1}{\gamma}\right] \\ \gamma & \text{if } t \in \left(\dfrac{1}{\gamma}, T\right] \end{cases} \qquad x \in (0,1)$$

$$v_3(t,x)=\begin{cases}\gamma x & \text{if } t\in\left[0,\dfrac{1}{\gamma}\right] \quad x\in(0,1)\\ 0 & \text{if } x\in(0,x_0] \quad t=\left(\dfrac{1}{\gamma},T\right]\\ \gamma & \text{if } x\in(x_0,1) \quad t\in\left(\dfrac{1}{\gamma},T\right]\end{cases}$$

where $x_0 \in (0,1)$. The non-uniqueness in v of the solution of problem (3.51) was proved. Finally, let us notice that $(v_4; \sigma)$ is also a solution of (3.51) where

$$v_4(t,x)=\begin{cases}\gamma x & \text{if } t\in\left[0,\dfrac{1}{\gamma}\right]\\ (t\gamma-1)x^2+(\gamma+1-t\gamma)x & \text{if } t\in\left(\dfrac{1}{\gamma},T\right].\end{cases} \quad x\in(0,1).\tag{3.54}$$

Since the functions (3.53) and (3.54) are continuous on $[0,T]\times(0,1)$ we proved the non-uniqueness in v of the solution of (3.51) even in the class of continuous functions.

Example 3.2 The following example shows that under the assumptions of Theorem 3.1 if $(u;\sigma)$ is a solution of the problem (3.40)–(3.46) then $\dot{u}=v$ can be a discontinuous function. Let $\Omega=(0,1)$, $\Gamma_1=\{0,1\}$, $\Gamma_2=\varnothing$, $\mathscr{A}=1$, $K=[-1,1]$, $T<8$, $g=0$, $\sigma_0=0$, $u_0=0$ and

$$b(t,x)=\begin{cases}2tx & \text{if } 0<x<\frac{1}{2}\\ -2t(1-x) & \text{if } \frac{1}{2}<x<1.\end{cases}\tag{3.55}$$

For all $t\in[0,T]$ we denote by $\tilde{\mathscr{H}}_{ad}(t)$ the set defined by

$$\tilde{\mathscr{H}}_{ad}(t)=\left\{\tau\in H^1(\Omega)\,\middle|\,\frac{d\tau}{dx}+b=0\right\}$$

and using (3.55) we have

$$\tau\in\tilde{\mathscr{H}}_{ad}(t)\Leftrightarrow\tau(t,x)=C_\tau(t)-\begin{cases}tx^2 & \text{if } 0<x\le\frac{1}{2}\\ t(1-x)^2 & \text{if } \frac{1}{2}<x<1\end{cases}\tag{3.56}$$

where C_τ is a function defined on $[0,T]$.

The elastic solution. The elastic solution of problems (3.40)–(3.46) is defined on $[0,T_0]$ (T_0 to be determined) and for every $t\in[0,T_0]$ it minimizes on $\tilde{\mathscr{H}}_{ad}(t)$ the energy function given by $E(\tau)=\frac{1}{2}\int_0^1|\tau(t,x)|^2\,dx$. Since from (3.56) we have $E(\tau)=\frac{1}{160}t^2-\frac{1}{12}C_\tau(t)+\frac{1}{2}C_\tau^2(t)$ for all t $\in[0,T_0]$ and $\tau\in\tilde{\mathscr{H}}_{ad}(t)$

it results that the elastic solution is given by

$$\sigma(t, x) = \frac{t}{12} - \begin{cases} tx^2 & \text{if } 0 < x \leq \frac{1}{2} \\ t(1-x)^2 & \text{if } \frac{1}{2} < x < 1 \end{cases} \quad \forall t \in [0, T_0]. \tag{3.57}$$

The time interval T_0 is determined by the condition $\sigma(t, x) \in K$ for all $(t, x) \in (0, T_0) \times (0, 1)$; since for all $t \in [0, T_0]$ we have $\min_{x \in [0,1]} \sigma(t, x) = \sigma(\frac{1}{2}, t) = t/6$ and $\max_{x \in [0,1]} \sigma(t, x) = t/12 < 1$ we deduce $T_0 = 6$ since at $t = T_0$ the point $x = \frac{1}{2}$ is plasticized.

The elastic–plastic solution. It is easy to see that the safety assumption (3.5) is satisfied if we take $\delta = 1 - T/8$ and

$$\chi(t, x) = \frac{t}{8} - \begin{cases} tx^2 & \text{if } 0 < x \leq \frac{1}{2} \\ t(1-x)^2 & \text{if } \frac{1}{2} < x < 1 \end{cases} \quad \forall t \in [0, T].$$

For $t \in [T_0, T]$ the solution in σ of problems (3.40)–(3.46) is given by

$$\sigma(t, x) = \frac{t}{4} - 1 - \begin{cases} tx^2 & \text{if } 0 < x \leq \frac{1}{2} \\ t(1-x)^2 & \text{if } \frac{1}{2} < x < 1. \end{cases} \tag{3.58}$$

Indeed, by the uniqueness in σ of the solution of (3.40)–(3.46) (consequence of Remark 3.2) it is sufficient to prove that

$$\int_0^1 \dot{\sigma}(\tau - \sigma) \, dx \geq 0 \qquad \forall \tau \in \mathscr{K} \cap \mathscr{H}_{ad}(t) \quad \text{and} \quad t \in [T_0, T] \tag{3.59}$$

where $\mathscr{K} = \{\tau \in H^1(\Omega) \mid |\tau(x)| \leq 1 \text{ for all } x \in \Omega\}$. Let $t \in [T_0, T]$ and $\tau \in \mathscr{K} \cap \mathscr{H}_{ad}(t)$. From (3.56) it results $t/4 - 1 \leq C_\tau(t) \leq 1$ and since by (3.58) we have $\int_0^1 \dot{\sigma}(\tau - \sigma) \, dx = (t/4 - 1 - C_\tau(t))(-\frac{1}{6})$ we deduce that (3.59) is satisfied.

It follows that the solution in σ of (3.40)–(3.46) is given by (3.57) if $t \in [0, T_0]$ and by (3.58) if $t \in [T_0, T]$. Since from (3.58) we deduce $\sigma(t, x) \in (-1, 1)$ for all $x \in (0, 1)$, $x \neq \frac{1}{2}$ and $t \in [T_0, T]$ we get that in $[T_0, T]$ only the point $x = \frac{1}{2}$ is plasticized. Hence in $(0, 1) \backslash \{\frac{1}{2}\}$ the solution in v is elastic; this means that

$$\frac{dv}{dx} = \dot{\sigma} \qquad \text{a.e. in } (0, 1) \quad \forall t \in [T_0, T]. \tag{3.60}$$

By (3.60) and (3.58) we obtain

$$v(t, x) = \frac{x}{4} - \frac{x^3}{3} + \begin{cases} \lambda_1(t) & \text{if } 0 < x < \frac{1}{2} \\ -x + x^2 + \lambda_2(t) & \text{if } \frac{1}{2} < x < 1 \end{cases} \quad \forall t \in [T_0, T], \tag{3.61}$$

where $\lambda_1(t)$ and $\lambda_2(t)$ are determined by the boundary conditions (see (3.45)): $v(0, t) = 0$, $v(1, t) = 0$ for all $t \in [T_0, T]$. It follows that $\lambda_1(t) = 0$, $\lambda_2(t) = \frac{1}{12}$

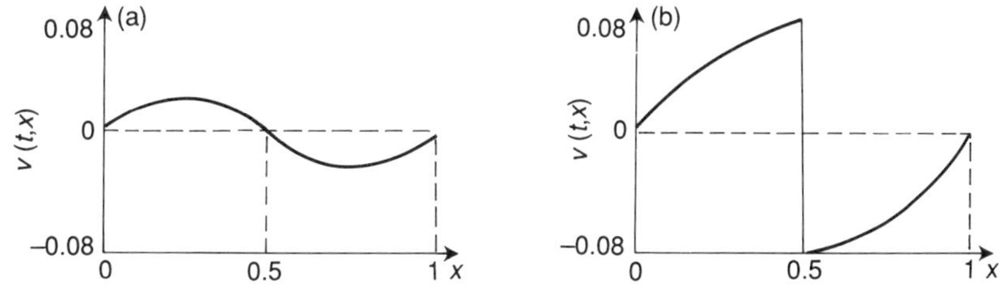

Fig. 3.1 The graphic representation of the velocity field with $T_0 = 6$ in Example 3.2 for (a) $t \in [0, T_0)$; (b) $t \in [T_0, T]$.

for all $t \in [T_0, T]$ and by (3.61) we deduce

$$v(t, x) = \frac{x}{4} - \frac{x^3}{3} + \begin{cases} 0 & \text{if } 0 < x < \frac{1}{2} \\ -x + x^2 + \frac{1}{12} & \text{if } \frac{1}{2} < x < 1 \end{cases} \tag{3.62}$$

for all $t \in [T_0, T]$. So, it results that the velocity field is discontinuous in $x = \frac{1}{2}$, for $t \in [T_0, T]$, as it can be seen in Figure 3.1.

4 A numerical approach

In this section we use an explicit Euler method, an internal approximation technique (possible a finite element one) for the displacement and strain fields and an external one for the stress field, in order to reduce the continuous problem to a recursive sequence of linear algebraic systems. The error is estimated (Theorem 4.1) over a finite time interval.

For large time intervals or for a large Lipschitz constant L of G (usually for metals) the error estimation obtained in Theorem 4.1 is not so useful. The following question arises: 'How large a time step can we choose in order to obtain numerical information about the long-term behaviour of the solution?' To throw some light on this problem only the viscoelastic case is considered. In this case the error is estimated over an infinite time interval provided that the time step τ is restricted to be less than τ_0, a value which depends on the material constants (Theorem 4.2).

It is not established that a critical value $\tau_{cr} > \tau_0$ for which the numerical solution is divergent if $\tau > \tau_{cr}$, but, however, Example 4.1 shows that such a critical value exists and hence the restriction on the time step is also a necessary condition in order to have a bound for the error for an infinite time interval.

Other 'pathological' examples for which Theorem 4.1 can be applied are successfully integrated over a large time interval using the time step given in Theorem 4.2.

For the sake of simplicity let us suppose in this section that G does not depend on time and (1.24)–(1.29) hold. Let us denote by $(\, , \,)_0 \colon V_1 \times V_1 \to \mathbb{R}$ the inner product given by

$$(u, v)_0 = \langle \mathscr{A}\varepsilon(u), \varepsilon(v) \rangle_{\mathscr{H}} \qquad \forall u, v \in V_1,$$

which generates an equivalent norm on V_1 denoted by $\| \cdot \|_0$.

4.1 Error estimates over a finite time interval

Let $T > 0$, $M \in \mathbb{N}$ and $\tau = T/M$ be the time step. We consider $V_h \subset V_1$, a finite-dimensional subspace of V_1 (constructed for instance using a finite element method), and let $(u_h^n, \sigma_h^n)_{n=\overline{0,M}}$ be the solution of the following recursive algebraic systems:

$$u_h^0 = \bar{u}_h^0 + \tilde{u}(0); \qquad \bar{u}_h^0 \in V_h, \ \sigma_h^0 \in \mathscr{H} \tag{4.1}$$

$$\bar{u}_h^{n+1} \in V_h; \quad (\bar{u}_h^{n+1}, v_h)_0 = (\bar{u}_h^n, v_h)_0 - \tau \langle G(\sigma_h^n, \varepsilon(u_h^n)), \varepsilon(v_h) \rangle_{\mathscr{H}} \quad \forall v_h \in V_h \tag{4.2}$$

$$u_h^{n+1} = \bar{u}_h^{n+1} + \tilde{u}((n+1)\tau) \tag{4.3}$$

$$\sigma_h^{n+1} = \sigma_h^n + \mathscr{A}\varepsilon(u_h^{n+1}) - \mathscr{A}\varepsilon(u_h^n) + \tau G(\sigma_h^n, \varepsilon(u_h^n)) \tag{4.4}$$

where $(\tilde{u}; \tilde{\sigma})$ is given by (1.40).

The couple $(\tilde{u}, \tilde{\sigma})$ is used only for the homogenization of the boundary conditions. It plays a technical role in the estimate of the error and is not used in practice.

The following theorem gives an upper bound of the distance between the exact solution $(u; \sigma)$ of (1.20)–(1.23) and the approximate one $(u_h^n; \sigma_h^n)_{n=\overline{0,M}}$.

Theorem 4.1 *Let* (1.24)–(1.29) *hold. Then for all* $n = \overline{0, M}$ *we have*

$$\begin{aligned} &\|u(n\tau) - u_h^n\|_{H_1} + \|\sigma(n\tau) - \sigma_h^n\|_{\mathscr{H}} \\ &\quad \le C \exp(CLT)[\tau(\dot{I}(T) + \dot{U}(T) + \dot{\Sigma}(T))(\exp(CLT) - 1) + S(T) \\ &\qquad + \|u_0 - u_h^0\|_{H_1} + \|\sigma_0 - \sigma_h^0\|_{\mathscr{H}}] \end{aligned} \tag{4.5}$$

where the constant C *depends only on* Ω, Γ_1, $\mathscr{A}$ *and*

$$\begin{aligned} &\dot{U}(T) = \|\dot{u}^*\|_{0,T,H_1}, \qquad \dot{\Sigma}(T) = \|\dot{\sigma}^*\|_{0,T,\mathscr{H}} \\ &\dot{I}(T) = C(\|\dot{b}\|_{0,T,H}) + \|\dot{g}\|_{0,T,H_\Gamma} + \|\dot{f}\|_{0,T,V_{\Gamma 1}}) \end{aligned} \tag{4.6}$$

$$S(T) = \sup_{t \in [0,T]} \left(\inf_{v_h \in V_h} \|u^*(t) - v_h\|_{H_1} \right) \tag{4.7}$$

with $u^* = u - \tilde{u}$, $\sigma^* = \sigma - \tilde{\sigma}$.

PROOF. In order to prove Theorem 4.1 we shall use Theorem A4.1 with the following notations: $X = \varepsilon(V_1)$, $Y = \mathscr{V}_2$ and $V = \mathscr{H}$, $X_h = \varepsilon(V_h)$. Let

$E \in B(V)$ be given by

$$(Er)(x) = \mathscr{A}(x)r(x) \qquad \text{a.e.} \quad x \in \Omega, \quad \forall r \in \mathscr{H}. \tag{4.8}$$

If we let $x(t) = \varepsilon(u^*(t)) \in X$, $y(t) = \sigma^*(t) \in Y$, then from (1.26), (1.27) we get (A.4.41), (A.4.42), where $x_0 = \varepsilon(u_0^*)$, $y_0 = \sigma_0^*$. Using this notation one can remark that if we let $x_h^n = \varepsilon(\bar{u}_h^n)$, $y_h^n = \bar{\sigma}_h^n = \sigma_h^n - \tilde{\sigma}(n\tau)$, then (4.1)–(4.4) are equivalent to (A.4.53)–(A.4.55). Hence we can use Theorem A.4.1 to obtain the error estimate (4.5).

4.2 Error estimates over an infinite time interval in the viscoelastic case

In this section the constitutive function G from (1.21) is supposed to be given by (2.1) and let us suppose that (2.2)–(2.5) hold. We shall also assume that

$$\begin{aligned} I &= \|b\|_{\boldsymbol{H}}^{\infty} + \|g\|_{\boldsymbol{H}_\Gamma}^{\infty} + \|f\|_{V_{\Gamma 1}'}^{\infty} < +\infty \\ \dot{I} &= \|\dot{b}\|_{\boldsymbol{H}}^{\infty} + \|\dot{g}\|_{\boldsymbol{H}}^{\infty} + \|\dot{f}\|_{V_{\Gamma 1}'}^{\infty} < +\infty \end{aligned} \tag{4.9}$$

From Theorem 2.1 and (4.9) one can deduce that

$$\dot{U} = \|\dot{u}^*\|_0^{\infty} < +\infty, \qquad \dot{\Sigma} = \|\dot{\sigma}^*\|_{\mathscr{H}}^{\infty} < +\infty, \tag{4.10}$$

where $u^* = u - \tilde{u}$, $\sigma^* = \sigma - \tilde{\sigma}$, $(\tilde{u}; \tilde{\sigma})$ is given by (1.40) and $(u; \sigma)$ is the solution of (1.20)–(1.23), (2.1).

Let $\tau > 0$ be the time step and let us consider the following recursive algebraic systems, which differ slightly from (4.1)–(4.4):

$$u_h^0 = \bar{u}_h^0 + \tilde{u}(0), \quad \bar{u}_h^0 \in V_h; \qquad \sigma_h^0 \in \mathscr{H} \tag{4.11}$$

$$\begin{aligned} \bar{u}_h^{n+1} \in V_h; \qquad (\bar{u}_h^{n+1}, v_h)_0 = (\bar{u}_h^n, v_h)_0 + \tau k[\langle f(n\tau), \gamma_1(v_h)\rangle_{V_{\Gamma 1}', V_{\Gamma 1}} \\ + \langle b(n\tau), v_h\rangle_{\boldsymbol{H}} - \langle F(\varepsilon(u_h^n)), \varepsilon(v_h)\rangle_{\mathscr{H}}] \\ \forall v_h \in V_h \end{aligned} \tag{4.12}$$

$$u_h^{n+1} = \bar{u}_h^{n+1} + \tilde{u}((n+1)\tau) \tag{4.13}$$

$$\sigma_h^{n+1} = (1 - \tau k)\sigma_h^n + \mathscr{A}\varepsilon(u_h^{n+1} - u_h^n) + \tau k F(\varepsilon(u_h^n)) \qquad \forall n \in \mathbb{N}. \tag{4.14}$$

Remark 4.1 *The sequence* $(u_h^n)_{n\in\mathbb{N}}$ can be computed from (4.11)–(4.13) without any computations performed on the sequence $(\sigma_h^n)_{n\in\mathbb{N}}$.

The following theorem gives an upper bound of the distance between the solution $(u; \sigma)$ of (1.20)–(1.23), (2.1) and the approximative solution $(u_h^n; \sigma_h^n)_{n\in\mathbb{N}}$ obtained by means of (4.11)–(4.14) for all $n \in \mathbb{N}$ (i.e. on the time interval $[0, +\infty)$) provided that the time step is small enough.

Theorem 4.2 *Let* $\tau_0 = \min(1/(2k), \alpha^2 a/(2kL_0^2 Q))$. *If* $0 < \tau < \tau_0$ *then*

$$q_1 = \tau k L_0^2 Q/(\alpha^2 a)(1 - \exp(-\tau ka/Q)) + \exp(-\tau ka/Q) < 1,$$
$$q_2 = \tau k(1 - \exp(-\tau k)) + \exp(-\tau k) < 1; \; q = \max(q_1, q_2) < 1. \quad (4.15)$$

Moreover for all $n \in \mathbb{N}$ *we have*

$$\|u(n\tau) - u_h^n\|_0 \leq q_1^n \|u_0 - u_h^0\|_0 + \tau C L_0/a(\dot{I} + \dot{U}) + C(L_0 S/a + \dot{S}/(kL_0)), \quad (4.16)$$

$$\|\sigma(n\tau) - \sigma_h^n\|_{\mathcal{H}} \leq q_2^n \|\sigma_0 - \sigma_h^0\|_{\mathcal{H}} + CL_0(1 - \exp(-\tau k))nq^n \|u_0 - u_h^0\|_0$$
$$+ \tau C[\dot{\Sigma} + L_0(L_0 + a)(\dot{U} + \dot{I})/a] + C(L_0^2 S/a + \dot{S}/k). \quad (4.17)$$

where C depends only on Ω, Γ_1, $\mathcal{A}$ *and* $S, \dot{S}$ *are given by*

$$S = \sup_{t \in \mathbb{R}_+} \left(\inf_{v_h \in V_h} \|u^*(t) - v_h\|_0 \right), \qquad \dot{S} = \sup_{t \in \mathbb{R}_+} \left(\inf_{v_h \in V_h} \|\dot{u}^*(t) - v_h\|_0 \right) \quad (4.18)$$

Remark 4.2 *As follows from* (4.16) *and* (4.17), *the error of the initial displacement* $\|u_0 - u_h^0\|_0$ *and initial stress* $\|\sigma_0 - \sigma_h^0\|_{\mathcal{H}}$ *vanishes in the estimate of the error at time* $n\tau$ *for a large n. This is a consequence of the asymptotic stability of* (1.20)–(1.23), (2.1) (*see Theorem* 2.1).

PROOF OF THEOREM 4.2. Let $u^* \in C^1(\mathbb{R}_+, V_1)$, $\sigma^* \in C^1(\mathbb{R}_+, \mathcal{V}_2)$ be the solution of (1.45), (1.46). If we multiply (1.45) by $\varepsilon(v)$, we use (2.1) and we integrate the result over Ω we get

$$(\dot{u}^*(t), v)_0 = k\langle f(t), \gamma_1 v\rangle_{V'_{\Gamma 1}, V_{\Gamma 1}} + k\langle b(t), v\rangle_H$$
$$- k\langle F(\varepsilon(u^*(t)) + \varepsilon(\tilde{u}(t)), \varepsilon(v)\rangle_{\mathcal{H}} \qquad \forall v \in V_1. \quad (4.19)$$

Let $J: \mathbb{R}_+ \times V_1 \to V_1$ be given by

$$(J(t, v), w)_0 = k\langle f(t), \gamma_1 w\rangle_{V'_{\Gamma 1}, V_{\Gamma 1}} + k\langle b(t), w\rangle_H - k\langle F(\varepsilon(v) + \varepsilon(\tilde{u}(t))), \varepsilon(w)\rangle_{\mathcal{H}}$$
$$\forall v, w \in V_1. \quad (4.20)$$

From (4.19), (4.20) we deduce that u^* is the solution of the following Cauchy problem:

$$u^*(0) = u_0^*, \qquad \dot{u}^*(t) = J(t, u^*(t)), \quad t > 0. \quad (4.21)$$

Since (2.2)–(2.5) hold, we have

$$(J(t, v_1) - J(t, v_2), v_1 - v_2)_0 \leq -(ka/Q)\|v_1 - v_2\|_0^2, \quad (4.22)$$

$$\|J(t_1, v_1) - J(t_2, v_2)\|_0 \leq kL_0/\alpha \cdot \|v_1 - v_2\|_0 + kL_0 C\dot{I}|t_1 - t_2|$$
$$\forall v_1, v_2 \in V_1, t, t_1, t_2 \in \mathbb{R}_+. \quad (4.23)$$

Let P_h: $V_1 \to V_h$ be the projector map and J_h: $\mathbb{R}_+ \times V_1 \to V_h$ given by $J_h = P_h J$ and let us remark that u_h^n given by (4.11)–(4.13) is the solution of the following recursive system:

$$\bar{u}_h^0 \in V_h, \qquad \bar{u}_h^{n+1} = \bar{u}_h^n + \tau J_h(n\tau, \bar{u}_h^n). \tag{4.24}$$

If we use now Lemma A.4.7 we get $q_1 < 1$ and (4.16). We also have

$$\|\dot{u}^*(n\tau) - (\bar{u}_h^{n+1} - \bar{u}_h^n)/\tau\|_0 \leq kL_0 C\|u(n\tau) - u_h^n\|_0 + \dot{S} \tag{4.25}$$

Let W: $\mathbb{R}_+ \times \mathscr{H} \to \mathscr{H}$ given by

$$W(t, \sigma) = -k\sigma + \mathscr{A}\varepsilon(\dot{u}^*(t) - k\tilde{\sigma}(t) + kF(\varepsilon(u(t))) \qquad \forall\sigma \in \mathscr{H} \quad \text{and} \quad t \in \mathbb{R}_+.$$

Bearing in mind that

$$\mathscr{A}\varepsilon(\dot{u}^*(t)) - k\tilde{\sigma}(t) + kF(\varepsilon(u(t))) = \dot{\sigma}^*(t) + k\sigma^*(t) \in \mathscr{V}_2,$$

we deduce that $W(t, \sigma) \in \mathscr{V}_2$ for all $\sigma \in \mathscr{V}_2$. Hence $\sigma^* \in C^1(\mathbb{R}_+, \mathscr{V}_2)$ is the solution of the following Cauchy problem:

$$\sigma^*(0) = \sigma_0^* \in \mathscr{V}_2, \qquad \dot{\sigma}^*(t) = W(t, \sigma^*(t)) \qquad \forall t > 0. \tag{4.26}$$

If for all $n \in \mathbb{N}$ we denote by $W_0(n\tau, \cdot)$: $\mathscr{H} \to \mathscr{H}$ the function defined by

$$W_0(n\tau, \sigma) = -k\sigma + \mathscr{A}\varepsilon(\bar{u}_h^{n+1} - \bar{u}_h^n)/\tau - k\tilde{\sigma}(n\tau) + kF(\varepsilon(u_h^n)) \qquad \forall\sigma \in \mathscr{H} \tag{4.27}$$

and we let $\bar{\sigma}_h^n = \sigma_h^n - \tilde{\sigma}(n\tau)$, then from (4.27) and (4.14) we obtain

$$\bar{\sigma}_h^{n+1} = \bar{\sigma}_h^n + \tau W_0(n\tau, \bar{\sigma}_h^n) \qquad \forall n \in \mathbb{N}. \tag{4.28}$$

From (4.27), (4.25), and (4.16) we have

$$\|W(n\tau, \sigma) - W_0(n\tau, \sigma)\|_{\mathscr{H}} \leq C(kL_0\|u^*(n\tau) - \bar{u}_h^n\|_0 + \dot{S}) \tag{4.29}$$

for all $\sigma \in \mathscr{H}$; we can also deduce

$$\|W(t_1, \sigma_1) - W(t_2, \sigma_2)\|_{\mathscr{H}} \leq k\|\sigma_1 - \sigma_2\|_{\mathscr{H}} + kL_0 C(\dot{U} + \dot{I})\cdot|t_1 - t_2| \tag{4.30}$$

$$\langle W(t, \sigma_1) - W(t, \sigma_2), \sigma_1 - \sigma_2\rangle_{\mathscr{H}} \leq -k\|\sigma_1 - \sigma_2\|_{\mathscr{H}}^2$$

$$\forall\sigma_1, \sigma_2 \in \mathscr{H}, t, t_1, t_2 \in \mathbb{R}_+. \tag{4.31}$$

From (4.28)–(4.31) and Lemma A.4.5 we can obtain (4.17).

4.3 Numerical examples

In order to illustrate the numerical method previously presented we shall give some one-dimensional numerical examples. In this section $\Omega = (0, 1) \subset \mathbb{R}$ and the following initial–boundary-value problem (which is a one-dimensional version of (1.20)–(1.23)) is considered:

$$\dot{\sigma}(t, x) = \mathscr{A}\dot{\varepsilon}(t, x) + G(\sigma(t, x), \varepsilon(t, x)) \tag{4.32}$$

$$\varepsilon(t, x) = \frac{\partial}{\partial x} u(t, x) \qquad \text{for } x \in (0, 1) \tag{4.33}$$

$$\frac{\partial}{\partial x} \sigma(t, x) + b(t, x) = 0 \tag{4.34}$$

$$u(1, t) = g(t), u(0, t) = 0 \tag{4.35}$$

or

$$\sigma(1, t) = f(t), u(0, t) = 0 \qquad \text{for } t > 0 \tag{4.35'}$$

$$u(0, x) = u_0(x), \quad \sigma(0, x) = \sigma_0(x) \qquad \text{for } x \in (0, 1). \tag{4.36}$$

Example 4.1 Let us consider the linear standard viscoelastic material, i.e. $G(\sigma, \varepsilon) = -k(\sigma - \mathscr{E}\varepsilon)$ with the initial data $u_0(x) = \varepsilon_0 x, \sigma_0(x) = \sigma_0$ for $x \in (0, 1)$. In order to imagine a creep experiment we put $f(t) = \sigma_0$ and $b(t, x) = 0$. In this case one easily integrates (4.32)–(4.36) to obtain that the solution of (4.32)–(4.36) is given by $\sigma(t, x) = \sigma_0$, $u(t, x) = \varepsilon(t)x$,

$$\varepsilon(t) = \varepsilon_0 \exp(-k\mathscr{E}\mathscr{A}^{-1}t) + (1 - \exp(-k\mathscr{E}\mathscr{A}^{-1}t))\sigma_0\mathscr{E}^{-1}.$$

If we choose V_h to be the finite element space constructed with polynomial functions of degree greater than or equal to 1 then we get that $u_h^n = u^n$ and $\sigma_h^n = \sigma^n = \sigma_0$ does not depend on h. Let $E(n, \tau)$ be the relative strain error

$$E(n, \tau) = |\varepsilon(u^n) - \varepsilon(n\tau)|/|\varepsilon(n\tau)|. \tag{4.37}$$

at iteration n for the time step τ.

Some numerical evaluations of $E(n, \tau)$ are presented in Figures 4.1 and 4.2 for $\mathscr{A} = 20$, $k = 10$, $\sigma_0 = 40$, $\mathscr{E} = 10$.

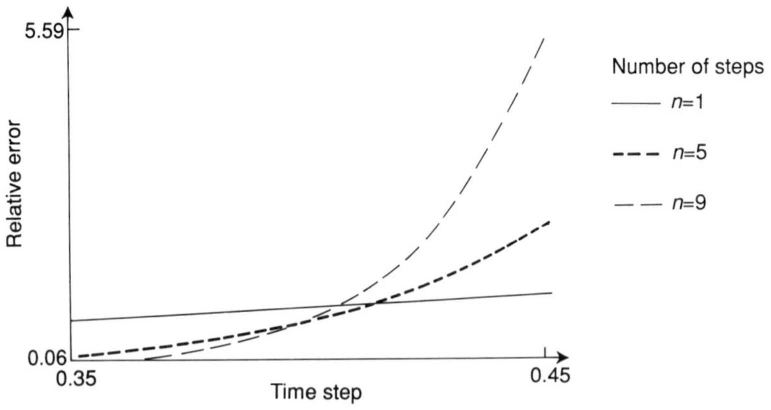

Fig. 4.1 The strain relative error in Example 4.1.

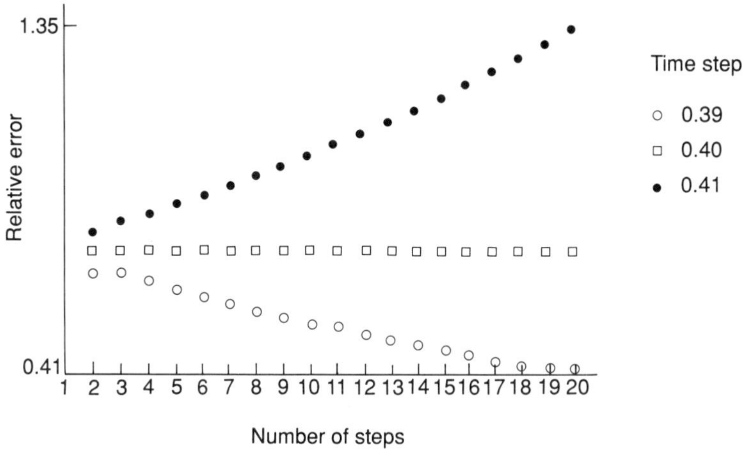

Fig. 4.2 The strain relative error in Example 4.1.

In Figure 4.1, after the first step the error is linear with respect to the time step τ but after nine steps, the error seems to have an exponential behaviour.

In Figure 4.2 we see that for $\tau = 0.39$ the error decreases as the number n of steps increases, but for $\tau = 0.4$ the error is almost constant for any n and for $\tau = 0.41$ the error quickly increases. This example suggests that there exists a critical time step τ_{cr} (in our case $\tau_{cr} = 0.4$) such that the error is bounded iff $\tau \leq \tau_{cr}$.

Example 4.2 In this example we shall consider a non-linear viscoelastic case (2.1) with F a non-monotone function given by (2.8) in Remark 2.2 with $E = 10.0$, $E_1 = 5.0$, $\gamma = 30.0$, $\Delta = 30$, $\alpha = 2.0$, $\beta = 4.0$. Let $b(t, x) = 0$, $f(t) = \sigma_0(x) = \sigma_0 = 15.0$, $\mathscr{A} = 20.0$ and $k = 10.0$; hence we have $\varepsilon_1 = 1.5$, $\varepsilon_2 = 3.0$, $\varepsilon_3 = 4.5$. In this example we choose $\varepsilon_0(x) = 3.0 - 0.075(x - 0.5)$, $u_0(x) = \int_0^x \varepsilon_0(s)\, ds$ in order to have $\varepsilon_0(x) \in A_3$ for $x < 0.5$ and $\varepsilon_0(x) \in A_1$ for $x > 0.5$ and $\varepsilon_0(x)$ to be very close to the unstable solution ε_2. The space V_h is the finite element space constructed with polynomial functions of degree 3 (i.e. $V_h \subset C^1(\bar{\Omega})$). The domain Ω was divided into 50 finite elements and the time step $\tau = 0.05$. In Figure 4.3, the computed solution $\varepsilon(u_h^n)$ is plotted and the results, which agree with the theoretical values presented in Remark 2.2, show that one half of the bar becomes shorter and the other one becomes longer, although the external data are constant in time and space and the initial data are 'almost' constant in space.

Example 4.3 Let us consider the elastic–viscoplastic case

$$G(\sigma, \varepsilon) = -\frac{1}{\mu}(\sigma - P_K \sigma)$$

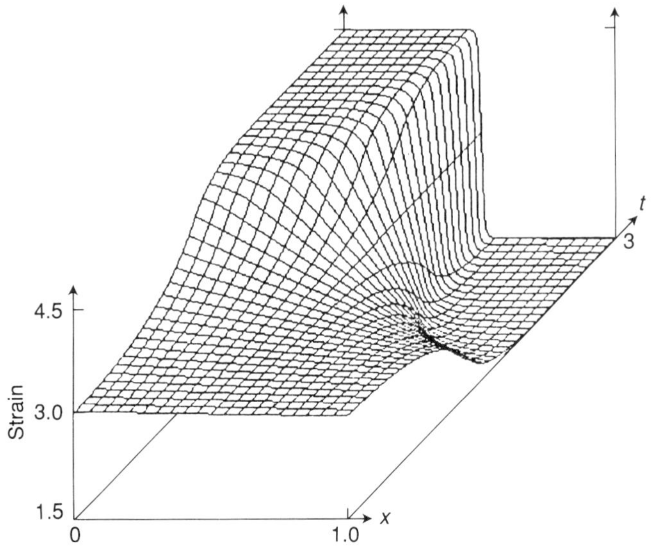

Fig. 4.3 The computed strain field $\varepsilon(u_h^n)$ from Example 4.2.

where P_K is the projector map on the plasticity convex $K = [-1, 1]$. Let $g(t) = 0$, $u_0(x) = 0$, $\sigma_0(x) = 0$, $\mathscr{A} = 1$, and b be given by (3.55). The elastic–perfectly plastic version of this example was considered in Example 3.2.

As follows from Theorem 3.1, the solution of the elastic–viscoplastic problem considered here approaches the solution of elastic–perfectly plastic problem given in Example 3.2 for small viscosity coefficient μ. In order to obtain in our case similar properties of the solution (described in Example 3.2) we choose a small $\mu = 0.01$. In this way one can consider the elastic–viscoplastic case as a penalized elastic–perfectly plastic problem. Let us remark that for $0 < t \leq T_0 = 6$ the solution of (4.32)–(4.36) is an elastic one given by (3.57) and

$$u(t, x) = \begin{cases} tx(\frac{1}{4} - x^2)/3 & \text{for } x < \frac{1}{2} \\ t(x-1)(\frac{1}{4} - (x-1)^2)/3 & \text{for } x > \frac{1}{2}. \end{cases} \tag{4.38}$$

For $t = T_0 = 6$ we have $\sigma(t, 0.5) = -1$, hence the point $x = 0.5$ is plasticized. The space V_h was constructed as in Example 4.2 and the time step is $\tau = 0.01$.

In Figure 4.4 the computed stress field σ_h^n is plotted. One can see that at $t = 6$ the point $x = 0.5$ is plasticized, and the stress function remains continuous for $t > 6$.

In Figures 4.5 and 4.6 the computed displacement field u_h^n and the velocity field $v_h^n = (u_h^{n+1} - u_h^n)/\tau$ are plotted. In these figures one can see that for $t > 6$ a 'discontinuity' appears in the displacement and velocity fields at $x = 0.5$, which is developing in time.

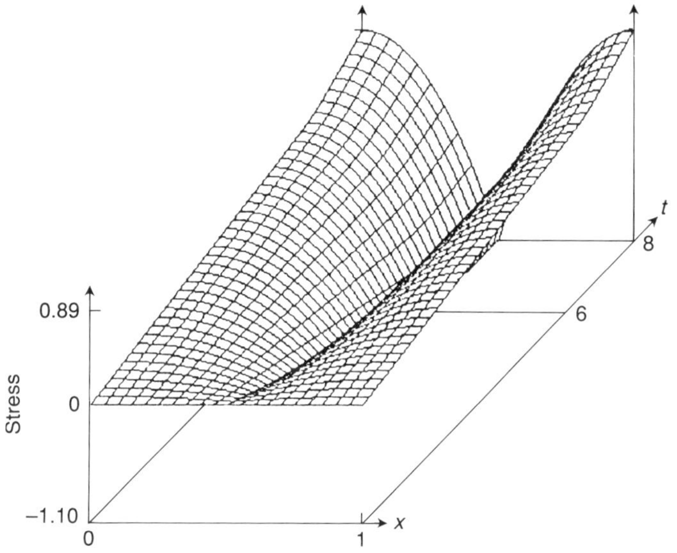

Fig. 4.4 The computed stress field σ_h^n from Example 4.3.

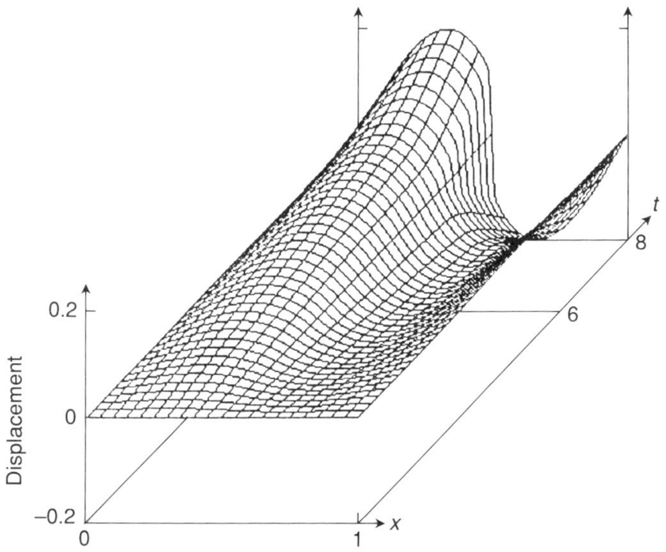

Fig. 4.5 The computed displacement field u_h^n from Example 4.3.

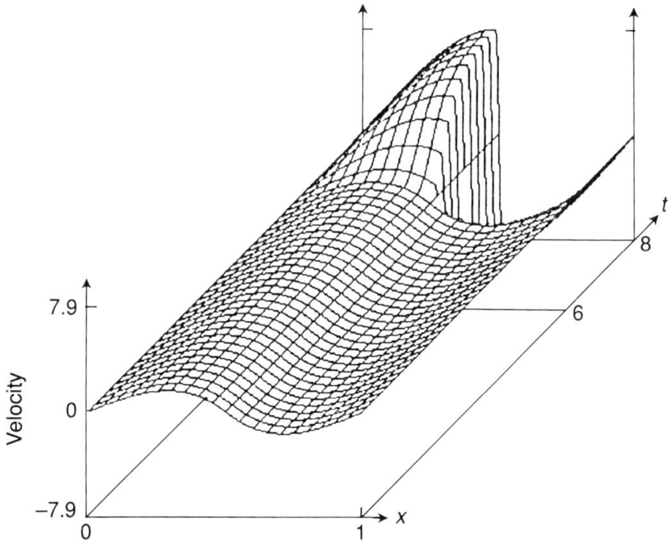

Fig. 4.6 The computed velocity field $v_h^n = (u_h^{n+1} - u_h^n)/\tau$ from Example 4.3.

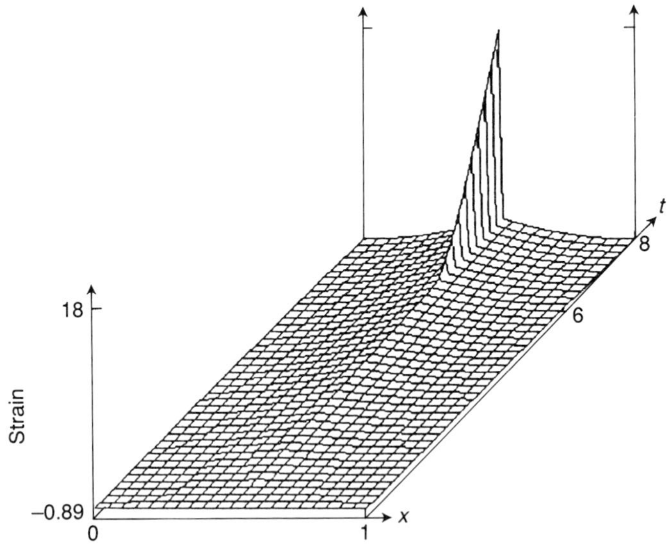

Fig. 4.7 The computed strain field $-\varepsilon(u_h^n)$ from Example 4.3.

In Figure 4.7 (see p. 88) the computed strain field $\varepsilon(u_h^n)$ is plotted and we remark that at the 'discontinuity' point $x = 0.5$ the strain increases quickly.

5 Quasistatic processes for rate-type viscoplastic materials with internal state variables

In this section we consider rate-type constitutive equations for which the plastic rate of deformation tensor depends also on an internal state variable and the evolution of this parameter is described by a constitutive equation. Some concrete examples of this internal state variable are presented in Section 5.1. In Section 5.2 the problem statement and the constitutive assumptions are presented and in Section 5.3 we give existence and uniqueness results (Theorem 5.1) and we study the dependence of the solution upon the input data (Theorem 5.2). In Section 5.4 a numerical approach to the problem is given.

5.1 Rate-type constitutive equations with internal state variables

We consider here constitutive equations for rate-type materials involving an internal state variable $\mathscr{H}$ of the form

$$\dot{\sigma} = \mathscr{A}\dot{\varepsilon} + G(\sigma, \varepsilon, \kappa) \tag{5.1}$$

in which $\mathscr{A}$ and G are constitutive functions (see also (1.2)). As was pointed out in Section 1.1 we can consider the equivalent form of (5.1) given by

$$\dot{\varepsilon} = \tilde{\mathscr{A}}\dot{\sigma} + \tilde{G}(\sigma, \varepsilon, \kappa). \tag{5.2}$$

Equation (5.2) underlines the viscoplastic (irreversible) strain rate defined by

$$\dot{\varepsilon}^I = \tilde{G}(\sigma, \varepsilon, \kappa). \tag{5.3}$$

A concrete example of function G in (5.2) can be obtained by taking

$$\tilde{G}(\sigma, \kappa) = \frac{1}{\mu}(\sigma - P_{K(\kappa)}\sigma) \tag{5.4}$$

where $P_{K(\tau)}$ is the projector map on the convex $K(\kappa)$ defined by the yield function $\kappa \to k(\kappa)$:

$$K(\kappa) = \{\sigma \in \mathscr{S}_3 | \sqrt{(\sigma_{\mathrm{II}})} \le k(\kappa)\}. \tag{5.5}$$

Using (1.17) and the notations used in Example 1.4 we can give the explicit form of (5.4):

$$\tilde{G}(\sigma, \kappa) = \begin{cases} 0 & \text{if } \sqrt{(\sigma_{\mathrm{II}})} \le k(\kappa) \\ \dfrac{1}{\mu}\left(\sigma - \dfrac{k(\kappa)}{\sqrt{(\sigma_{\mathrm{II}})}}\right) & \text{if } \sqrt{(\sigma_{\mathrm{II}})} > k(\kappa). \end{cases} \tag{5.6}$$

There exists a large scope in the choice of the internal state variables, authors having individual preferences varying even from paper to paper (for references in the field see Cristescu and Suliciu (1982), Ch. VI). Some of the internal state variables considered by many authors are the plastic strain, a number of tensor variables that take into account the spatial display of dislocations and the work-hardening of the material. A major and still remaining open problem in viscoplasticity concerns the way of establishing the evolution equations for the internal state variables. Here we suppose that κ is a vector-valued function which satisfies the equation

$$\dot{\kappa} = \varphi(\sigma, \varepsilon, \kappa) \tag{5.7}$$

in which $\varphi\colon \mathscr{S}_N \times \mathscr{S}_N \times \mathbb{R}^m \to \mathbb{R}^m$ is a given function.

A simple example of internal state variable, for which an evolution equation of the form (5.7) with $m = 1$ holds, can be obtained by considering

$$\varphi(\sigma, \varepsilon, \kappa) = \Phi(|\dot{\varepsilon}^I|) = \Phi(\tilde{G}(\sigma, \varepsilon, \kappa)) \tag{5.8}$$

where $\Phi\colon \mathbb{R}_+ \to \mathbb{R}$ is a given function. The parameter κ defined in this way is called the *strain-hardening parameter*. If we consider $\Phi(r) = \frac{2}{3}r$, then κ given by (5.7) and (5.8) represents the irreversible equivalent strain.

Another example of internal state variable can be obtained by considering in (5.7) the function φ given by

$$\varphi(\sigma, \varepsilon, \kappa) = \sigma \cdot \dot{\varepsilon}^I = \sigma \cdot \tilde{G}(\sigma, \varepsilon, \kappa). \tag{5.9}$$

Since $\sigma \cdot \dot{\varepsilon}^I$ is the irreversible stress power, the parameter κ defined in this way is called the *work-hardening parameter*.

Concrete examples for viscoplastic models involving hardening parameters can be found in Cristescu (1989) for rock-like materials (see also Section 6.1 of this chapter).

Remark 5.1 *Some authors (see for instance Laborde,* 1979) *used instead of* (5.7) *the following equation for* κ:

$$\kappa(t) = f\left(\int_0^t \varphi(\sigma(s), \varepsilon(s), \kappa(s))\right) \mathrm{d}s \qquad \forall t \geq 0. \tag{5.10}$$

It is not difficult to see that after a substitution (5.10) *can be written as* (5.7). *Indeed, if we put* $\bar{G}(\sigma, \varepsilon, \kappa) = G(\sigma, \varepsilon, f(\kappa))$, $\bar{\varphi}(\sigma, \varepsilon, \kappa) = \varphi(\sigma, \varepsilon, f(\kappa))$, then we observe that if $\sigma, \varepsilon, \bar{\kappa}$ satisfies $\dot{\sigma} = \mathscr{A}\dot{\varepsilon} + \bar{G}(\sigma, \varepsilon, \bar{\kappa})$ and $\dot{\bar{\kappa}} = \bar{\varphi}(\sigma, \varepsilon, \bar{\kappa})$, *then* (5.1) *and* (5.10) *hold for* $\kappa = f(\bar{\kappa})$.

In the next section we consider a quasistatic problem for materials of the form (5.1), (5.7) for which we present similar results to those for problems (1.20)–(1.23). For more generality we suppose that the functions G and φ in (5.1), (5.7) depend also on t by virtue of the temperature.

Though all the results of the next sections hold for $m \geq 1$, for simplicity, we shall consider in the following only the case $m = 1$.

5.2 Problem statement

Using the same notations as in Section 1, we consider the following mixed problem:

Find the displacement function $u: \mathbb{R}_+ \times \Omega \to \mathbb{R}^N$, the stress function $\sigma: \mathbb{R}_+ \times \Omega \to \mathscr{S}_N$ and the parameter $\kappa: \mathbb{R}_+ \times \Omega \to \mathbb{R}$ such that

$$\operatorname{Div} \sigma(t) + b(t) = 0 \tag{5.11}$$

$$\dot{\sigma}(t) = \mathscr{A}\varepsilon(\dot{u}(t)) + G(t, \sigma(t), \varepsilon(u(t)), \kappa(t)) \tag{5.12}$$

$$\dot{\kappa}(t) = \varphi(t, \sigma(t), \varepsilon(u(t)), \kappa(t)) \qquad \text{in } \Omega, \tag{5.13}$$

$$u(t) = g(t) \quad \text{on } \Gamma_1, \quad \sigma(t)\nu = f(t) \quad \text{on } \Gamma_2, \qquad \text{for } t > 0 \tag{5.14}$$

$$u(0) = u_0, \quad \sigma(0) = \sigma_0, \quad \kappa(0) = \kappa_0 \qquad \text{in } \Omega. \tag{5.15}$$

The meaning of problems (5.11)–(5.15) is similar to that of problem (1.20)–(1.23), except that the constitutive equation (1.21) has been replaced by the more general constitutive equation (5.12); for the unknown κ the evolution equation (5.13) has been introduced (φ given) and the initial value of κ has been assigned in (5.15). As in Section 1, the constitutive functions G and φ may depend on time by means of the temperature field.

In the study of problems (5.11)–(5.15) the following assumptions are made:

$G: \Omega \times \mathbb{R}_+ \times \mathscr{S}_N \times \mathscr{S}_N \times \mathbb{R} \to \mathscr{S}_N$ has the following properties:

(a) there exist $L > 0$ and $\theta \in C(\mathbb{R}_+, H)$ such that

$$|G(x, t_1, \sigma_1, \varepsilon_1, \kappa_1) - G(x, t_2, \sigma_2, \varepsilon_2, \kappa_2)|$$
$$\leq L(|\theta(x, t_1) - \theta(x, t_2)| + |\sigma_1 - \sigma_2| + |\varepsilon_1 - \varepsilon_2| + |\kappa_1 - \kappa_2|)$$
$$\forall t_1, t_2 \in \mathbb{R}_+, \sigma_1, \sigma_2, \varepsilon_1, \varepsilon_2 \in \mathscr{S}_N, \kappa_1, \kappa_2 \in \mathbb{R}, \text{ a.e. in } \Omega \tag{5.16}$$

(b) $x \to G(x, t, \sigma\, \varepsilon, \kappa)$ is a measurable function with respect to the Lebesgue measure on Ω, for all $t \in \mathbb{R}_+$, $\sigma, \varepsilon \in \mathscr{S}_N$ and $\kappa \in \mathbb{R}$

(c) $x \to G(x, t, 0_N, 0_N, 0) \in \mathscr{H}$ for all $t \in \mathbb{R}_+$;

$\varphi: \Omega \times \mathbb{R}_+ \times \mathscr{S}_N \times \mathscr{S}_N \times \mathbb{R} \to \mathbb{R}$ has the following properties:

(a) there exist $\tilde{L} > 0$ and $\psi \in C(\mathbb{R}_+, H)$ such that

$$|\varphi(x, t_1, \sigma_1, \varepsilon_1, \kappa_1) - \varphi(x, t_2, \sigma_2, \varepsilon_2, \kappa_2)|$$
$$\leq \tilde{L}(|\psi(x, t_1) - \psi(x, t_2)| + |\sigma_1 - \sigma_2| + |\varepsilon_1 - \varepsilon_2| + |\kappa_1 - \kappa_2|)$$
$$\forall t_1, t_2 \in \mathbb{R}_+, \sigma_1, \sigma_2, \varepsilon_1, \varepsilon_2 \in \mathscr{S}_N, \kappa_1, \kappa_2 \in \mathbb{R}, \text{ a.e. in } \Omega; \tag{5.17}$$

(b) $x \to \varphi(x, t, \sigma, \varepsilon, \kappa)$ is a measurable function with respect to the Lebesgue measure on Ω, for all $t \in \mathbb{R}_+$, $\sigma, \varepsilon \in \mathscr{S}_N$ and $\kappa \in \mathbb{R}$

(c) $x \to \varphi(x, t, 0_N, 0_N, 0) \in H \ \forall t \in \mathbb{R}_+$;

$$\kappa_0 \in H. \tag{5.18}$$

Regarding (5.16) and (5.17) we can make similar comments to those of Remark 1.3(1), (2). Moreover, let us observe the following.

Remark 5.2 (1) *The constitutive assumption* (5.16) *is satisfied for the function* $\tilde{G}$ *in* (5.4) *if the yield function* $\kappa \to k(\kappa)$: $\mathbb{R} \to \mathbb{R}_+$ *is a Lipschitz function. Indeed, since* (5.16*b,c*) *are obvious, we shall give the proof of* (5.16*a*). *For this, let* σ_1, σ_2, ε_1, $\varepsilon_2 \in \mathscr{S}_N$ *and* $\kappa_1, \kappa_2 \in \mathbb{R}$. *From* (5.4) *we have*

$$|\tilde{G}(x, \sigma_1, \varepsilon_1, \kappa_1) - \tilde{G}(x, \sigma_2, \varepsilon_2, \kappa_2)|$$
$$\leq C(|\sigma_1 - \sigma_2| + |P_{K(\kappa_1)}\sigma_1 - P_{K(\kappa_2)}\sigma_2|)$$
$$\leq C(|\sigma_1 - \sigma_2| + |P_{K(\kappa_1)}\sigma_1 - P_{K(\kappa_1)}\sigma_2| + |P_{K(\kappa_1)}\sigma_2 - P_{K(\kappa_2)}\sigma_2|).$$

Since the projector map is a non-expensive operator, we get

$$|\tilde{G}(x, \sigma_1, \varepsilon_1, \kappa_1) - \tilde{G}(x, \sigma_2, \varepsilon_2, \kappa_2)|$$
$$\leq C(2|\sigma_1 - \sigma_2| + |P_{K(\kappa_1)}\sigma_2 - P_{K(\kappa_2)}\sigma_2|). \quad (5.19)$$

If $k(\kappa_1) = k(\kappa_2)$, (5.5) *implies* $P_{K(\kappa_1)}\sigma_2 = P_{K(\kappa_2)}\sigma_2$ *hence* (5.16*a*) *results from* (5.19). *Then let* $k(\kappa_1) \neq k(\kappa_2)$ *and let us suppose for instance* $k(\kappa_1) < k(\kappa_2)$. *There exist three possibilities, as follows:*

(1) $\sqrt{\sigma_{2_{\rm II}}} \leq k(\kappa_1) < k(\kappa_2)$; *in this case from* (5.5) *and Lemma* 1.1 *we get* $|P_{K(\kappa_1)}\sigma_2 - P_{K(\kappa_2)}\sigma_2| = 0$;

(2) $k(\kappa_1) < k(\kappa_2) < \sqrt{\sigma_{2_{\rm II}}}$; *in this case from* (5.5) *and Lemma* 1.1 *we get*

$$|P_{K(\kappa_1)}\sigma_2 - P_{K(\kappa_2)}\sigma_2| = |((k(\kappa_2) - k(\kappa_1))/\sqrt{\sigma_{2_{\rm II}}})\sigma_2'|$$
$$= \sqrt{2}|k(\kappa_1) - k(\kappa_2)| \leq L|\kappa_1 - \kappa_2|$$

since k *is a Lipschitz function;*

(3) $k(\kappa_1) < \sqrt{\sigma_{2_{\rm II}}} \leq k(\kappa_2)$; *in this case, from* (5.5) *and Lemma* 1.1 *we get*

$$|P_{K(\kappa_1)}\sigma_2 - P_{K(\kappa_2)}\sigma_2| = \left|\sigma_2 - \frac{\operatorname{tr}\sigma_2}{3} I_3 - \frac{k(\kappa_1)}{\sqrt{\sigma_{2_{\rm II}}}}\sigma_2'\right|$$
$$= |\sigma_2' - (k(\kappa_1)/\sqrt{\sigma_{2_{\rm II}}})\sigma_2'|$$
$$= |\sigma_2'| \cdot |(\sqrt{\sigma_{2_{\rm II}}} - k(\kappa_1)/\sqrt{\sigma_{2_{\rm II}}}|$$
$$= \sqrt{2}|k(\kappa_1) - \sqrt{\sigma_{2_{\rm II}}}|$$
$$\leq 2(k(\kappa_2) - k(\kappa_1))$$
$$= \sqrt{2}|k(\kappa_1) - k(\kappa_2)|$$
$$\leq L|\kappa_1 - \kappa_2| \qquad \textit{as in } (2).$$

So, in any case we have $|P_{K(\kappa_1)}\sigma_2 - P_{K(\kappa_2)}\sigma_2| \le L|\sigma_1 - \sigma_2|$ *hence by* (5.19) *we get* (5.16*a*).

(2) *The constitutive assumption* (5.16) *is satisfied for the function* φ *given by* (5.8) *if* $\tilde{G}$ *satisfies* (5.16) *and* $\phi: \mathbb{R}_+ \to \mathbb{R}_+$ *is a Lipschitz function such that* $\phi(0) = 0$.

5.3 Existence, uniqueness, and continuous dependence of the solution

For the problem (5.11)–(5.15) we have the following existence and uniqueness result:

Theorem 5.1 *Suppose that* (1.24), (1.26)–(1.29), (5.16)–(5.18) *hold. Then, there exists a unique solution of the problem* (5.11)–(5.15) *which satisfies* (1.30), (1.31) *and* $\kappa \in C^1(\mathbb{R}_+, H)$.

PROOF. The same technique as in the proof of Theorem 1.1 is used. More precisely, after the same homogenization of the boundary data given by Lemma 1.3 and using the same arguments as in the proof of Lemma 1.4, (5.11)–(5.15) reduces to the study of the following Cauchy problem on X:

$$\dot{x}(t) = A(t, x(t)) \qquad \forall t > 0 \tag{5.20}$$

$$x(0) = x_0, \tag{5.21}$$

for which the following notations are assumed:

$X = V_1 \times \mathcal{V}_2 \times H$ endowed with the inner product

$$\langle x, y\rangle_X = \langle \mathscr{A}\varepsilon(u), \varepsilon(v)\rangle_{\mathscr{H}} + \langle \mathscr{A}^{-1}\sigma, \tau\rangle_{\mathscr{H}} + \langle \kappa, \eta\rangle_H$$

for all $x = (u; \sigma; \kappa) \in X$, $y = (v; \tau; \eta) \in X$.

$A: \mathbb{R}_+ \times X \to X$ is the operator defined by Riesz's representation theorem from the equality

$$\begin{aligned}\langle A(t, x), y\rangle_X = &-G(t, \sigma + \tilde{\sigma}(t), \varepsilon(u) + \varepsilon(\tilde{u}(t)), \kappa(t)), \varepsilon(v)\rangle_{\mathscr{H}} \\ &+ \langle \mathscr{A}^{-1}G(t, \sigma + \tilde{\sigma}(t), \varepsilon(u) + \varepsilon(\tilde{u}(t)), \kappa(t)), \varepsilon(v)\rangle_{\mathscr{H}} \\ &+ \langle \varphi(t, \sigma + \tilde{\sigma}(t), \varepsilon(u) + \varepsilon(\tilde{u}(t)), \kappa(t)), \eta\rangle_H\end{aligned} \tag{5.22}$$

for all $x = (u; \sigma; \eta) \in X$, $y = (v; \tau; \eta) \in X$ and $t \in \mathbb{R}_+$ with $(\tilde{u}; \tilde{\sigma})$ given by (1.40).

Finally $u^* = u - \tilde{u}$, $\sigma^* = \sigma - \tilde{\sigma}$ and

$$x = (u^*; \sigma^*; \kappa), \qquad x_0 = (u_0^*; \sigma_0^*; \kappa_0) = (u_0 - \tilde{u}(0); \sigma_0 - \tilde{\sigma}(0); \kappa_0).$$

The existence and uniqueness of the solution of (5.20)–(5.21) follows also from Theorem A3.2 since A defined by (5.22) satisfies all the requirements in Lemma 1.5.

The same technique as in the proof of Theorem 1.2 can be used in order to study the continuous dependence of the solution in problems (5.11)–(5.15).

So, with minor adjustments in the proof of Theorem 1.2 we get the following result:

Theorem 5.2 *Let* (1.24), (5.16), (5.17) *hold and let* $(u_i; \sigma_i; \kappa_i)$ *be the solution of* (5.11)–(5.15) *for the data* b_i, g_i, f_i, u_{0i}, σ_{0i}, κ_{0i}, $i = 1, 2$ *which satisfies* (1.26)–(1.29), (5.18). *Then, for all* $T > 0$ *there exists* $C > 0$ *which depends only on* Ω, Γ_1, T, $\mathscr{A}$, G, *and* φ *such that*

$$\begin{aligned}\|u_1 - u_2\|_{j,T,H_1} &+ \|\sigma_1 - \sigma_2\|_{j,T,\mathscr{H}_2} + \|\kappa_1 - \kappa_2\|_{j,T,H} \\ &\leq C(\|u_{01} - u_{02}\|_{H_1} + \|\sigma_{01} - \sigma_{02}\|_{\mathscr{H}_2} + \|\kappa_{01} - \kappa_{02}\|_H \\ &\quad + \|b_1 - b_2\|_{j,T,H} + \|g_1 - g_2\|_{j,T,H_\Gamma} + \|f_1 - f_2\|_{j,T,V'_{\Gamma 1}}), \qquad j = 0, 1. \end{aligned} \tag{5.23}$$

From (5.23) it is easy to deduce the following result:

Corollary 5.1 *Let the hypotheses of Theorem* 5.2 *hold. If* $b_1 = b_2$, $f_1 = f_2$, $g_1 = g_2$, *then*

$$\begin{aligned}\|u_1 - u_2\|_{j,T,H_1} &+ \|\sigma_1 - \sigma_2\|_{j,T,\mathscr{H}_2} + \|\kappa_1 - \kappa_2\|_{j,T,H} \\ &\leq C(\|u_{01} - u_{02}\|_{H_1} + \|\sigma_{01} - \sigma_{02}\|_{\mathscr{H}_2} + \|\kappa_{01} - \kappa_{02}\|_H), \qquad j = 0, 1. \end{aligned} \tag{5.24}$$

The concepts of stability, finite time stability and asymptotic stability for a solution of the problems (5.11)–(5.15) can be formulated in the same way as was made precise in Section 1.4. Morever, we have:

Remark 5.3 *The inequality* (5.24) *implies the finite time stability of every solution of problem* (5.11)–(5.15). *As in the case of the problem* (1.20)–(1.23), *generally speaking stability does not hold.*

5.4 A numerical approach

In this section we shall suppose that θ, ψ from (5.16), (5.17) are more regulate:

$$\theta, \psi \in C^1(\mathbb{R}_+, L^2(\Omega)). \tag{5.25}$$

Let $T > 0$, $M \in \mathbb{N}$ and $\tau = T/M$ be the time step. Let us consider $V_h \subset V_1$ a finite-dimensional subspace constructed using for instance the finite-element method and let $(u_h^n, \sigma_h^n, \kappa_h^n)$ be the solution of the following recursive algebraic systems:

$$u_h^0 = \bar{u}_h^0 + \tilde{u}(0), \quad \bar{u}_h^0 \in V_h, \quad \sigma_h^0 \in \mathscr{H}, \quad \kappa_h^0 \in H \tag{5.26}$$

$$\begin{aligned}&\bar{u}_h^{n+1} \in V_h, \langle \mathscr{A}\varepsilon(\bar{u}_h^{n+1}), \varepsilon(v_h)\rangle_{\mathscr{H}} \\ &\quad = \langle \mathscr{A}\varepsilon(\bar{u}_h^n), \varepsilon(v_h)\rangle_{\mathscr{H}} - \tau\langle G(n\tau, \sigma_h^n, \varepsilon(u_h^n), \kappa_h^n), \varepsilon(v_h)\rangle_{\mathscr{H}} \qquad \forall v_h \in V_h \end{aligned} \tag{5.27}$$

$$u_h^{n+1} = \bar{u}_h^{n+1} + \tilde{u}((n+1)\tau) \tag{5.28}$$

$$\sigma_h^{n+1} = \sigma_h^n + \mathscr{A}\varepsilon(u_h^{n+1} - u_h^n) + \tau G(n\tau, \sigma_h^n, \varepsilon(u_h^n), \kappa_h^n) \tag{5.29}$$

$$\kappa_h^{n+1} = \kappa_h^n + \tau\varphi(n\tau, \sigma_h^n, \varepsilon(u_h^n), \mathscr{H}_h^n) \tag{5.30}$$

for all $n = \overline{0, M-1}$ where $(\tilde{u}, \tilde{\sigma})$ are given by (1.40).

The couple $(\tilde{u}, \tilde{\sigma})$ is used only for the homogenization of the boundary conditions. It plays a technical role in the estimates of the error and it is not used in practice.

The following theorem gives an error estimate for the numerical scheme (5.26)–(5.30).

Theorem 5.3 *Let* (5.16)–(5.18), (1.26)–(1.29), (1.24), (5.25) *hold. Then for all* $n = \overline{0, M}$ *we have*

$$\begin{aligned} &\|u(n\tau) - u_h^n\|_{H_1} + \|\sigma(n\tau) - \sigma_h^n\|_{\mathscr{H}} + \|\kappa(n\tau) - \kappa_h^n\|_H \\ &\quad \leq C\exp(CL_0T)[\tau(\dot{I}_1(T) + \dot{U}(T) + \dot{\Sigma}(T) + \dot{\Phi}(T)(\exp(CL_0T) - 1) \\ &\qquad + S(T) + \|u_0 - u_h^0\|_{H_1} + \|\sigma_0 - \sigma_h^0\|_{\mathscr{H}} + \|\kappa_0 - \kappa_h^0\|_H], \end{aligned} \tag{5.31}$$

where $C > 0$ *depends only on* Ω, Γ_1, $\mathscr{A}$, *and* $L_0 = \max(L, \tilde{L})$. *Here we have used the notation*

$$\dot{I}_1(T) = \dot{I}(T) + \|\dot{\psi}\|_{0,T,H} \tag{5.32}$$

$$\dot{\Phi}(T) = \|\dot{\kappa}\|_{0,T,H} \tag{5.33}$$

and $\dot{I}(T)$, $\dot{U}(T)$, $\dot{\Sigma}(T)$, *and* $S(T)$ *are given by* (4.6), (4.7).

PROOF. In order to prove this theorem we shall use Theorem A.4.1 with the following notations: $V = \mathscr{H} \times H$, $X = \varepsilon(V_1) \times \{0\}$, $Y = \mathscr{V}_2 \times H$. One can easily deduce that $V = X \oplus Y$. If we define $E \in B(V)$ by

$$(E(r_1, r_2))(x) = (\mathscr{A}(x)r_1(x), r_2(x)) \qquad \text{a.e. on } \Omega$$

for all $r_1 \in \mathscr{H}$, $r_2 \in H$ and define $B\colon \mathbb{R}_+ \times (\mathscr{H} \times H) \times (\mathscr{H} \times H) \to \mathscr{H} \times H$ by

$$B(t, r_1, r_2) = (G(t, \tilde{\sigma}(t) + \eta_2, \varepsilon(\tilde{u}(t)) + \eta_1, \kappa_2); \varphi(t, \tilde{\sigma}(t) + \eta_2, \varepsilon(\tilde{u}(t)) + \eta_1, \kappa_2)) \tag{5.34}$$

for all $t \in \mathbb{R}_+$, $r_i = (\eta_i, \kappa_i) \in \mathscr{H} \times H$, $i = 1, 2$ then we can deduce (A.4.50).

Let us denote by $x = (\varepsilon(u^*), 0)$, $y = (\sigma^*, \kappa)$ (see the notations given in the proof of Theorem 5.1) and let us remark that (5.11)–(5.15) is equivalent with (A.4.41)–(A.4.42). If we let $X_h = \varepsilon(V_h) \times \{0\}$, $x_h^n = (\varepsilon(\bar{u}_h^n), 0)$, $y_h^n = (\bar{\sigma}_h^n, \kappa_h^n)$, where $\bar{\sigma}_h^n = \sigma_h^n - \tilde{\sigma}(n\tau)$, then one can deduce that (5.26)–(5.30) is equivalent with (A.4.53)–(A.4.55). The error estimate (5.31) easily follows from Theorem A.4.1.

6 An application to a mining engineering problem

In this paragraph we study the creep of coal in a long wall, where self-advancing mechanized roof supports are used. In order to reduce the complexity, we consider the problem to be a plane strain one. A schematic cross-section is shown in Figure 6.1; GF is the face of the long wall with D_1 the solid coal in front of the face; $ABEF$ is the floor while MN is the proof, $BCDE$ is the steel base of the self-advancing support unit. D_3 is the mined-out area which is not back-filled so that the roof strata are allowed to cave in behind the retracting pillar line.

In the case studied, the material below the worked coal strata consists of an argillaceous coal of a much softer nature (domain D_2). The problem to be studied is what would happen if the work were not advancing for a certain period of time:

- how would the stress distribution evolve in time?
- how important is the creep of the coal and that of the argillaceous coal?

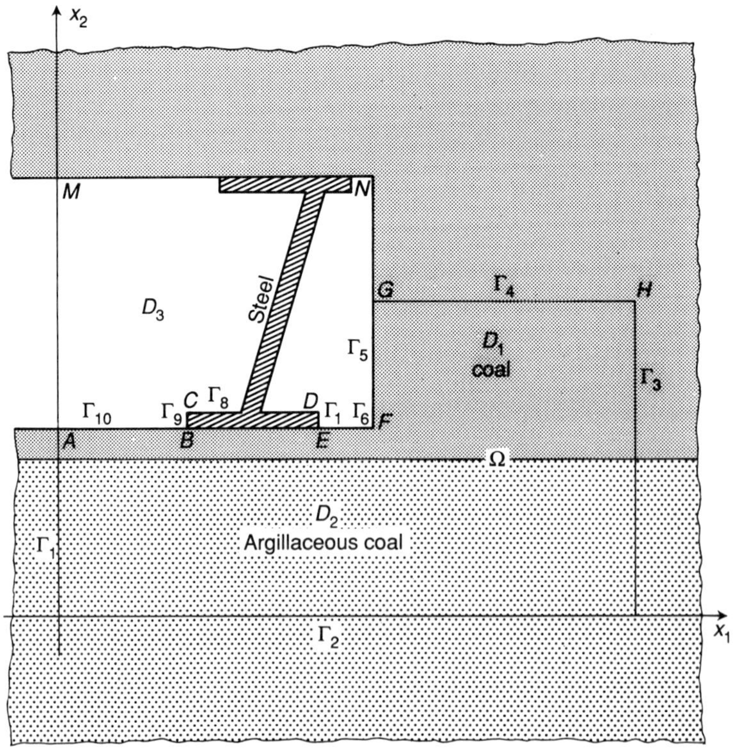

Fig. 6.1 Schematic cross section of the mining domain.

- how thick must the coal strata be under the footing in order to resist the pressure exerted by the support?
- might damage eventually be caused at the face of the long wall?

We try to answer all these questions in the present section.

6.1 Constitutive assumptions and material constants

For both coal and argillaceous coal we use an elastic–viscoplastic constitutive equations proposed by Cristescu (1987, 1989) which is of the form (5.1) or (5.2). The elastic rate of deformation component is related to the stress via a linear elastic isotropic relationship

$$\dot{\varepsilon}^E = \tilde{\mathscr{A}}\dot{\sigma} = \left(\frac{1}{3K} - \frac{1}{2G}\right)\frac{\operatorname{tr}(\dot{\sigma})}{3} I_3 + \frac{1}{2G}\dot{\sigma}, \tag{6.1}$$

where K is the bulk modulus and G the shear modulus. We have also

$$\dot{\sigma} = \mathscr{A}\dot{\varepsilon}^E = \lambda \operatorname{tr}(\dot{\varepsilon}^E) + 2\mu\dot{\varepsilon}^E \tag{6.2}$$

with the Lamé constants $\lambda = (3K - 2G)/3$ and $\mu = G$. The complete associated constitutive equation is written in the form

$$\dot{\varepsilon} = \left(\frac{1}{9K} - \frac{1}{6G}\right)\operatorname{tr}(\dot{\sigma}) I_3 + \frac{1}{2G}\dot{\sigma} + k\left\langle 1 - \frac{\kappa}{H(\sigma)}\right\rangle \frac{\partial H}{\partial \sigma}(\sigma) \tag{6.3}$$

where H is the viscoplastic potential, which depends on the mean stress $\sigma^{\mathrm{m}} = (\frac{1}{3})\operatorname{tr}(\sigma)$ and on the equivalent stress $\sigma^{\mathrm{e}} = \sqrt{(\frac{2}{3})}|\sigma'|$ where $\sigma' = \sigma - \sigma^{\mathrm{m}} I_3$. Finally k is a viscosity parameter and $\langle A \rangle = \frac{1}{2}(|A| + A)$ for all $A \in \mathbb{R}$.

Note that if we let

$$\tilde{G}(\sigma, \varepsilon, \kappa) = k\left\langle 1 - \frac{\kappa}{H(\sigma)}\right\rangle \frac{\partial H}{\partial \sigma}(\sigma),$$

then (6.3) becomes (5.2). We can also write (6.3) as follows:

$$\dot{\sigma} = \lambda \operatorname{tr}(\dot{\varepsilon}) I_3 + 2\mu\dot{\varepsilon} - k\left\langle 1 - \frac{\kappa}{H(\sigma)}\right\rangle \left[\lambda \operatorname{tr}\left(\frac{\partial H}{\partial \sigma}\right) I_3 + 2\mu \frac{\partial H}{\partial \sigma}(\sigma)\right], \tag{6.4}$$

which is a constitutive equation of the form (5.1) with

$$G(\sigma, \varepsilon, \mathscr{H}) = -k\left\langle 1 - \frac{\kappa}{H(\sigma)}\right\rangle \left[\lambda \operatorname{tr}\left(\frac{\partial H}{\partial \sigma}\right) I_3 + 2\mu \frac{\partial H}{\partial \sigma}(\sigma)\right].$$

The work hardening parameter κ is chosen as independent variable, i.e. the irreversible stress work per unit volume:

$$\kappa(t) = \int_0^t \operatorname{tr}(\sigma(\tau)\cdot\dot{\varepsilon}^I(\tau))\, d\tau. \tag{6.5}$$

From the above formula we obtain for the stress power

$$\dot{\kappa} = k\left\langle 1 - \frac{\mathscr{H}}{H(\sigma)}\right\rangle \frac{\partial H}{\partial \sigma}(\sigma)\cdot\sigma. \tag{6.6}$$

Note that (6.5) is of the form (5.7) with

$$\varphi(\sigma, \varepsilon, \kappa) = k\left\langle 1 - \frac{\kappa}{H(\sigma)}\right\rangle \frac{\partial H}{\partial \sigma}(\sigma)\cdot\sigma.$$

For coal the constitutive constants and functions (for the associated constitutive equation) are

$$K = 16\,850 \text{ kgf/cm}^2, \qquad G = 3790 \text{ kgf/cm}^2, \qquad k = 468 \times 10^{-6}\,\text{s}^{-1},$$

$$H(\sigma^m, \sigma^e) = \left[a_0 - a_1\left(\frac{\sigma^m}{\alpha} - \frac{\sigma^e}{\alpha}\right)\right]\left(\frac{\sigma^e}{\alpha}\right)^2 + \left[b_0 - b_1\left(\frac{\sigma^m}{\alpha} - \frac{\sigma^E}{\alpha}\right)\right]\left(\frac{\sigma^e}{\alpha}\right)$$

$$+\begin{cases} c\left(1 - \cos\left(\omega\dfrac{\sigma^m}{\alpha}\right)\right) & \text{for } 0 \le \sigma^m \le \sigma_1 \\ 2c & \text{for } \quad \sigma^m \ge \sigma_1 \end{cases}$$

where $a_0 = 2.477 \times 10^{-5}$ kgf/cm^2, $a_1 = 3.533 \times 10^{-7}$ kgf/cm^2, $b_0 = 5.347 \times 10^{-3}$ kgf/cm^2, $b_1 = 6.651 \times 10^{-4}$ kgf/cm^2, $c = 0.095$ kgf/cm^2, $\omega = 7.2°$, $\alpha = 1$ kgf/cm^2, $\sigma_1 = 25$ kgf/cm^2. For argillaceous coal we have $K = 4074$ kgf/cm^2, $G = 1235$ kgf/cm^2, $k = 8.24 \times 10^{-6}\,\text{s}^{-1}$,

$$H(\sigma^m, \sigma^e) = \frac{1}{d_1 + d_2\dfrac{\sigma^m}{\alpha}}\left(\frac{\sigma^e}{\alpha}\right)^2 + b\left(\frac{\sigma^m}{\alpha}\right)$$

$$+\begin{cases} c_0 \sin\left(\omega\dfrac{\sigma^m}{\alpha} + \varphi\right) + c_1 & \text{for } 0 \le \sigma^m \le \sigma_2 \\ c_0 + c_1 & \text{for } \quad \sigma^m > \sigma_2 \end{cases}$$

where $d_1 = 11\,505\cdot 68(\text{kgf/cm}^2)^{-1}$, $d_2 = 133.312(\text{kgf/cm}^2)^{-1}$, $b = 0.00333$ kgf/cm^2, $c_0 = c_1 = 0.046$ kgf/cm^2, $\alpha = 1$ kgf/cm^2, $\sigma_2 = 28.125$ kgf/cm^2, $\omega = 6.4°$, $\varphi = -90°$.

Finally for the steel we consider

$$E = 2\,100\,000 \text{ kgf/cm}^2, \qquad \nu = 0.27, \qquad G = 810\,000 \text{ kgf/cm}^2$$

and it will be assumed that the steel base remains in an elastic state.

6.2 Boundary conditions and initial data

Let us first describe in more detail the mathematical formulation of the general problem (5.11)–(5.15) which will be used here. The geometry of the problem is shown in Figures 6.1 and 6.2. Due to the symmetries already mentioned we consider the problem to be a plane strain one, i.e. $u = (u_1, u_2, 0)$ and $u_i = u_i(x_1, x_2)$, $i = 1, 2$ and therefore $\varepsilon_{33} = \varepsilon_{13} = \varepsilon_{23} = 0$. We carry the computations in the polygonal domain Ω shown in Figures 6.1 and 6.2, and included in $\mathbb{R}^2$ (i.e. $N = 2$). The particular dimensions considered are $l_1 = 150$ cm, $l_c = 30$ cm, $l_a = 120$ cm, $l_1 = l_c + l_a$, $l_3 = 300$ cm, $l_4 = 300$ cm, $l_5 = 120$ cm, $l_6 = 65$ cm, $l_7 = 15$ cm, $l_8 = 155$ cm, $l_{10} = 150$ cm. The partition of boundary Γ from (5.11)–(5.15) corresponds in this concrete problem as follows: Γ_1 with Γ_2 and Γ_2 with $\Gamma \backslash \Gamma_2$.

For the primary stress we make the standard assumption

$$\sigma^p = \begin{Vmatrix} \sigma_h & 0 & 0 \\ 0 & \sigma_v & 0 \\ 0 & 0 & \sigma_h \end{Vmatrix}$$

with $\sigma_v = \sigma_{22}^p = -18$ kgf/cm^2, $\sigma_h = \sigma_{11}^p = \sigma_{33}^p = 0.3\sigma_v$.

The boundary conditions which are of the form (5.14) are schematically presented in Figure 6.2. On the boundary Γ_2 (which corresponds with Γ_1 in (5.14)) we prescribe the displacement

$$u(t) = 0.$$

Along Γ_1 a linear distribution of normal stress is assumed

$$\sigma(t, x_2)v = f_1(x_2) = \left(\frac{l_1 - x_2}{l_1} \sigma_h, 0 \right).$$

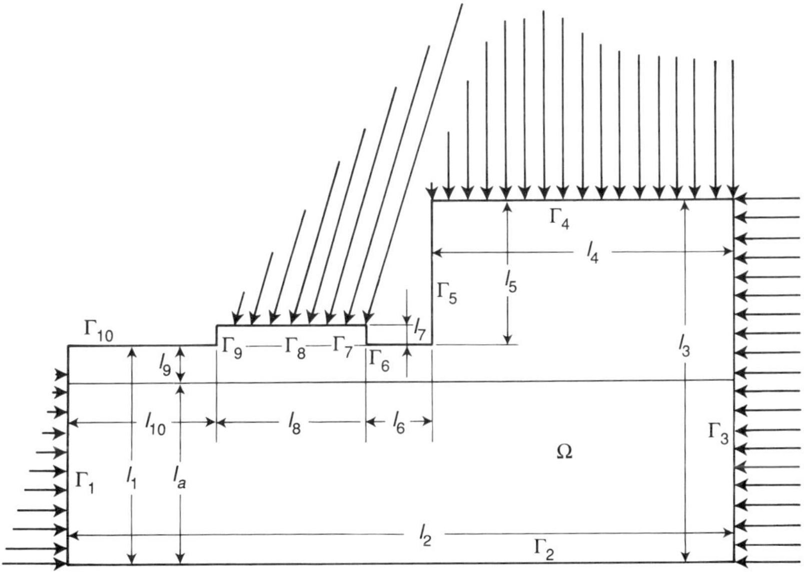

Fig. 6.2 Schematical representation of the boundary conditions.

On $\Gamma_5 \cup \Gamma_6 \cup \Gamma_7 \cup \Gamma_9 \cup \Gamma_{10}$ the boundary is stress free, i.e. $\sigma(t)\nu = f_2 = (0, 0)$. Along Γ_8 a linear stress distribution was assumed (suggested by field measurements):

$$\sigma(\mathrm{t}, x_1)\nu = f_3(x_1) = \left(\beta \frac{x_1 - l_{10}}{l_8} \tan(17°); \beta \frac{x_1 - l_{10}}{l_8}\right),$$

where $\beta = 41.5$ kgf/cm^2, the pressure exerted by the foot of the support. On Γ_3 the assumed boundary condition is

$$\sigma(t, x_2)\nu = (\sigma_h, 0) = f_4$$

while on Γ_4

$$\sigma(t, x_1)\nu = (0, g(x_1)) = f_5(x_1)$$

where g has the graphical representation shown in Figure 6.2.

From the various possible stresses along Γ_4 we have chosen one suggested by field measurements.

The initial data $(u_0, \sigma_0, \kappa_0)$ from (5.15) are quite difficult to formulate. For this reason we have chosen as initial data the elastic solution which would be obtained if the excavation were performed instantaneously. Recall that according to the constitutive equations used the 'instantaneous' responses of both coals are elastic.

Since the reference configuration is stressed before excavation, in order to determine the elastic solution the previously formulated boundary conditions must be modified accordingly. Thus instead of the force f acting on $\Gamma\backslash\Gamma_2$ we will consider the force $\hat{f} = f - \sigma^P\nu$ acting on $\Gamma\backslash\Gamma_2$.

Therefore the initial data must satisfy the equations

$$\operatorname{Div} \hat{\sigma} = 0 \tag{6.7}$$

$$\hat{\sigma} = \mathscr{A}\varepsilon(u_0) \tag{6.8}$$

$$u = 0 \quad \text{on } \Gamma_2, \qquad \hat{\sigma}\nu = \hat{f} \quad \text{on } \Gamma\backslash\Gamma_2. \tag{6.9}$$

In this way we get u_0, $\sigma_0 = \hat{\sigma}_0 + \sigma^P$ and we put $\kappa_0 = H(\sigma_0)$.

Since we shall use the numerical approach (5.26)–(5.30) we need the approximation of the initial data u_h^0, σ_h^0, κ_h^0. From (6.7)–(6.9) we deduce the following variational equation in the finite element space V_h:

$$\int_\Omega \mathscr{A}\varepsilon(u_h^0)\cdot\varepsilon(v)\,\mathrm{d}x = \int_{\Gamma\backslash\Gamma_2} \hat{f}v\,\mathrm{d}s \qquad \forall v \text{ in } V_h.$$

From here we get u_h^0 and further $\hat{\sigma}_h^0 = \mathscr{A}\varepsilon(u_h^0)$, and therefore $\sigma_h^0 = \hat{\sigma}_h^0 + \sigma^P$. With this procedure we obtain the initial data u_h^0, σ_h^0, and $\kappa_h^0 = \kappa_0$.

6.3 Numerical results

In order to construct the space V_h, the finite element method was used. The two-dimensional domain Ω was divided into 867 quadrilateral elements.

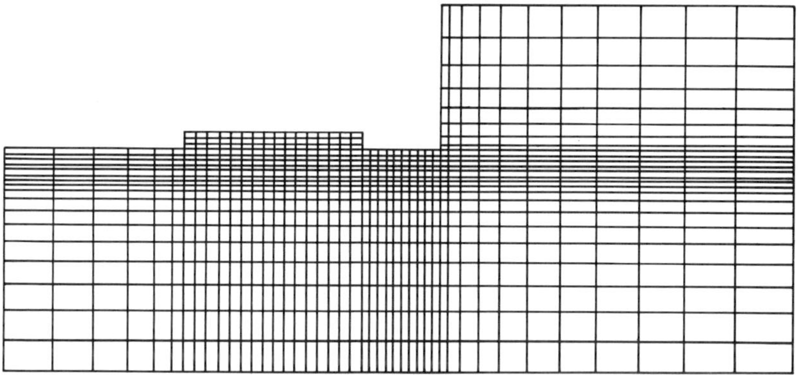

Fig. 6.3 The mesh pattern used in the finite element discretization of the domain Ω.

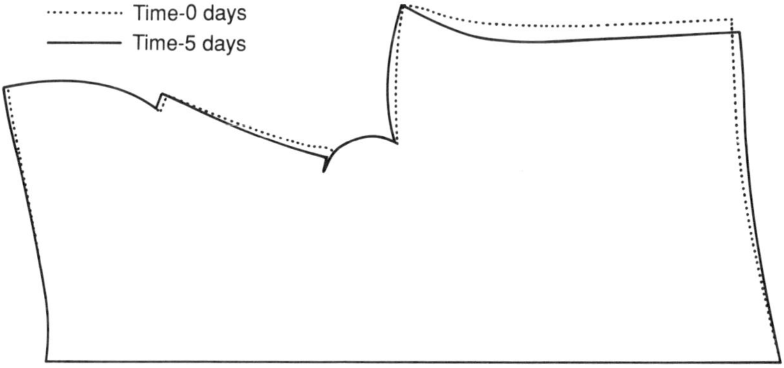

Fig. 6.4 The deformed boundary of Ω at the initial time (dashed line) and after five days (solid line).

The mesh pattern used in the finite element discretization is presented in Figure 6.3.

The evolution of the deformed boundary of Ω is shown in Figure 6.4. Note that the steel base of the self-advancing support unit is sinking in time (it advances towards Γ_2) and the face *GF* of the long wall is bulging in time (it advances towards the mined-out area). For a better visualization of the evolution of the boundary, the displacement fields were exaggerated with respect to the length of the domain Ω.

The computed field of the mean stress after five days is shown in Figure 6.5. Note that at the point *E* the mean stress is reaching its maximum value and on *EF* we have negative mean stress, i.e. the central part of the line *EF* is subjected to tractions.

The computed field of the equivalent stress after five days is shown in

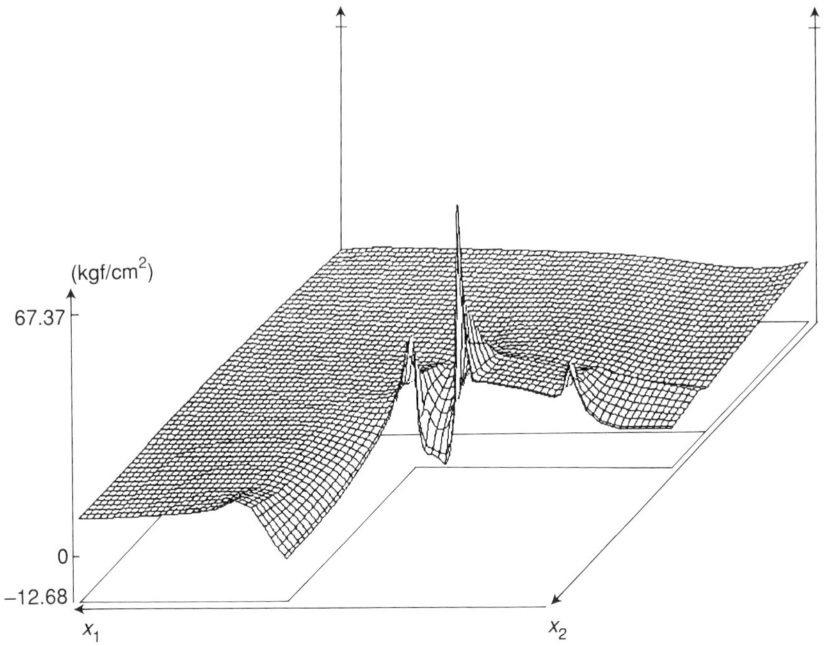

Fig. 6.5 The computed field on the mean stress after five days.

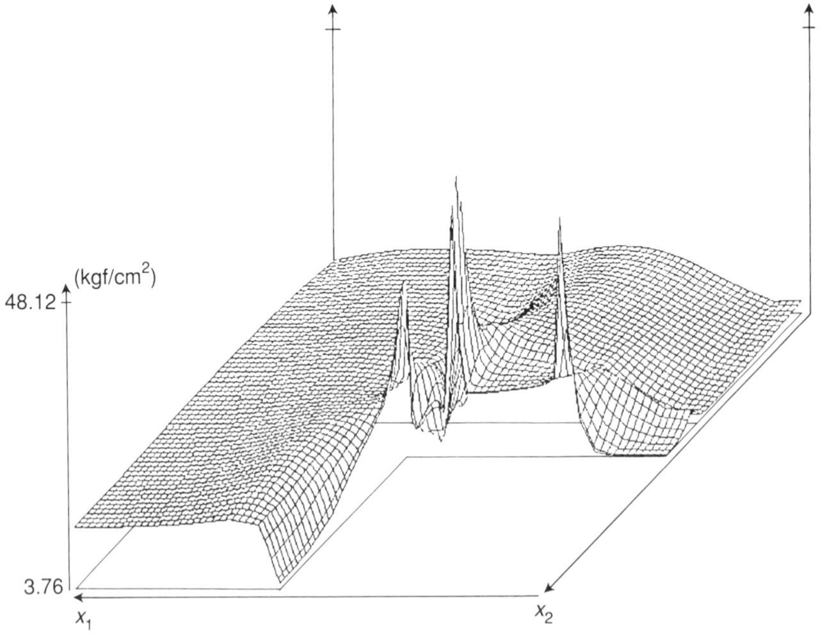

Fig. 6.6 The computed field of the equivalent stress after five days.

Figure 6.6 on page 102. Note that the greatest values of the equivalent stress are obtained in the points B, E, and F, where a significant shearing is expected.

Figures 6.7–10 show the evolution of the equivalent and mean stress in different points of the domain Ω. Note that both are almost constant after 3–4 days.

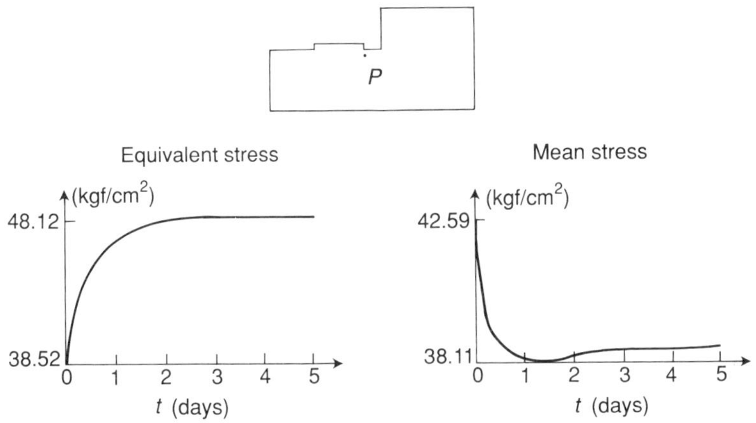

Fig. 6.7 The evolution of the equivalent stress σ^e and of the mean stress σ^m at the point P.

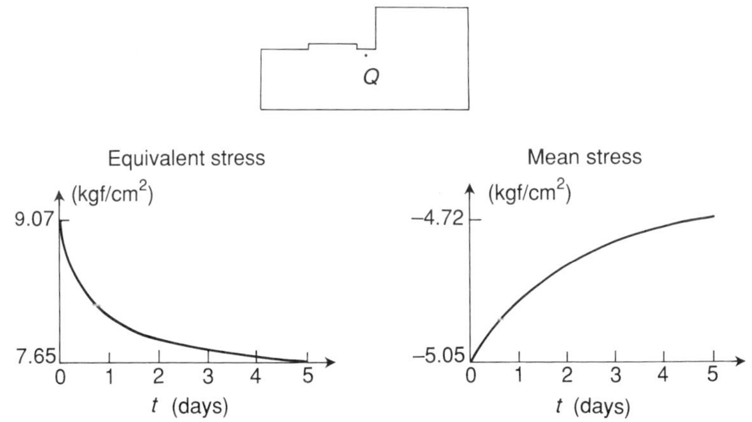

Fig. 6.8 The evolution of the equivalent stress σ and of the mean stress σ^m at the point Q.

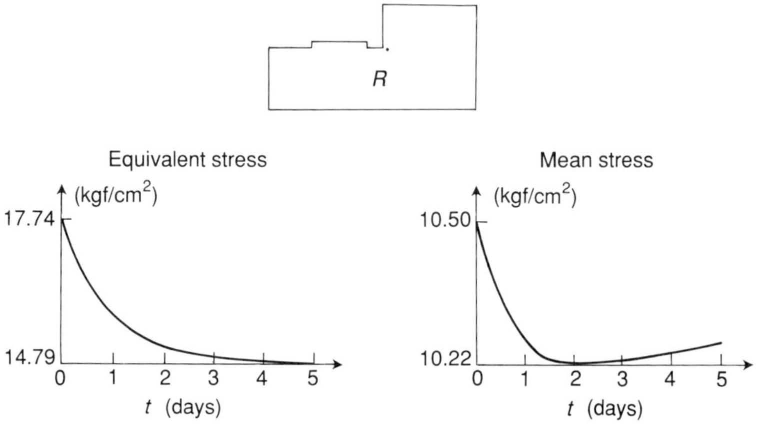

Fig. 6.9 The evolution of the equivalent stress σ^e and of the mean stress σ^m at the point R.

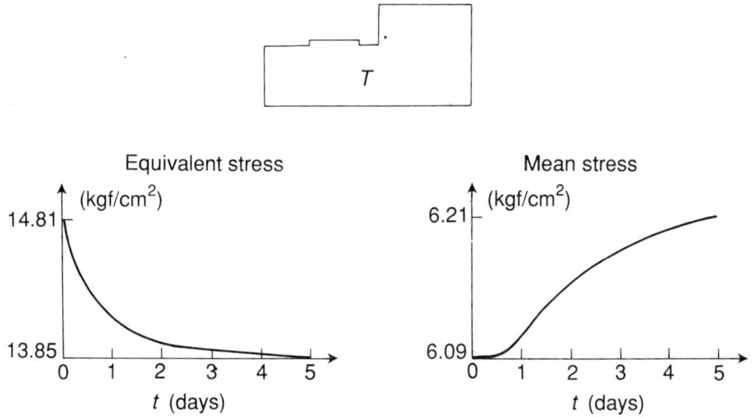

Fig. 6.10 The evolution of the equivalent stress σ^m and of the mean stress σ^m at the point T.

6.4 Failure

With respect to instantaneous failure we accept a classical criterion formulated in terms of stresses

$$|\sigma_i| \leq \sigma_t, \quad i = 1, 2, 3 \qquad \text{if } \sigma_i \leq 0$$
$$\sigma^e \leq a + b\sigma^m \qquad \text{if } \sigma^m > 0$$

where σ_i are principal stresses, σ_t is the uniaxial tensile strength, and σ^e and σ^m are defined in Section 6.2.

For coal we have $\sigma_t = 17\ \text{kgf/cm}^2$, $a = 10\ \text{kgf/cm}^2$, $b = 0.47$, while for argillaceous coal $\sigma_t = 4\ \text{kgf/cm}^2$, $a = 10\ \text{kgf/cm}^2$, $b = 0.37$. This criterion is a classical Mohr–Mises kind of criterion which is coupled with limit tension cut-off planes.

The numerical example has been carried out for $l_c = 15$ cm and $l_c = 20$ cm. It has been found that for both cases the stratum of coal under Γ_6 boundary is subjected to tensile stress which produce failure across the whole thickness. Thus if we would like to have under the self-advancing support a coal stratum which may deform but not fail, this stratum has to be at least 25–30 cm thick. A too thick layer would result in loss of extracted coal; it was decided that a thickness of 30 cm would be safe enough to ensure a smooth advancement of the self-advancing support.

What concerns the long-term failure which is to be considered, if for some reason the self-advancing support does not advance for some time, we use energetic criteria stating that long-term failure by creep-dilatation will occur if the energy of microcracking surpasses a certain limit value, specific for that particular material (see Cristescu, 1989). Shortly, we must estimate the value of

$$W_v(t) = \frac{1}{3}\int_0^t (\operatorname{tr}\sigma(\tau))(\operatorname{tr}\dot{\varepsilon}^I(\tau))\,d\tau$$

during creep. For coal, this limit value for $W_v(t)$ ranges between $0.5 \times 10^3\ \mathrm{J\,m^{-3}}$ and $10^5\ \mathrm{J\,m^{-3}}$. In these preliminary calculations we have found that long-term damage at the face of the long wall is not possible. However, if we take into account that the excavation is not really instantaneous, and that a material point still well ahead of the long wall is subjected to creep deformation during the processes of advancement of the long wall, well before its position reaches its present one (at R, say), then damage by creep-dilatation is possible after several hours just near the long-wall surface, mainly in the neighbourhood of point R. This subject matter needs further consideration.

Generally, if the whole history of long-wall advancement were considered, more significant deformations are to be expected due to creep than those described above.

Bibliographical notes

The general form (1.1) of a quasilinear rate-type constitutive equation was proposed by Cristescu (1963, 1964) in the one-dimensional case. The semilinear equation (1.2) was proposed by Perzyna (1963) in the three-dimensional case. Various results and mechanical interpretations concerning constitutive equations (1.1), (1.2) (or some particular forms of them) may be found for instance in the papers of Geiringer and Freudenthal (1958), Cristescu (1967), Cristescu and Suliciu (1982), Gurtin *et al.* (1984), Suliciu (1984), Podio-Guidugli and Suliciu (1984).

Quasistatic problems of the form (1.20)–(1.23) were studied by many authors, with different methods. So, for f constant in time and (1.21) replaced by (1.8), (1.7), an existence and uniqueness result was given by Duvaut and Lions (1972), Ch. 5, using a regularization method and monotony arguments. In the case of the constitutive equation (1.8) in which $\tilde{G}$ depends only on σ, existence and uniqueness results were given by Suquet (1981*a*,*b*) using a time-discretization technique and also monotony arguments. A similar technique as in the proof of Theorem 1.1, based on the equivalence between (1.20)–(1.23) and an ordinary differential equation in a Hilbert space was used by Suquet (1982, Ch. 3) in the case in which G does not depend on ε.

Other existence results for materials of the form (1.2) in which G depends only on σ were given by Anzellotti and Gianquinta (1980), Anzellotti (1983), Djaoua and Suquet (1984) (the case of Norton–Hoff materials); regularity results with respect to spatial variables for such types of problems were obtained by Suquet (1982, Ch. 3).

The results presented in Sections 1.2–1.4 and 2.1–2.4 were written following Ionescu and Sofonea (1988). A slightly different technique in the proof of Theorem 1.1 can be found in the work of Sofonea (1988). A similar technique presented in proof of Theorem 2.2 can be found in the work of Suquet (1982, Ch. 3). For other results concerning periodic solutions in the study of viscoelastic or viscoplastic materials we refer the reader to Halphen (1978) and Wesfreid (1980). The elastic problem (2.21)–(2.23) was considered by many authors with different assumptions on the function F (see for instance Fichera (1972), Duvaut and Lions (1972, Ch. 3), Mazilu and Sburlan (1973) and others).

The evolution problem for elastic–plastic solids has developed remarkably in the last twenty years. Duvaut and Lions (1972, Ch. 5) used the constructive theory of partial differential equations including Galerkin approximations, regularization, and penalization to obtain the existence and uniqueness of the stress field. Moreau (1975) built the geometrical theory of 'catching up by a moving convex set' and proved directly the existence and the uniqueness of the stress history. Other authors were interested in the evolution problem for the velocity (or displacement field): Nayroles (1970), Johnson (1976), Suquet (1981*b*), Le Tallec (1990). The main difficulty of this problem is the regularity of the strain tensor which was avoided by means of a weak formulation. The results presented in Section 3 were obtained by Suquet (1981*a*, 1982). In these papers the existence of a velocity field history for the quasistatic evolution of a perfectly plastic body is proved in the more general case when K depends on time and on spatial coordinates. Moreover, the technique presented in the papers mentioned above justifies the numerous regularizations currently used to approximate the perfectly plastic law (see Theorem 3.1). The 'safety assumption' (3.5) was introduced by Moreau (1975) and Johnson (1976). For details, history and complete references on quasistatic processes in perfect plasticity we refer the reader to the work of Suquet (1982, Ch. 3).

Section 4 was written following the paper of Ionescu (1990*a*). The explicit Euler method was also applied by Geymonat and Raous (1977) for a linear viscoelastic quasistatic problem. In order to obtain numerical stability, the time step is also restricted to be less than a critical value.

Concrete constitutive equations of the form (5.2), (5.7) or (5.8) were proposed by Cristescu (1985, 1987, 1989) in order to describe the behaviour of rock-like materials. Other references on rate-type constitutive equations involving a hardening parameter or other internal state variables can be found in the papers of Nečas and Kratochvil (1973) and John (1974).

The results from Section 5.3 were presented following the ideas of Sofonea (1988), (1989*c*) and the results of Section 5.4 belong to Ionescu (1990*b*).

Section 6 was written following the paper of Cristescu *et al.* (1991).

4

DYNAMIC PROCESSES FOR RATE-TYPE ELASTIC–VISCOPLASTIC MATERIALS

This chapter deals with dynamic problems for semilinear rate-type constitutive materials. The existence and uniqueness of the solution is given using arguments of linear continuous semigroups in Hilbert spaces. In the viscoelastic case an energy bound of the solution is given and it is proved that linear elasticity is a proper asymptotic theory for viscoelastic materials. An approach to perfect plasticity is also studied, by a passage to the limit. The case of rate-type materials involving internal state variables is also discussed and some monotony and fixed point methods in the study of dynamic problems are presented. Finally the evolution of the perturbations of homogeneous simple shear is analysed and the link with strain localization is given.

1 Discussion of a dynamic elastic–viscoplastic problem

In this section we study an initial and boundary-value problem describing dynamic processes for semilinear rate-type elastic–viscoplastic materials. Constitutive assumptions and assumptions on the data are given; the existence and uniqueness of the solution is proved by reducing the mechanical problem to a semilinear hyperbolic equation in a Hilbert space. The continuous dependence of the solution upon the input data is also discussed.

We start this section by presenting the mechanical problem investigated and the assumptions on the data.

1.1 Problem statement

Let Ω be a domain in $\mathbb{R}^N$ ($N = 1, 2, 3$) with a smooth boundary $\partial\Omega = \Gamma$ and let Γ_1 be an open subset of Γ. Let $\Gamma_2 = \Gamma\backslash\bar{\Gamma}_1$ and let us consider the following mixed problem:

Find the displacement function u: $[0, T] \times \Omega \to \mathbb{R}^N$ and the stress function σ: $[0, T] \times \Omega \to \mathscr{S}_N$ such that

$$\rho\ddot{u}(t) = \text{Div}\,\sigma(t) + \rho b(t) \tag{1.1}$$

$$\dot{\sigma}(t) = \mathscr{A}\varepsilon(\dot{u}(t)) + G(t, \sigma(t), \varepsilon(u(t))) \qquad \text{in } \Omega, \tag{1.2}$$

$$u(t) = g(t) \text{ on } \Gamma_1, \quad \sigma(t)\nu = f(t) \text{ on } \Gamma_2, \qquad t \in (0, T] \tag{1.3}$$

$$u(0) = u_0, \quad \dot{u}(0) = v_0, \quad \sigma(0) = \sigma_0 \qquad \text{in } \Omega. \tag{1.4}$$

In (1.1) $\rho: \Omega \to \mathbb{R}_+$ is the density of the mass and $b: [0, T] \times \Omega \to \mathbb{R}^N$ is the given body force (Section 2.4, Chapter 1). Equation (1.2) represents a semilinear rate-type elastic–viscoplastic constitutive law in which $\mathscr{A}$ is a fourth-order tensor and G is a constitutive function which depends usually on time by means of the temperature field (see Section 1.1 in Chapter 3). The functions u_0, v_0, σ_0 are the initial data and f, g are the boundary data.

In the study of problems (1.1)–(1.4) the following assumptions are made:

$$\mathscr{A} \text{ satisfies (1.24) in Chapter 3} \tag{1.5}$$

$$G: \Omega \times [0, T] \times \mathscr{S}_N \times \mathscr{S}_N \to \mathscr{S}_N \text{ satisfies (1.25) in Chapter 3}$$
$$\text{on } [0, T] \text{ with } \theta \in W^{1,1}(0, T, H) \tag{1.6}$$

$$\rho \in L^\infty(\Omega), \qquad \rho(x) \geq c_1 > 0 \quad \text{a.e. in } \Omega \tag{1.7}$$

$$b \in W^{1,1}(0, T, \boldsymbol{H}) \tag{1.8}$$

$$u_0, v_0 \in \boldsymbol{H}_1, \qquad \sigma_0 \in \mathscr{H}_2 \tag{1.9}$$

$$u_0 = g(0), \qquad v_0 = \dot{g}(0) \quad \text{on } \Gamma_1, \qquad \sigma_0 \nu = f(0) \quad \text{on } \Gamma_2 \tag{1.10}$$

there exists $\tilde{u} \in W^{3,1}(0, T, \boldsymbol{H}) \cap W^{2,1}(0, T, \boldsymbol{H}_1)$ such that

$$\tilde{u}(t) = g(t) \qquad \text{on } \Gamma_1 \quad \forall t \in [0, T] \tag{1.11}$$

there exists $\tilde{\sigma} \in W^{2,1}(0, T, \mathscr{H}) \cap W^{1,1}(0, T, \mathscr{H}_2)$ such that

$$\tilde{\sigma}(t)\nu = f(t) \text{ on } \Gamma_2 \text{ for all } t \in [0, T]. \tag{1.12}$$

Remark 1.1 (1) *For the mechanical and mathematical interpretations of* (1.5), (1.6) *see Remark* 1.3 *in Chapter* 3.

(2) *Assumption* (1.10) *ensures the compatibility between the boundary and the initial data in order to have no discontinuities (shocks) at the initial time.*

(3) *If we want to give assumptions on the boundary data f, g in order to have* (1.11), (1.12) *we can make a stronger assumption, namely*

$$g \in W^{3,1}(0, T, \boldsymbol{H}_\Gamma), \qquad f \in W^{2,1}(0, T, V'_{\Gamma_1}). \tag{1.13}$$

Indeed let us suppose that (1.13) holds; using Remark 3.1 in Chapter 2 in the case $X_1 = H_\Gamma$, $X_2 = \boldsymbol{H}_1$, $A = z_1$ (see (2.29) and (2.30) of Chapter 2), $k = 3$, $p = 1$ we get that there exists $\tilde{u} \in W^{3,1}(0, T, \boldsymbol{H}_1)$ such that $\tilde{u}(t) = g(t)$ on Γ for all $t \in [0, T]$; using again Remark 3.1 in Chapter 2 in the case $X_1 = V'_{\Gamma_1}$, $X_2 = \mathscr{H}_2$, $A = z_{2,\Gamma_1}$ (see Lemma 2.2 of Chapter 2), $k = 2$, $p = 1$ we get that there exists $\tilde{\sigma} \in W^{2,1}(0, T, \mathscr{H}_2)$ such that $\tilde{\sigma}(t)\nu = f(t)$ on Γ_2 for all $t \in [0, T]$. Hence (1.13) implies (1.11) and (1.12).

1.2 An existence and uniqueness result

In this section an existence and uniqueness result for the problem (1.1)–(1.4) is proved. The technique used here is based on the equivalence between the studied problem and a semilinear evolution equation in a Hilbert space. To be more specific the mechanical problems (1.1)–(1.4) is reduced to a Lipschitz perturbation of a linear hyperbolic evolution equation and semigroup techniques are used.

The main result of this section is the following theorem:

Theorem 1.1 *Let* (1.5)–(1.12) *hold. Then there exists a unique solution of the problem* (1.1)–(1.4) *such that*

$$u \in W^{2,\infty}(0, T, \boldsymbol{H}) \cap W^{1,\infty}(0, T, \boldsymbol{H}_1) \tag{1.14}$$

$$\sigma \in W^{1,\infty}(0, T, \mathcal{H}) \cap L^{\infty}(0, T, \mathcal{H}_2). \tag{1.15}$$

In order to prove Theorem 1.1 we need some preliminary results. Let us homogenize the boundary conditions (1.3) by denoting $u^* = u - \tilde{u}$, $v^* = \dot{u} - \dot{\tilde{u}}$, $\sigma^* = \sigma - \tilde{\sigma}$ and $u_0^* = u_0 - \tilde{u}(0)$, $v_0^* = v_0 - \dot{\tilde{u}}(0)$, $\sigma_0^* = \sigma_0 - \tilde{\sigma}(0)$. From (1.10) we notice that $u_0^*, v_0^* \in \boldsymbol{V}_1$, $\sigma_0^* \in \mathcal{V}_1$ hence we can easily deduce the following lemma:

Lemma 1.1 *The couple* $(u; \sigma)$ *is a solution of* (1.1)–(1.4) *such that* (1.14), (1.15) *hold iff* $u^* \in W^{1,\infty}(0, T, \boldsymbol{V}_1)$, $v^* \in W^{1,\infty}(0, T, \boldsymbol{H}) \cap L^{\infty}(0, T, \boldsymbol{V}_1)$ *and* $\sigma^* \in W^{1,\infty}(0, T, \mathcal{H}) \cap L^{\infty}(0, T, \boldsymbol{V}_1)$ *is the solution of the following problem:*

$$\dot{u}^*(t) = v^*(t) \tag{1.16}$$

$$\dot{v}^*(t) = \rho^{-1} \operatorname{Div} \sigma^*(t) + a(t) \tag{1.17}$$

$$\dot{\sigma}^*(t) = \mathcal{A}\varepsilon(v^*(t)) + H(t, \sigma^*(t), \varepsilon(u^*(t))) \tag{1.18}$$

$$u^*(0) = u_0^*, \qquad v^*(0) = v_0^*, \qquad \sigma^*(0) = \sigma_0^* \tag{1.19}$$

where $a \in W^{1,1}(0, T, \boldsymbol{H})$ *and* $H: \Omega \times [0, T] \times \mathcal{S}_N \times \mathcal{S}_N \to \mathcal{S}_N$ *are given by*

$$a(t) = b(t) - \ddot{\tilde{u}}(t) + \rho^{-1} \operatorname{Div} \tilde{\sigma}(t) \tag{1.20}$$

$$H(t, \tau, \sigma) = G(t, \tilde{\sigma}(t) + \tau, \varepsilon(\tilde{u}(t)) + \sigma) + \mathcal{A}\varepsilon(\dot{\tilde{u}}(t)) - \dot{\tilde{\sigma}}(t) \tag{1.21}$$

for all $t \in [0, T]$, $\tau, \sigma \in \mathcal{S}_N$.

Let us consider the real Hilbert space $X = \boldsymbol{V}_1 \times \boldsymbol{H} \times \mathcal{H}$ and let $D(A) = \boldsymbol{V}_1 \times \boldsymbol{V}_1 \times \mathcal{V}_1 \subset X$, $A: D(A) \subset X \to X$ be given by

$$A(u; v; \sigma) = (v; \rho^{-1} \operatorname{Div} \sigma; \mathcal{A}\varepsilon(v)) \quad \text{for all } u, v \in \boldsymbol{V}_1, \sigma \in \mathcal{V}_1. \tag{1.22}$$

With the above notations we have:

Lemma 1.2 *The operator* A *is the infinitesimal generator of a* C_0 *semigroup* $(S(t))_{t \geq 0}$ *of bounded linear operators in* X.

In order to prove this lemma let us consider $Y = \boldsymbol{H} \times \mathcal{H}$ with the inner product

$$\langle (v_1; \tau_1), (v_2; \tau_2) \rangle_Y = \langle \rho v_1, v_2 \rangle_{\boldsymbol{H}} + \langle \mathcal{A}^{-1} \tau_1, \tau_2 \rangle_{\mathcal{H}} \tag{1.23}$$

which generates an equivalent norm on Y. Let $D(B) = \boldsymbol{V}_1 \times \mathcal{V}_1 \subset Y$ and $B\colon D(B) \subset Y \to Y$ be given by

$$B(v; \sigma) = (\rho^{-1} \operatorname{Div} \sigma; \mathcal{A}\varepsilon(v)) \quad \text{for all } (v; \sigma) \in \boldsymbol{V}_1 \times \mathcal{V}_1. \tag{1.24}$$

Lemma 1.3 *If we consider in Y the inner product given by* (1.23) *then $B^* = -B$, where B^* denotes the adjoint operator of B.*

PROOF. Let $(v; \sigma) \in D(B^*)$ and $(\bar{v}; \bar{\sigma}) = B^*(v; \sigma)$. For all $(u; \tau) \in D(B) = \boldsymbol{V}_1 \times \mathcal{V}_1$ we have

$$\langle \operatorname{Div} \tau, v \rangle_{\boldsymbol{H}} + \langle \varepsilon(u), \sigma \rangle_{\mathcal{H}} = \langle \rho u, \bar{v} \rangle_{\boldsymbol{H}} + \langle \mathcal{A}^{-1} \tau, \bar{\sigma} \rangle_{\mathcal{H}}. \tag{1.25}$$

If we put $\tau = 0$ in (1.25) we get $\langle \varepsilon(u), \sigma \rangle_{\mathcal{H}} = \langle u, \rho \bar{v} \rangle_{\boldsymbol{H}}$ for all $u \in D$. Hence $\operatorname{Div} \sigma = -\rho \bar{v}$, i.e. $\sigma \in \mathcal{H}_2$. We also have $\langle \varepsilon(u), \sigma \rangle_{\mathcal{H}} + \langle \operatorname{Div} \sigma, u \rangle_{\boldsymbol{H}} = 0$ for all $u \in \boldsymbol{V}_1$; hence $\sigma \in \mathcal{V}_1$. If we now put $u = 0$ in (1.25) we get $\langle \operatorname{Div} \tau, v \rangle_{\boldsymbol{H}} = \langle \tau, \mathcal{A}^{-1} \bar{\sigma} \rangle_{\mathcal{H}}$ for all $\tau \in \mathcal{D}$. So we get $\varepsilon(v) \in \mathcal{H}$ and from (2.26) of Chapter 2 we obtain $v \in \boldsymbol{H}_1$ and $\varepsilon(v) = -\mathcal{A}^{-1} \bar{\sigma}$. We can also deduce

$$\langle \operatorname{Div} \tau, v \rangle_{\boldsymbol{H}} + \langle \varepsilon(v), \tau \rangle_{\mathcal{H}} = 0 \qquad \forall \tau \in \mathcal{V}_1. \tag{1.26}$$

If $v \notin \boldsymbol{V}_1$ then $\gamma_1(v) \notin \boldsymbol{V}_{\Gamma_1}$ and there exists $f \in \boldsymbol{H}'_\Gamma$ such that $f = 0$ on $\boldsymbol{V}_{\Gamma_1}$ and $\langle f, \gamma_1 v \rangle_{\boldsymbol{H}'_\Gamma, \boldsymbol{H}_\Gamma} \neq 0$. From (2.45) in Chapter 2 we find that there exists $\tau \in \mathcal{H}_2$ such that $\gamma_2 \tau = f$. Since $\gamma_2 \tau = 0$ on $\boldsymbol{V}_{\Gamma_1}$ we have $\tau \in \mathcal{V}_1$ and from (1.26) we deduce $\langle \gamma_2 \tau, \gamma_1 v \rangle_{\boldsymbol{H}'_\Gamma, \boldsymbol{H}_\Gamma} = 0$, a contradiction. Hence $v \in \boldsymbol{V}_1$ and we have $B^*(v; \sigma) = -(\rho^{-1} \operatorname{Div} \sigma; \mathcal{A}\varepsilon(v)) = -B(v; \sigma)$. We have just proved that $D(B^*) \subset D(B)$ and $B^* = -B$ on $D(B^*)$. A simple calculation can prove that $D(B) \subset D(B^*)$ and the lemma is proved.

PROOF OF LEMMA 1.2. Let us denote by $(T(t))_{t \geq 0} \subset B(Y)$ the C_0-semigroup generated by B (see Lemma 1.3 and Lemma A3.4). Let $T_1(t) \in B(Y, \boldsymbol{H})$, $T_2(t) \in B(Y, \mathcal{H})$ such that $T(t)(v; \sigma) = (T_1(t)(v; \sigma); T_2(t)(v; \sigma)) \in \boldsymbol{H} \times \mathcal{H}$ for all $t \geq 0$. We shall denote by $S(t)$ the operator

$$S(t)(u; v; \sigma) = \left(u + \int_0^t T_1(s)(v; \sigma)\, \mathrm{d}s; T(t)(v; \sigma) \right) \tag{1.27}$$

for all $(u; v; \sigma) \in \boldsymbol{V}_1 \times \boldsymbol{H}_1 \times \mathcal{H} = X$ and $t \geq 0$. If we have in mind that $[D(B)]$ endowed with the graph norm of B is isomorphic with $\boldsymbol{V}_1 \times \mathcal{V}_1$ (we use (1.5), (1.7) together with (2.28) and (2.43) from Chapter 2) and the fact that $y \to \int_0^t T(s) y \, \mathrm{d}s \in B(Y, [D(B)])$ for all $t \geq 0$ (see Lemma A3.3) we obtain that $y \to \int_0^t T_1(s) y \, \mathrm{d}s \in B(Y, \boldsymbol{V}_1)$, hence $S(t) \in B(X)$ for all $t \geq 0$. Since B is the generator of $(T(t))_{t \geq 0}$ one can easily check that A is the infinitesimal generator of the semigroup $(S(t))_{t \geq 0}$.

PROOF OF THEOREM 1.1. Let $x(t) = (u^*(t); v^*(t); \sigma^*(t))$, $x_0 = (u_0^*; v_0^*; \sigma_0^*)$ and let $h: [0, T] \times X \to X$ be given by

$$h(t, (u; v; \sigma)) = (0; a(t); H(t, \sigma, \varepsilon(u)) \tag{1.28}$$

where $a(t)$ and $H(t, \cdot, \cdot)$ are given by (1.20), (1.21). Using this notation note that (1.16)–(1.19) can be written as

$$\dot{x}(t) = Ax(t) + h(t, x(t)), \qquad t > 0, \tag{1.29}$$

$$x(0) = x_0. \tag{1.30}$$

We shall use Theorem A3.9 in order to prove the existence and the uniqueness of the solution of (1.29), (1.30) and from Lemma 1.1 the statement of Theorem 1.1 will follow.

Since from Lemma 1.2 we have that A is the infinitesimal generator of a C_0-semigroup of linear operators on X we have only to notice that $x_0 \in D(A)$ (we use (1.9) and (1.10)) and to check (A3.39). Indeed, for all $(u_i; \sigma_i; x_i) \in X$, $t_i \in [0, T]$, $i = 1, 2$, from (1.5), (1.6) we have

$$\begin{aligned}
&\|h(t_1, (u_1; v_1; \sigma_1)) - h(t_2, (u_2; v_2; \sigma_2))\|_X \\
&\le C[\|a(t_1) - a(t_2)\|_H + \|\dot{\tilde{\sigma}}(t_1) - \dot{\tilde{\sigma}}(t_2)\|_{\mathscr{H}} + \|\theta(t_1) - \theta(t_2)\|_H \\
&\qquad + \|\varepsilon(\tilde{u}(t_1)) - \varepsilon(\tilde{u}(t_2))\|_{\mathscr{H}} + L(\|\varepsilon(\tilde{u}(t_1)) - \varepsilon(\tilde{u}(t_2))\|_{\mathscr{H}} + \|\tilde{\sigma}(t_1) - \tilde{\sigma}(t_2)\|_{\mathscr{H}} \\
&\qquad + \|\sigma_1 - \sigma_2\|_{\mathscr{H}} + \|\varepsilon(u_1) - \varepsilon(u_2)\|_{\mathscr{H}})] \\
&\le C\Bigg[\int_{t_1}^{t_2} [\|\dot{b}(s) - \dddot{\tilde{u}}(s) + \rho^{-1} \operatorname{Div} \dot{\tilde{\sigma}}(s)\|_H + \|\ddot{\tilde{\sigma}}(s)\|_{\mathscr{H}} + \|\ddot{\tilde{u}}(s)\|_{H_1} \\
&\qquad + \|\dot{\theta}(s)\|_H + L(\|\dot{\tilde{u}}(s)\|_{H_1} + \|\ddot{\tilde{\sigma}}(s)\|_{\mathscr{H}})] \, ds \\
&\qquad + L(\|u_1 - u_2\|_{H_1} + \|\sigma_1 - \sigma_2\|_{\mathscr{H}})\Bigg].
\end{aligned}$$

So, having in mind that $[D(A)]$ endowed with the graph norm of A is isomorphic with $V_1 \times V_1 \times \mathscr{V}_1$, from Theorem A3.9 we get there exists a unique solution of (1.16)–(1.19) $u^* \in W^{1,\infty}(0, T, V_1)$, $v^* \in W^{1,\infty}(0, T, H) \cap L^\infty(0, T, V_1)$, $\sigma^* \in W^{1,\infty}(0, T, \mathscr{H}) \cap L^\infty(0, T, \mathscr{V}_1)$.

1.3 The dependence of the solutions upon the input data

In this section the difference between two solutions of the problem (1.1)–(1.4) for two different input data is evaluated over a finite time interval. In this way the continuous dependence of the solution upon all input data is obtained (Theorem 1.2), hence the finite time stability holds (Corollary 1.1).

Theorem 1.2 *Let* (1.5)–(1.7) *hold and let* $(u_i; \sigma_i)$ *be the solutions of* (1.1)–(1.4) *for the data* b_i, g_i, f_i, u_{0i}, v_{0i}, $i = 1, 2$, *which satisfies* (1.8)–(1.10), (1.13).

Then there exists $C > 0$ which depends on Ω, Γ_1, $\mathscr{A}$, ρ, L and T such that

$$\|u_1 - u_2\|_{0,T,\boldsymbol{H}_1} + \|\dot{u}_1 - \dot{u}_2\|_{0,T,\boldsymbol{H}} + \|\sigma_1 - \sigma_2\|_{0,T,\mathscr{H}}$$
$$\leq C[\|u_{01} - u_{02}\|_{\boldsymbol{H}_1} + \|v_{01} - v_{02}\|_{\boldsymbol{H}} + \|\sigma_{01} - \sigma_{02}\|_{\mathscr{H}}$$
$$+ \|b_1 - b_2\|_{0,T,\boldsymbol{H}} + \|g_1 - g_2\|_{2,T,\boldsymbol{H}_\Gamma} + \|f_1 - f_2\|_{1,T,\boldsymbol{V}'_{\Gamma 1}}]. \quad (1.31)$$

PROOF. Let $\tilde{u}_i \in W^{3,1}(0, T, \boldsymbol{H}_1)$, $\tilde{\sigma}_i \in W^{2,1}(0, T, \mathscr{H}_2)$ which are given in Remark 1.1, $i = 1, 2$. Hence (1.11), (1.12) hold and from (3.14) of Chapter 2 we have that there exists $C > 0$ depending only on Ω and Γ_1 such that

$$\|\ddot{\tilde{u}}_1 - \ddot{\tilde{u}}_2\|_{0,T,\boldsymbol{H}} \leq C\|\ddot{g}_1 - \ddot{g}_2\|_{0,T,\boldsymbol{H}_\Gamma} \quad (1.32)$$

$$\|\varepsilon(\tilde{u}_1) - \varepsilon(\tilde{u}_2)\|_{0,T,\mathscr{H}} \leq C\|g_1 - g_2\|_{0,T,\boldsymbol{H}_\Gamma} \quad (1.33)$$

$$\|\varepsilon(\dot{\tilde{u}}_1) - \varepsilon(\dot{\tilde{u}}_2)\|_{0,T,\mathscr{H}} \leq C\|\dot{g}_1 - \dot{g}_2\|_{0,T,\boldsymbol{H}_\Gamma} \quad (1.34)$$

$$\|\mathrm{Div}\, \tilde{\sigma}_1 - \mathrm{Div}\, \tilde{\sigma}_2\|_{0,T,\boldsymbol{H}} \leq C\|f_1 - f_2\|_{0,T,\boldsymbol{V}'_{\Gamma 1}} \quad (1.35)$$

$$\|\tilde{\sigma}_1 - \tilde{\sigma}_2\|_{0,T,\mathscr{H}} \leq C\|f_1 - f_2\|_{0,T,\boldsymbol{V}'_{\Gamma 1}} \quad (1.36)$$

$$\|\dot{\tilde{\sigma}} - \dot{\tilde{\sigma}}_2\|_{0,T,\mathscr{H}} \leq C\|\dot{f}_1 - \dot{f}_2\|_{0,T,\boldsymbol{V}'_{\Gamma 1}}. \quad (1.37)$$

Let $u_i^* = u_i - \tilde{u}_i$, $v_i^* = \dot{u}_i - \dot{\tilde{u}}_i$, $\sigma_i^* = \sigma_i - \tilde{\sigma}_i$, $u_{0i}^* = u_{0i} - \tilde{u}_i(0)$, $v_{0i}^* = v_{0i} - \dot{\tilde{u}}_i(0)$, $\sigma_{0i}^* = \sigma_{0i} - \tilde{\sigma}_i(0)$, $i = 1, 2$ then from Lemma 1.1 we get that $(u_i^*, v_i^*; \sigma_i^*)$ are solutions for (1.16)–(1.19) in which $a(t)$, $H(t, \cdot, \cdot)$ are replaced by $a_i(t)$, $H_i(t, \cdot, \cdot)$ given by (1.20), (1.21) for $\tilde{u}_i, \tilde{\sigma}_i$. From (1.32)–(1.37) and (1.5), (1.6) we also have

$$\|a_1 - a_2\|_{0,T,\boldsymbol{H}} \leq C(\|b_1 - b_2\|_{0,T,\boldsymbol{H}} + \|\ddot{g}_1 - \ddot{g}_2\|_{0,T,\boldsymbol{H}_\Gamma} + \|f_1 - f_2\|_{0,T,\boldsymbol{V}'_{\Gamma 1}}), \quad (1.38)$$

$$\|H_1(t, \sigma_1, \tau_1) - H_2(t, \sigma_2, \tau_2)\|_{\mathscr{H}}$$
$$\leq C[Q\|\dot{g}_1 - \dot{g}_2\|_{0,T,\boldsymbol{H}_\Gamma} + \|\dot{f}_1 - \dot{f}_2\|_{0,T,\boldsymbol{V}'_{\Gamma 1}}$$
$$+ L(\|g_1 - g_2\|_{0,T,\boldsymbol{H}_\Gamma} + \|f_1 - f_2\|_{0,T,\boldsymbol{V}'_{\Gamma 1}}$$
$$+ \|\sigma_1 - \sigma_2\|_{\mathscr{H}} + \|\tau_1 - \tau_2\|_{\mathscr{H}})]. \quad (1.39)$$

If we denote by h_i: $[0, T] \times X \to X$ the function given by (1.28) in which H and a are replaced by H_i and a_i, $i = 1, 2$, we obtain that $x_i = (u_i^*; v_i^*; \sigma_i^*)$ is a solution for the Cauchy problem

$$\dot{x}_i(t) = Ax_i(t) + h_i(t, x_i(t)), \qquad t \in [0, T] \quad (1.40)$$

$$x_i(0) = x_{0i} \quad (1.41)$$

where $x_{0i} = (u_{0i}^*, v_{0i}^*; \sigma_{0i}^*)$, $i = 1, 2$ and A is given by (1.22). From (1.38) and (1.39) we deduce

$$\|h_1(t, y_1) - h_2(t, y_2)\|_X \leq \bar{C}(\|b_1 - b_2\|_{0,T,\boldsymbol{H}} + \|g_1 - g_2\|_{2,T,\boldsymbol{H}_\Gamma}$$
$$+ \|f_1 - f_2\|_{1,T,\boldsymbol{V}'_{\Gamma 1}} + \|y_1 - y_2\|_X) \quad (1.42)$$

for all $t \in [0, T]$, $y_1, y_2 \in X$, where $\bar{C}$ depends only on Ω, Γ_1, $\mathscr{A}$, ρ, and L. Since $(S(t))_{t \geq 0}$ is the semigroup generated by A, from (A3.38) and (1.40), (1.41) we get

$$x_i(t) = S(t)x_{0i} + \int_0^t S(t-s)h_i(s, x_i(s))\,ds \qquad \forall t \in [0, T], \quad i = 1, 2. \quad (1.43)$$

Let $x = x_1 - x_2$, $x_0 = x_{01} - x_{02}$; then from (1.43) we obtain

$$x(t) = S(t)x_0 + \int_0^t S(t-s)(h_1(s, x_1(s)) - h_2(s, x_2(s)))\,ds \qquad \forall t \in [0, T]. \quad (1.44)$$

Let $N(T)$ be such that $\|S(t)\|_{B(X)} \leq N(T)$ for all $t \in [0, T]$. From (1.42) and (1.44) we obtain

$$\|x(t)\|_X \leq N(T)\|x_0\|_X + \bar{C}N(T)T(\|b_1 - b_2\|_{0,T,\boldsymbol{H}} + \|g_1 - g_2\|_{2,T,\boldsymbol{H}_\Gamma}$$
$$+ \|f_1 - f_2\|_{1,T,V_{\Gamma 1}}) + \bar{C}N(T)\int_0^t \|x(s)\|_X\,ds \qquad \forall t \in [0, T]. \quad (1.45)$$

Using Corollary A4.1, after some algebra (1.45) implies (1.31).

Using the same technique as in the proof of Theorem 1.2 we can easily deduce the following result:

Corollary 1.1 *Let* (1.5)–(1.8), (1.11), (1.12) *hold and let* $(u_i; \sigma_i)$ *be the solution of* (1.1)–(1.4) *for the initial data* $u_{0i}, v_{0i}, \sigma_{0i}$, $i = 1, 2$, *which satisfies* (1.9), (1.10). *Then there exists* $C > 0$ *which depends only on* Ω, Γ_1, $\mathscr{A}$, ρ, L, *and* T *such that*

$$\|u_1 - u_2\|_{0,T,\boldsymbol{H}_1} + \|\dot{u}_1 - \dot{u}_2\|_{0,T,\boldsymbol{H}} + \|\sigma_1 - \sigma_2\|_{0,T,\mathscr{H}}$$
$$\leq C(\|u_{01} - u_{02}\|_{\boldsymbol{H}_1} + \|v_{01} - v_{02}\|_{\boldsymbol{H}} + \|\sigma_{01} - \sigma_{02}\|_{\mathscr{H}}). \quad (1.46)$$

Subsequently we give the definition of stability and finite time stability for the problem (1.1)–(1.4). A solution $(u; \sigma)$ of (1.1)–(1.4) will be called:

(i) *stable* if there exists $m: \mathbb{R}_+ \to \mathbb{R}_+$ a continuous increasing function with $m(0) = 0$ such that

$$\|u(t) - u_1(t)\|_{\boldsymbol{H}_1} + \|\dot{u}(t) - \dot{u}_1(t)\|_{\boldsymbol{H}} + \|\sigma(t) - \sigma_1(t)\|$$
$$\leq m(\|u_0 - u_{01}\|_{\boldsymbol{H}_1} + \|v_0 - v_{01}\|_{\boldsymbol{H}} + \|\sigma_0 - \sigma_{01}\|_{\mathscr{H}}) \quad (1.47)$$

for all $t \in \mathbb{R}_+$ and all $u_{01}, v_{01}, \sigma_{01}$ satisfying (1.9), (1.10) where $(u_1; \sigma_1)$ is the solution of (1.1)–(1.4) for the initial data $u_{01}, v_{01}, \sigma_{01}$;

(ii) *finite time stable* if (1.47) holds for all finite time intervals.

Remark 1.2 *From* (1.46) *we deduce the finite time stability of every solution of* (1.1)–(1.4).

1.4 Weak solutions

The assumptions made in Section 1.1 are too strong to include the case of discontinuous boundary data. The purpose of this section is to prove the existence and uniqueness of weak solutions for problems (1.1)–(1.4) using assumptions weak enough to allow time discontinuities of boundary data and space discontinuities of initial data for the velocity and stress fields. After the variational (weak) formulation of the problem, the main result is stated. The proof is based on the extension of the semilinear hyperbolic problem considered in Section 1.2 to a larger suitable space. An abstract result (Theorem A3.5) which gives a construction technique for extended generators of C^0-semigroups of linear contractions is used. In what follows we shall consider the spaces $\boldsymbol{H}$ and $\mathscr{H}$ with the inner products

$$((u_1, u_2))_{\boldsymbol{H}} = \langle \rho u_1, u_2 \rangle_{\boldsymbol{H}}, \qquad ((\tau_1, \tau_2))_{\mathscr{H}} = \langle \mathscr{A}^{-1}\tau_1, \tau_2 \rangle_{\mathscr{H}}$$

for all $u_i \in \boldsymbol{H}$, $\tau_i \in \mathscr{H}$, $i = 1, 2$, which generates equivalent norms on $\boldsymbol{H}$ and $\mathscr{H}$ respectively. We shall also identify $\boldsymbol{H}$ and $\mathscr{H}$ with their duals (i.e. $\boldsymbol{H} = \boldsymbol{H}'$, $\mathscr{H} = \mathscr{H}'$) hence the following continuous inclusions hold (see section A1.3):

$$V_1 \subset \boldsymbol{H} \subset V_1'; \qquad \mathscr{V}_1 \subset \mathscr{H} \subset \mathscr{V}_1'.$$

If the solution (u, σ) of (1.1)–(1.4) is smooth enough, for instance if it has the regularity obtained in Theorem 1.1, then the following weak formulation of (1.1)–(1.4) can be deduced:

$$\dot{u}(t) = v(t) \tag{1.48}$$

$$\frac{\mathrm{d}}{\mathrm{d}t}((v(t), w))_{\boldsymbol{H}} = -\langle \sigma(t), \varepsilon(w) \rangle_{\mathscr{H}} + \langle f(t), w \rangle_{\boldsymbol{H}_\Gamma', \boldsymbol{H}_\Gamma} + \langle b(t), w \rangle_{\boldsymbol{H}} \tag{1.49}$$

$$\begin{aligned} \frac{\mathrm{d}}{\mathrm{d}t}((\sigma(t), \tau))_{\mathscr{H}} = {} & -\langle v(t), \operatorname{Div} \tau \rangle_{\mathscr{H}} + \langle \gamma_2(\tau), \dot{g}(t) \rangle_{\boldsymbol{H}_\Gamma', \boldsymbol{H}_\Gamma} \\ & + ((G(t, \sigma(t), \varepsilon(u(t))), \tau))_{\mathscr{H}} \qquad \forall w \in \boldsymbol{V}_1, \quad \tau \in \mathscr{V}_1 \end{aligned} \tag{1.50}$$

$$u(t) = g(t) \qquad \text{on } \Gamma_1, \quad t \in (0, T] \tag{1.51}$$

$$u(0) = u_0, \qquad v(0) = v_0, \qquad \sigma(0) = \sigma_0. \tag{1.52}$$

In order to prove an existence and uniqueness result for the problem (1.48)–(1.52) the following assumptions will be used:

$$b \in L^1(0, T, \boldsymbol{H}) \tag{1.53}$$

there exists $\tilde{u} \in W^{2,1}(0, T, \boldsymbol{H}) \cap W^{1,1}(0, T, \boldsymbol{H}_1)$ such that

$$\tilde{u}(t) = g(t) \qquad \text{on } \Gamma_1 \quad \forall t \in [0, T] \tag{1.54}$$

there exists $\tilde{\sigma} \in W^{1,1}(0, T, \mathscr{H}) \cap L^1(0, T, \mathscr{H}_2)$ such that

$$\tilde{\sigma}(t)\nu = f(t) \qquad \text{on } \Gamma_2 \quad \forall t \in [0, T] \tag{1.55}$$

$$u_0 \in \boldsymbol{H}_1, \qquad v_0 \in \boldsymbol{H}, \qquad \sigma_0 \in \mathscr{H} \tag{1.56}$$

$$u_0 = g(0) \qquad \text{on } \Gamma_1. \tag{1.57}$$

Remark 1.3 *Assumptions* (1.54), (1.55) *are weak enough to permit time discontinuities in the boundary conditions for velocity and stress fields. Indeed, let us consider* $N = 1$, $T = 2$, $\Omega = (0, 1)$, $\Gamma_2 = \{0\}$ *and* $f(t) = 0$ *for* $t < 1$, $f(t) = 1$ *for* $t > 1$. *In this case we can take* $\tilde{\sigma}(t, x) = 1$ *for* $x \geq 2(1 - t)$,

$$\tilde{\sigma}(t, x) = (x - 1 + t)/(1 - t)$$

for $1 - t < x < 2(1 - t)$ *and* $\tilde{\sigma}(t, x) = 0$ *for* $x < 1 - t$, *where* $x \in (0, 1)$, $t \in (0, 2)$. *One can easily verify that* $\tilde{\sigma} \in W^{1,1}(0, T, \boldsymbol{H}) \cap L^1(0, T, \boldsymbol{H}_1)$ *and* $\tilde{\sigma}(t, 0) = f(t)$.

The main result of this section is the following:

Theorem 1.3 *Let* (1.5)–(1.7) *and* (1.53)–(1.57) *hold. Then there exists a unique solution* (u, v, σ) *of* (1.48)–(1.52) *such that*

$$u \in C^1([0, T], \boldsymbol{H}) \cap C^0([0, T], \boldsymbol{H}_1) \tag{1.58}$$

$$v \in W^{1,1}(0, T, \boldsymbol{V}_1') \cap C^0([0, T], \boldsymbol{H}) \tag{1.59}$$

$$\sigma \in W^{1,1}(0, T, \mathscr{V}_1') \cap C^0([0, T], \mathscr{H}). \tag{1.60}$$

Moreover, if we suppose

$$b, \ddot{u}, \operatorname{Div} \tilde{\sigma} \in C^0([0, T], \boldsymbol{V}_1') \tag{1.61}$$

$$\varepsilon(\dot{\ddot{u}}), \dot{\tilde{\sigma}} \in C^0([0, T], \mathscr{V}_1') \tag{1.62}$$

then we have

$$v \in C^1([0, T], \boldsymbol{V}_1'), \qquad \sigma \in C^1([0, T], \mathscr{V}_1'). \tag{1.63}$$

Remark 1.4 *The regularity* (1.59)–(1.60) *permits us to deduce that*

$$\frac{\mathrm{d}}{\mathrm{d}t}\langle \rho v(t), w \rangle_{\boldsymbol{H}} = \frac{\mathrm{d}}{\mathrm{d}t}((v(t), w))_{\boldsymbol{H}} = \langle \dot{v}(t), w \rangle_{\boldsymbol{V}_1', \boldsymbol{V}_1} \tag{1.64}$$

$$\frac{\mathrm{d}}{\mathrm{d}t}\langle \mathscr{A}^{-1}\sigma(t), \tau \rangle_{\mathscr{H}} = \frac{\mathrm{d}}{\mathrm{d}t}((\sigma(t), \tau))_{\mathscr{H}} = \langle \dot{\sigma}(t), \tau \rangle_{\mathscr{V}_1', \mathscr{V}_1}. \tag{1.65}$$

PROOF OF THEOREM 1.3. Let $Y = \boldsymbol{H} \times \mathscr{H}$, $\mathrm{D(B)} = \boldsymbol{V}_1 \times \mathscr{V}_1$ and $B: D(B) \subset Y \to Y$ be given by (1.24). From Lemma 1.3 we have $B^* = -B$, hence $D(B^*) = \boldsymbol{V}_1 \times \mathscr{V}_1$ and B is the generator of a C^0-semigroup $(T(t))_{t \geq 0}$ acting on Y. If we put $T(t) = ((T_1(t); T_2(t))$ and S given by (1.27) then we get from Lemma 1.2 that $S(t)$ is a C^0-semigroup on $X = \boldsymbol{V}_1 \times \boldsymbol{H} \times \mathscr{H}$. Its generator is A: $D(A) \subset X \to X$ given by (1.22).

If we make use of Theorem A3.5 we get the existence of $\tilde{B}$: $Y \subset V'_1 \to \boldsymbol{V}'_1$ defined by

$$\langle \tilde{B}(v; \sigma), (u; \tau) \rangle_{Y', Y} = -\langle v, \operatorname{Div} \tau \rangle_H - \langle \sigma, \varepsilon(u) \rangle_{\mathscr{H}} \tag{1.66}$$

for all $u \in \boldsymbol{V}_1$, $v \in \boldsymbol{H}$, $\tau \in \mathscr{V}_1$, $\sigma \in \mathscr{H}$. If we put $\tilde{T}(t) = (\tilde{T}_1(t), \tilde{T}_2(t))$ and

$$\tilde{S}(t)(u; v; \sigma) = \left(u + \int_0^t \tilde{T}_1(s)(v; \sigma) \, \mathrm{d}s, \, \tilde{T}(t)(v; \sigma) \right), \tag{1.67}$$

then we obtain that $(\tilde{S}(t))_{t \geq 0}$ is a C^0-semigroup acting on $\tilde{X} = \boldsymbol{H} \times \boldsymbol{V}'_1 \times \mathscr{V}'_1$ and its generator is $\tilde{A}$: $D(\tilde{A}) \subset X \to X$, $D(\tilde{A}) = \boldsymbol{H} \times \boldsymbol{H} \times \mathscr{H}$ given by

$$\tilde{A}(u; v; \sigma) = (v; \tilde{B}(v; \sigma)) \tag{1.68}$$

If we have in mind that $\tilde{B}|_{D(B)} = B$, $\tilde{T}(t)|_Y = T(t)$ we get

$$\tilde{A}|_{D(A)} = A, \qquad \tilde{S}(t)|_X = S(t) \tag{1.69}$$

Let F: $[0, T] \times X \to X$ defined by

$$F(t, u, v, \sigma) = (0; 0; G(t, \tilde{\sigma}(t) + \sigma, \varepsilon(\tilde{u}(t)) + \varepsilon(u))) \tag{1.70}$$

for all $u \in \boldsymbol{V}_1$, $v \in \boldsymbol{H}$, $\sigma \in \mathscr{H}$ and $h \in L^1(0, T, X)$ given by

$$h(t) = (0; a(t); \mathscr{A}\varepsilon(\dot{\tilde{u}}(t)) - \dot{\tilde{\sigma}}(t)), \tag{1.71}$$

where a is given by (1.20). From (1.6) we deduce that F is continuous and it is uniformly Lipschitz continuous in the second argument with respect to the first argument.

Let $(u; v; \sigma)$ be a solution of (1.48)–(1.52) and let us denote by $u^* = u - \tilde{u}$, $v^* = v - \dot{\tilde{u}}$, $\sigma^* = \sigma - \tilde{\sigma}$, $u_0^* = u_0 - \tilde{u}(0)$, $v_0^* = v_0 - \dot{\tilde{u}}(0)$, $\sigma_0^* = \sigma_0 - \tilde{\sigma}(0)$. If we put $x = (u^*; v^*; \sigma^*)$, $x_0 = (u_0^*; v_0^*; \sigma_0^*) \in X$ then it can be easily proved that $(u; v; \sigma)$ is a solution of (1.48)–(1.52) with the regularity (1.58)–(1.60) if and only if $x \in C^0([0, T], X) \cap W^{1,1}(0, T, \tilde{X})$ is a solution of the following Cauchy problem:

$$\dot{x}(t) = \tilde{A}x(t) + F(t, x(t)) + h(t) \qquad t \in (0, T] \tag{1.72}$$

$$x(0) = x_0. \tag{1.73}$$

Hence it only remains to prove the existence and uniqueness for the problem (1.72)–(1.73). To do this we consider the regular Cauchy problem

$$\dot{x}(t) = Ax(t) + F(t, x(t)) + h(t) \qquad t \in (0, T] \tag{1.74}$$

$$x(0) = x_0. \tag{1.75}$$

on X. Since F is Lipschitz continuous on X and $h \in L^1(0, T, X)$, it follows (see Theorem A3.8(ii) and (A3.28)) that there exists a unique mild solution of (1.74)–(1.75), i.e. there exists $x \in C^0([0, T], X)$ such that

$$x(t) = S(t)x_0 + \int_0^t S(t-s)(F(s, x(s)) + h(s))\, \mathrm{d}s. \tag{1.76}$$

If we denote by $j(t) = F(t, x(t)) + h(t)$ and we make use of (1.69), then we obtain

$$x(t) = \tilde{S}(t)x_0 + \int_0^t \tilde{S}(t-s)j(s)\, \mathrm{d}s. \tag{1.77}$$

Since $j \in L^1(0, T, [D(\tilde{A})])$, the relation (1.77) shows that $x \in W^{1,1}(0, T, \tilde{X})$ is the solution of (1.72)–(1.73).

If in addition we suppose (1.61)–(1.62), then we have $j \in L^1(0, T, [D(\tilde{A})]) \cap C^0([0, T], X)$; hence from Lemma A3.6, it follows that $x \in C^1([0, T], \tilde{X})$.

2 The behaviour of the solution in the viscoelastic case

In this section the behaviour of $(u; \sigma)$, the solution of (1.1)–(1.4), will be studied in the viscoelastic case, i.e. the constitutive function G will be supposed to be given by

$$G(\sigma, \varepsilon) = -k(\sigma - F(\varepsilon)) \qquad \forall \sigma, \varepsilon \in \mathscr{S}_N, \tag{2.1}$$

where $k > 0$ is a viscosity constant and $F: \Omega \times \mathscr{S}_N \to \mathscr{S}_N$ is a constitutive function which gives the equilibrium curve $\sigma = F(\varepsilon)$. In order to satisfy (1.6) we assume that

$$F \text{ satisfies (2.2)–(2.5) in Chapter 3.} \tag{2.2}$$

Some remarks concerning this assumption are given in the first part of Section 2 from Chapter 3.

The results stated in this section hold also in a more general case:

$$G(\sigma, \varepsilon) = -\mathscr{K}(\sigma - F(\varepsilon)) \qquad \forall \sigma, \varepsilon \in \mathscr{S}_N, \tag{2.3}$$

where $\mathscr{K}$ is a fourth-order tensor but for the sake of simplicity only (2.1) will be considered.

2.1 The energy function

We say that the constitutive equation

$$\dot{\sigma} = \mathscr{A}\dot{\varepsilon} + G(\sigma, \varepsilon) \tag{2.4}$$

has an *energy function* $\psi: \mathscr{S}_N \times \mathscr{S}_N \to \mathbb{R}$ if the second law of thermodynamics

(in the isothermal case)

$$\sigma(t)\cdot\dot{\varepsilon}(t) - \rho\frac{\mathrm{d}}{\mathrm{d}t}\psi(\sigma(t), \varepsilon(t)) \geq 0 \tag{2.5}$$

holds along any curve $(\sigma(t); \varepsilon(t))_{t\geq 0}$ which satisfies (2.4). As was proved in Mihailescu–Suliciu and Suliciu (1979), Gurtin *et al.* (1984), the constitutive law (2.4) has a smooth energy function iff

$$\frac{\partial\psi}{\partial\varepsilon}(\sigma, \varepsilon) + \mathscr{A}\frac{\partial\psi}{\partial\sigma}(\sigma, \varepsilon) = \rho^{-1}\sigma \tag{2.6}$$

$$\frac{\partial\psi}{\partial\sigma}(\sigma, \varepsilon)\cdot G(\sigma, \varepsilon) \leq 0 \tag{2.7}$$

for all $\sigma, \varepsilon \in \mathscr{S}_N$.

In order to prove that in the viscoelastic case an energy function exists we shall suppose for the sake of simplicity that the material is homogeneous, i.e.

$$\mathscr{A}, \rho \text{ and } F \text{ do not depend on } x \in \Omega. \tag{2.8}$$

We also assume that

$$F \text{ is of class } C^1, \qquad F(0) = 0 \tag{2.9}$$

and $\partial F/\partial\varepsilon$ is symmetric, i.e.

$$\frac{\partial F}{\partial\varepsilon}(\varepsilon)\tau_1\cdot\tau_2 = \frac{\partial F}{\partial\varepsilon}(\varepsilon)\tau_2\cdot\tau_1 \qquad \forall\varepsilon, \tau_1, \tau_2 \in \mathscr{S}_N. \tag{2.10}$$

The following assumption

$$\left(\mathscr{A} - \frac{\partial F}{\partial\varepsilon}(\varepsilon)\right)\tau\cdot\tau \geq \beta|\tau|^2 \qquad \forall\varepsilon, \tau \in \mathscr{S}_N \tag{2.11}$$

with $\beta > 0$ expresses the fact (which agrees with the experimental data—see Cristescu and Suliciu (1982), p. 41) that the instantaneous curve $\sigma = \mathscr{A}\varepsilon$ lies over the equilibrium curve (or relaxation boundary) $\sigma = F(\varepsilon)$ and $\mathscr{A} > \partial F/\partial\varepsilon$.

The main result of this section is the following theorem:

Theorem 2.1 *Let* (1.5), (2.1), (2.2), (2.8)–(2.11) *hold. Then there exists an energy function* ψ *of class* C^2 *such that* (2.6), (2.7) *hold and there exist* C_1, C_2 *(which depend only on* $\mathscr{A}$, L_0, a, *and* β*) such that*

$$C_1(|\sigma|^2 + |\varepsilon|^2) \leq \rho\psi(\sigma, \varepsilon) \leq C_2(|\sigma|^2 + |\varepsilon|^2) \qquad \forall\sigma, \varepsilon \in \mathscr{S}_N. \tag{2.12}$$

PROOF. Let us remark that from (2.2) and (2.9) one can easily obtain

$$a|\tau|^2 \leq \frac{\partial F}{\partial\varepsilon}(\varepsilon)\tau\cdot\tau \leq L_0|\tau|^2 \qquad \forall\varepsilon, \tau \in \mathscr{S}_N. \tag{2.13}$$

If we denote by $H: \mathscr{S}_N \to \mathscr{S}_N$ the function given by

$$H(\tau) = \mathscr{A}\tau - F(\tau) \qquad \forall \tau \in \mathscr{S}_N, \tag{2.14}$$

then from (2.2) we get that H is a Lipschitz function and from (2.11) we obtain

$$\frac{\partial H}{\partial \tau}(\varepsilon)\tau \cdot \tau \geq \beta |\tau|^2 \qquad \forall \varepsilon, \tau \in \mathscr{S}_N; \tag{2.15}$$

hence (see Remark A2.1) H is a strongly monotone operator. Using Theorem A2.4 we deduce that H is one to one and surjective and H^{-1} is also a strongly monotone and Lipschitz operator. If we denote by

$$R(\varepsilon) = -(-H)^{-1}(\varepsilon) - \mathscr{A}^{-1}\varepsilon \qquad \forall \varepsilon \in \mathscr{S}_N \tag{2.16}$$

and we have in mind that $\partial H(\varepsilon)/\partial\varepsilon = \mathscr{A} - \partial F(\varepsilon)/\partial\varepsilon$ then from (2.11), (2.13) and using Lemma A4.17 we deduce that there exist $d_1, d_2 > 0$ such that

$$d_1|\tau|^2 \leq \frac{\partial R}{\partial \varepsilon}(\varepsilon)\tau \cdot \tau \leq d_2|\tau|^2 \qquad \forall \varepsilon, \tau \in \mathscr{S}_N. \tag{2.17}$$

Since $\partial R/\partial\varepsilon$ is continuous, symmetric and positively defined, from Theorem A2.5 and Lemma A2.10(iii) we can deduce that there exists a uniformly convex function $\varphi: \mathscr{S}_N \to \mathbb{R}$ such that

$$\frac{\partial \varphi}{\partial \varepsilon}(\varepsilon) = R(\varepsilon), \qquad \varphi(0) = 0. \tag{2.18}$$

If we have in mind that

$$\varphi(\varepsilon) = \int_0^1 R(t\varepsilon)\cdot\varepsilon \,\mathrm{d}t = \int_0^1 \int_0^t \frac{\partial R}{\partial \varepsilon}(s\varepsilon)\varepsilon\cdot\varepsilon \,\mathrm{d}s\,\mathrm{d}t$$

then from (2.17) we deduce

$$\frac{d_1}{2}|\varepsilon|^2 \leq \varphi(\varepsilon) \leq \frac{d_2}{2}|\varepsilon|^2 \qquad \forall \varepsilon \in \mathscr{S}_N. \tag{2.19}$$

Let $\psi: \mathscr{S}_N \times \mathscr{S}_N \to \mathbb{R}$ be given by

$$\psi(\sigma, \varepsilon) = \rho^{-1}[\tfrac{1}{2}\mathscr{A}^{-1}\sigma\cdot\sigma + \varphi(\sigma - \mathscr{A}\varepsilon)] \qquad \forall \sigma, \varepsilon \in \mathscr{S}_N. \tag{2.20}$$

From (2.19) and (1.5) we deduce

$$\tfrac{1}{2}(\alpha|\sigma|^2 + d_1|\sigma - \mathscr{A}\varepsilon|^2) \leq \rho\psi(\sigma, \varepsilon) \leq \tfrac{1}{2}(Q|\sigma|^2 + d_2|\sigma - \mathscr{A}\varepsilon|^2) \qquad \forall \sigma, \varepsilon \in \mathscr{S}_N.$$

Since $(\sigma; \varepsilon) \to |\sigma|^2 + d_i|\sigma - \mathscr{A}\varepsilon|^2$, $i = 1, 2$ is a norm on $\mathscr{S}_N \times \mathscr{S}_N$ which is a finite-dimensional space, from the above inequalities we get (2.12).

Let us now prove (2.6) and (2.7). For this, we observe that after some algebra we get

$$\rho \frac{\partial \psi}{\partial \sigma}(\sigma, \varepsilon) = \varepsilon - (-H)^{-1}(\sigma - \mathscr{A}\varepsilon)$$

and

$$\rho \frac{\partial \psi}{\partial \varepsilon}(\sigma, \varepsilon) = \sigma - \mathscr{A}\varepsilon + \mathscr{A}(-H)^{-1}(\sigma - \mathscr{A}\varepsilon),$$

and (2.6) easily follows. If we compute

$$\begin{aligned}\rho \frac{\partial \psi}{\partial \sigma}(\sigma, \varepsilon)\cdot G(\sigma, \varepsilon) &= -k\rho(\varepsilon - (-H)^{-1}(\sigma - \mathscr{A}\varepsilon))\cdot(\sigma - F(\varepsilon)) \\ &= -k\rho(\varepsilon - (-H)^{-1}(\sigma - \mathscr{A}\varepsilon))\cdot(\sigma - \mathscr{A}\varepsilon + H(\varepsilon))\end{aligned}$$

and if we let $\tau \in \mathscr{S}_N$ be such that $\sigma - \mathscr{A}\varepsilon = -H(\tau)$, then we get

$$\rho \frac{\partial \psi}{\partial \sigma}(\sigma, \varepsilon)\cdot G(\sigma, \varepsilon) = -k\rho(H(\varepsilon) - H(\tau))\cdot(\varepsilon - \tau).$$

If we bear in mind that H is monotone (a consequence of (2.15)) from the above equality we get (2.7).

2.2 An energy bound for isolated bodies

In this section we study the energy boundness of the solutions of (1.1)–(1.4), (2.1) in the isolated body case. A decisive role will be played by the energy function constructed in Section 2.1 (Theorem 2.1).

We say that the viscoelastic body is *isolated* if

$$b \equiv 0, \qquad g \equiv 0, \quad \text{and} \quad f \equiv 0. \tag{2.21}$$

Hence from Theorem 1.1 we find that in this case there exists a unique solution $(u; \sigma)$ of (1.1)–(1.4) such that $u \in C^1(\mathbb{R}_+, \boldsymbol{H}) \cap C(\mathbb{R}_+, \boldsymbol{H}_1)$, $\sigma \in C(\mathbb{R}_+, \mathscr{H}_2)$ and (1.14), (1.15) hold for all $T > 0$.

Let e: $\boldsymbol{H} \times \mathscr{H} \times \mathscr{H} \to \mathbb{R}$ be the total energy function given by

$$e(v, \varepsilon, \sigma) = \tfrac{1}{2}\rho\|v\|_{\boldsymbol{H}}^2 + \int_\Omega \rho\psi(\sigma, \varepsilon)\,\mathrm{d}x \tag{2.22}$$

for all $v \in \boldsymbol{H}$, $\varepsilon, \sigma \in \mathscr{H}$, where ψ is constructed in Theorem 2.1. From (2.12), there exist C_1, C_2 which depend only on $\mathscr{A}$ and F such that

$$\begin{aligned}&\rho C_1(\|v\|_{\boldsymbol{H}}^2 + \|\varepsilon\|_{\mathscr{H}}^2 + \|\sigma\|_{\mathscr{H}}^2) \\ &\quad \le e(v, \varepsilon, \sigma) \le \rho C_2(\|v\|_{\boldsymbol{H}}^2 + \|\varepsilon\|_{\mathscr{H}}^2 + \|\sigma\|_{\mathscr{H}}^2) \qquad \forall v \in \boldsymbol{H}, \quad \varepsilon, \sigma \in \mathscr{H}.\end{aligned} \tag{2.23}$$

The main result of this section is the following:

Theorem 2.2 *Let (1.5), (1.7), (1.9), (1.10), (2.2), (2.8)–(2.11), and (2.21) hold. If* $(u;\sigma)$ *is the solution of (1.1)–(1.4), (2.1) then for all* $t \geq 0$ *we have*

$$e(\dot{u}(t), \varepsilon(u(t)), \sigma(t)) \leq e(v_0, \varepsilon(u_0), \sigma_0). \tag{2.24}$$

Moreover, if we suppose in addition that

$$\text{meas } \Gamma_1 > 0, \tag{2.25}$$

then there exists $C > 0$ *which depends only on* Ω, Γ_1, $\mathscr{A}$, *and* F *such that for all* $t \geq 0$

$$\|u(t)\|_{\boldsymbol{H}_1} + \|\dot{u}(t)\|_{\boldsymbol{H}} + \|\sigma(t)\|_{\mathscr{H}} \leq C[\|u_0\|_{\boldsymbol{H}_1} + \|v_0\|_{\boldsymbol{H}} + \|\sigma_0\|_{\mathscr{H}}]. \tag{2.26}$$

Remark 2.1 *From (2.26) one can deduce that the steady solution* $u \equiv 0$, $\sigma \equiv 0$ *of (1.1)–(1.4) is stable (for the definition of stability for a solution of (1.1)–(1.4) see the last part of Section 1.3).*

PROOF OF THEOREM 2.1. Let us compute

$$\frac{\mathrm{d}}{\mathrm{d}t} e(\dot{u}(t), \varepsilon(u(t)), \sigma(t)) = \langle \rho\ddot{u}(t), \dot{u}(t)\rangle_{\boldsymbol{H}} + \left\langle \rho \frac{\partial\psi}{\partial\varepsilon}(\sigma(t), \varepsilon(u(t))), \varepsilon(\dot{u}(t)) \right\rangle_{\mathscr{H}}$$
$$+ \left\langle \rho \frac{\partial\psi}{\partial\sigma}(\sigma(t), \varepsilon(u(t))), \dot{\sigma}(t) \right\rangle_{\mathscr{H}}.$$

If we use (1.2), (2.6) and we have in mind that $\langle \sigma(t), \varepsilon(\dot{u}(t))\rangle_{\mathscr{H}} = -\langle \text{Div } \sigma(t), \dot{u}(t)\rangle_{\boldsymbol{H}} = -\langle \rho\ddot{u}(t), u(t)\rangle_{\boldsymbol{H}}$, then we get

$$\frac{\mathrm{d}}{\mathrm{d}t} e(\dot{u}(t), \varepsilon(u(t)), \sigma(t)) = \left\langle \rho \frac{\partial\psi}{\partial\sigma}(\sigma(t), \varepsilon(u(t))), G(\sigma(t), \varepsilon(u(t)) \right\rangle_{\mathscr{H}}.$$

From (2.7) we deduce

$$\frac{\mathrm{d}}{\mathrm{d}t} e(u(t), \varepsilon(u(t)), \sigma(t)) \leq 0;$$

hence (2.24) holds. Then inequality (2.26) easily follows from (2.24) and Korn's inequality (2.32) of Chapter 2.

2.3 An approach to linear elasticity

In the papers of Suliciu (1984) and Podio-Guidugli and Suliciu (1984) it is proved that for isolated bodies the following inequality holds:

$$\int_0^t \|\sigma(s) - F(\varepsilon(u(s)))\|_{\mathscr{H}}^2 \,\mathrm{d}s \leq C/k \qquad \forall t \geq 0.$$

This inequality shows that for large values of the viscosity constant k the solution of the viscoelastic problem (1.1)–(1.4), (2.1) almost obeys an elastic law. That is why one can expect that the study of the asymptotic behaviour of the viscoelastic solution with respect to k will show that elasticity is a proper asymptotic theory for viscoelastic materials. This fact has already been proved in Section 2.3 of Chapter 3 for the quasistatic case. The dynamic case of this singular perturbation problem is studied in this section. Using the energy function constructed in Section 2.1 and assuming that F is linear, it is proved that the solution of the linear viscoelastic problem converges to the solution of a linear elastic problem for a large k.

Let us consider the linear viscoelastic material (linear standard material)

$$G(\sigma, \varepsilon) = -k(\sigma - \mathscr{E}\varepsilon) \tag{2.27}$$

where $\mathscr{E}$ is a fourth-order tensor such that

$$\mathscr{E}_{ijkl} \in L^\infty(\Omega),\ \mathscr{E}(x)\varepsilon\cdot\tau = \mathscr{E}(x)\tau\cdot\varepsilon \qquad \forall\tau, \varepsilon \in \mathscr{S}_N \quad \text{a.e. in } \Omega \tag{2.28}$$

$$\mathscr{E}(x)\tau\cdot\tau \geq c|\tau|^2 \qquad \forall\tau \in \mathscr{S}_N \quad \text{a.e. in } \Omega\ (c > 0) \tag{2.29}$$

and $k > 0$ is a viscosity constant.

In order to satisfy (2.11) we shall suppose that

$$(\mathscr{A}(x) - \mathscr{E}(x))\tau\cdot\tau \geq d|\tau|^2 \qquad \forall\tau \in \mathscr{S}_N \quad \text{a.e. in } \Omega \text{ with } d > 0. \tag{2.30}$$

Let us now consider the following linear elastic problem: Find the displacement function $\hat{u}$: $[0, T] \times \Omega \to \mathbb{R}^N$ and the stress function $\hat{\sigma}$: $[0, T] \times \Omega \to \mathscr{S}_N$ such that

$$\rho\ddot{\hat{u}}(t) = \text{Div } \hat{\sigma}(t) + b(t) \tag{2.31}$$

$$\hat{\sigma}(t) = \mathscr{E}\varepsilon(\hat{u}(t)) \qquad \text{in } \Omega \tag{2.32}$$

$$\hat{u}(t) = g(t) \quad \text{on } \Gamma_1, \quad \hat{\sigma}(t)\nu = f(t) \quad \text{on } \Gamma_2 \qquad \forall t \in [0, T] \tag{2.33}$$

$$\hat{u}(0) = u_0, \qquad \dot{\hat{u}}(0) = v_0. \tag{2.34}$$

The following lemma ensures the existence and the uniqueness of the solution for the problems (2.31)–(2.34):

Lemma 2.1 *Let* (1.7)–(1.12), (2.28), (2.29) *hold with* $\sigma_0 = \mathscr{E}\varepsilon(u_0)$. *Then there exists a unique pair of functions* $(\hat{u}; \hat{\sigma})$ *forming a solution of* (2.31)–(2.34) *such that*

$$\hat{u} \in W^{2,\infty}(0, T, \boldsymbol{H}) \cap W^{1,\infty}(0, T, \boldsymbol{H}_1) \tag{2.35}$$

$$\hat{\sigma} \in W^{1,\infty}(0, T, \mathscr{H}) \cap L^\infty(0, T, \mathscr{H}_2) \tag{2.36}$$

PROOF. If we take the time derivative in (2.32) we get

$$\dot{\hat{\sigma}}(t) = \mathscr{E}\varepsilon(\dot{\hat{u}}(t)). \tag{2.37}$$

Hence (2.37) is exactly (1.2), in which we replace $\mathscr{A}$ by $\mathscr{E}$ and $G \equiv 0$. If we bear in mind that

$$\hat{\sigma}(0) = \sigma_0, \tag{2.38}$$

then from Theorem 1.1 we deduce that there exists a unique pair of functions $(\hat{u}; \hat{\sigma})$ forming a solution for the problems (2.31), (2.37), (2.33), (2.34), (2.38), such that (2.35) and (2.36) hold. Since $\mathrm{d}(\hat{\sigma}(t) - \mathscr{E}\varepsilon(\hat{u}(t)))/\mathrm{d}t = 0$ (see (2.37)) we get

$$\hat{\sigma}(t) - \mathscr{E}\varepsilon(\hat{u}(t)) = \sigma_0 - \mathscr{E}\varepsilon(u_0) = 0,$$

and (2.32) follows.

The main result of this section is the following:

Theorem 2.3 *Let* (1.5), (1.7)–(1.12), (2.25), *and* (2.28)–(2.30) *hold with* $\sigma_0 = \mathscr{E}\varepsilon(u_0)$. *If we denote by* $(u_k; \sigma_k)$ $(k > 0)$ *the solution of the problem* (1.1)–(1.4), (2.27), *then for all* $t \in [0, T]$ *we have that* $u_k(t) \to \hat{u}(t)$ *in* $\boldsymbol{H}_1$, $\dot{u}_k(t) \to \dot{\hat{u}}(t)$ *in* $\boldsymbol{H}$ *and* $\sigma_k(t) \to \hat{\sigma}(t)$ *in* $\mathscr{H}$ *for* $k \to +\infty$, *where* $(\hat{u}; \hat{\sigma})$ *is the solution of* (2.31)–(2.34).

PROOF. Let us denote by $\bar{u}_k = u_k - \hat{u}$, $\bar{v}_k = \dot{u}_k - \dot{\hat{u}}$, $\bar{\sigma}_k = \sigma_k - \hat{\sigma}$. Note that $\bar{u}_k \in W^{1,\infty}(0, T, \boldsymbol{V}_1)$, $v_k \in W^{1,\infty}(0, T, \boldsymbol{H}) \cap L^\infty(0, T, \boldsymbol{V}_1)$, $\bar{\sigma}_k \in W^{1,\infty}(0, T, \mathscr{H}) \cap L^\infty(0, T, \mathscr{V}_1)$ is the solution of the following problem:

$$\dot{\bar{u}}_k(t) = \bar{v}_k(t) \tag{2.39}$$

$$\rho\dot{\bar{v}}_k(t) = \operatorname{Div} \bar{\sigma}_k(t) \tag{2.40}$$

$$\dot{\bar{\sigma}}_k(t) = \mathscr{A}\varepsilon(\bar{v}_k(t)) - k(\bar{\sigma}_k(t) - \mathscr{E}\varepsilon(\bar{u}_k(t))) + h(t) \tag{2.41}$$

$$\bar{u}_k(0) = \bar{v}_k(0) = 0, \qquad \bar{\sigma}_k(0) = 0 \tag{2.42}$$

where $h \in L^\infty(0, T, \mathscr{H})$ is given by

$$h(t) = -\dot{\hat{\sigma}}(t) + \mathscr{A}\varepsilon(\dot{\hat{u}}(t)) \qquad \forall t \in [0, T].$$

Let $X = \boldsymbol{V}_1 \times \boldsymbol{H} \times \mathscr{H}$ and A be given by (1.22) and let us consider the operator $P: X \to X$ defined by

$$P(u; v; \sigma) = (0; 0; \sigma - \mathscr{E}\varepsilon(u)) \qquad \forall (u; v; \sigma) \in X. \tag{2.43}$$

Let $D_k: D(A) \subset X \to X$, $D_k = A - kP$ and bear in mind that $P \in B(X)$; then from Lemma 1.2 and Theorem A3.6 we can deduce that D_k is the infinitesimal generator of a C_0 semigroup denoted by $(U_k(t))_{t \geq 0}$. We shall now construct an energetic norm in X such that $(U_k(t))_{t \geq 0}$ is a contractions semigroup. Having in mind (2.30) from Lemma A4.17 one can easily deduce that $\mathscr{B}$ given by

$$\mathscr{B} = (\mathscr{A} - \mathscr{E})^{-1} - \mathscr{A}^{-1} \tag{2.44}$$

is symmetric and positively defined, i.e. there exists $\gamma = cd/(|\mathscr{A}||\mathscr{A} - \mathscr{E}|^2) > 0$ such that

$$\mathscr{B}(x)\tau\cdot\tau \geq \gamma|\tau|^2 \qquad \forall x \in \Omega, \quad \tau \in \mathscr{S}_N. \tag{2.45}$$

Let us consider in X the following energetic inner product:

$$\langle(u_1; v_1; \sigma_1), (u_2; v_2; \sigma_2)\rangle_E = \langle\rho v_1, v_2\rangle_{\boldsymbol{H}} + \langle\mathscr{A}^{-1}\sigma_1, \sigma_2\rangle_{\mathscr{H}} + \langle\mathscr{B}(\sigma_1 - \mathscr{A}\varepsilon(u_1)), \sigma_2 - \mathscr{A}\varepsilon(u_2)\rangle_{\mathscr{H}} \tag{2.46}$$

for all $u_i \in \boldsymbol{V}_i$, $v_i \in \boldsymbol{H}$, $\sigma_i \in \mathscr{H}$, $i = 1, 2$. This inner product generates the energetic norm $\|\cdot\|_E$ given by

$$\|(u; v; \sigma)\|_E^2 = \langle\rho v, v\rangle_{\boldsymbol{H}} + \langle\mathscr{A}^{-1}\sigma, \sigma\rangle_{\mathscr{H}} + \langle\mathscr{B}(\sigma - \mathscr{A}\varepsilon(u)), \sigma - \mathscr{A}\varepsilon(u)\rangle_{\mathscr{H}}, \tag{2.47}$$

which is exactly the total energy e given by (2.22) in our case. In order to prove that $\|\cdot\|_E$ is an equivalent norm in X we mention that there exists $c > 0$ such that $\|x\|_E \leq C\|x\|_X$ for all $x \in X$. Let $x_n = (u_n; v_n; \sigma_n) \in X$ be a Cauchy sequence in the energetic norm $\|\cdot\|_E$. From (2.47) we deduce that v_n, σ_n, and $\sigma_n - \mathscr{A}\varepsilon(u_n)$ are Cauchy sequences in $\boldsymbol{H}$ and $\mathscr{H}$ respectively. Hence there exist $v \in \boldsymbol{H}$, $\sigma \in \mathscr{H}$ such that $v_n \to v$ in $\boldsymbol{H}$ and $\sigma_n \to \sigma$ in $\mathscr{H}$. Since $\varepsilon(u_n)$ is a Cauchy sequence in $\mathscr{H}$ from Korn's inequality ((2.32) in Chapter 2), we find that u_n is a Cauchy sequence in $\boldsymbol{V}_1$, hence there exists $u \in \boldsymbol{V}_1$ such that $u_n \to u$ in $\boldsymbol{H}_1$. We have just proved that X is complete in the energy norm $\|\cdot\|_E$. Using the closed graph theorem one can easily deduce now that $\|\cdot\|_E$ is an equivalent norm on X.

After some algebra we obtain that for all $u, v \in \boldsymbol{V}_1$, $\sigma \in \mathscr{V}_1$, we have

$$\langle D_k(u; v; \sigma), (u; v; \sigma)\rangle_E = -k\langle(\mathscr{A} - \mathscr{E})^{-1}(\sigma - \mathscr{E}\varepsilon(u)), \sigma - \mathscr{E}\varepsilon(u)\rangle_{\mathscr{H}} \leq 0. \tag{2.48}$$

Having in mind that

$$\frac{1}{2}\frac{\mathrm{d}}{\mathrm{d}t}\|U_k(t)x\|_E^2 = \langle D_k U_k(t)x, U_k(t)x\rangle_E \leq 0,$$

we obtain $\|U_k(t)x\|_E \leq \|x\|_E$ for all $x \in D(A)$ and $t \geq 0$. Using the density of $D(A)$ in X, we get

$$\|U_k(t)x\|_E \leq \|x\|_E \qquad \forall x \in X, \quad t \geq 0. \tag{2.49}$$

Note now that if we define $l \in L^\infty(0, T, X)$ by

$$l(t) = (0; 0; h(t)) \qquad \forall t \in [0, T], \tag{2.50}$$

and let $x_k(t) = (\bar{u}_k(t); \bar{v}_k(t); \bar{\sigma}_k(t))$, then from (2.39)–(2.42) we deduce

$$\dot{x}_k(t) = D_k x_k(t) + l(t) \qquad \text{for } t \in [0, T], \quad x_k(0) = 0. \tag{2.51}$$

Hence we get

$$x_k(t) = \int_0^t U_k(t-s)l(s)\,\mathrm{d}s \qquad \forall t \in [0, T]. \tag{2.52}$$

Let us suppose for the moment that $l \in C^1([0, T]), X)$. From (2.49) and (2.52) we have that $\|x_k(t)\|_E \le \int_0^t \|l(s)\|_E\,\mathrm{d}s$; hence we have just obtained that

$$\bar{u}_k \text{ is bounded in } L^\infty(0, T, V_1) \tag{2.53}$$

$$\bar{\sigma}_k \text{ is bounded in } L^\infty(0, T, \mathscr{H}). \tag{2.54}$$

Bearing in mind that $\dot{x}_k(t) = U_k(t)l(0) + \int_0^t U_k(t-s)\dot{l}(s)\,\mathrm{d}s$, we obtain

$$\|\dot{x}_k(t)\|_E \le \|l(0)\|_E + \int_0^T \|\dot{l}(s)\|_E\,\mathrm{d}s;$$

hence we have

$$\dot{\bar{u}}_k = \bar{v}_k \text{ is bounded in } L^\infty(0, T, V_1) \tag{2.55}$$

$$\dot{\bar{v}}_k = \rho^{-1}\operatorname{Div}\bar{\sigma}_k \text{ is bounded in } L^\infty(0, T, \boldsymbol{H}) \tag{2.56}$$

$$\dot{\bar{\sigma}}_k \text{ is bounded in } L^\infty(0, T, \mathscr{H}). \tag{2.57}$$

From (2.51) we obtain $\|D_k x_k(t)\|_E \le \|\dot{x}_k(t)\|_E + \|l(t)\|_E < +\infty$. If we take the energetic scalar product in (2.51) with $x_k(t)$ we get

$$-\langle D_k x_k(t), x_k(t)\rangle_E \le (\|\dot{x}_k(t)\|_E + \|l(t)\|_E)\|x_k(t)\|_E \le C$$

and from (2.48) we obtain

$$\|\bar{\sigma}_k(t) - \mathscr{E}\varepsilon(u_k(t))\|_{\mathscr{H}} \le C/k \qquad \forall t \in [0, T], \quad k > 0. \tag{2.58}$$

From the *a priori* estimates (2.53)–(2.57) we deduce there exist

$$\bar{u} \in W^{1,\infty}(0, T, V_1), ,$$

$$\bar{v} \in W^{1,\infty}(0, T, \boldsymbol{H}) \cap L^\infty(0, T, V_1) \quad \text{and} \quad \bar{\sigma} \in W^{1,\infty}(0, T, \mathscr{H}) \cap L^\infty(0, T, \mathscr{V}_1)$$

such that

$$\bar{u}_k \to u, \quad \dot{\bar{u}} \to \dot{\bar{u}} \qquad \text{weakly* in } L^\infty(0, T, V_1) \tag{2.59}$$

$$\bar{v}_k \to \bar{v}, \quad \dot{\bar{v}}_k \to \dot{\bar{v}} \qquad \text{weakly* in } L^\infty(0, T, \boldsymbol{H}) \tag{2.60}$$

$$\bar{\sigma}_k \to \bar{\sigma}, \quad \dot{\bar{\sigma}}_k \to \dot{\bar{\sigma}} \qquad \text{weakly* in } L^\infty(0, T, \mathscr{H}) \tag{2.61}$$

$$\bar{v}_k \to \bar{v} \qquad \text{weakly* in } L^\infty(0, T, V_1) \tag{2.62}$$

$$\bar{\sigma}_k \to \bar{\sigma} \qquad \text{weakly* in } L^\infty(0, T, \mathscr{V}_1) \tag{2.63}$$

when $k \to +\infty$ (if necessary we can pass to a subsequence in (2.59)–(2.63)).

From (2.39), (2.40), and (2.59), (2.60), (2.63) we get

$$\dot{\bar{u}}(t) = \bar{v}(t) \tag{2.64}$$

$$\dot{\bar{v}}(t) = \rho^{-1} \operatorname{Div} \bar{\sigma}(t) \qquad \text{a.e. on } (0, T), \tag{2.65}$$

and from (2.58) we have

$$\bar{\sigma}(t) = \mathscr{E}\varepsilon(\bar{u}(t)) \qquad \forall t \in [0, T]. \tag{2.66}$$

Using now (2.59)–(2.61), we get that $\bar{u}_k(0) \to \bar{u}(0)$ weakly in $\boldsymbol{V}_1$, $\bar{v}_k(0) \to \bar{v}(0)$ *weakly in* $\boldsymbol{H}$ and $\bar{\sigma}_k(0) \to \bar{\sigma}(0)$ weakly in $\mathscr{H}$. From (2.42) we deduce that

$$\bar{u}(0) = 0, \qquad \bar{v}(0) = 0. \tag{2.67}$$

If we make use of the uniqueness result of Lemma 2.1, we deduce that $\bar{u} \equiv 0, \bar{v} \equiv 0, \bar{\sigma} \equiv 0$ because these constitute the unique solution of (2.64)–(2.67); hence

$$x_k \to 0 \qquad \text{weakly* in } L^\infty(0, T, X) \tag{2.68}$$

when $k \to +\infty$. Again using (2.48) from (2.51) we obtain

$$\frac{1}{2}\frac{\mathrm{d}}{\mathrm{d}t}\|x_k(t)\|_E^2 = \langle D_k x_k(t), x_k(t)\rangle_E + \langle l(t), x_k(t)\rangle_E \leq \langle l(t), x_k(t)\rangle_E \qquad \text{a.e. on } (0, T),$$

and after an integration we get

$$\tfrac{1}{2}\|x_k(s)\|_E^2 \leq \int_0^s \langle l(t), x_k(t)\rangle_E \,\mathrm{d}t \qquad \forall s \in [0, T].$$

Using (2.68) we have just proved that for all $l \in C^1(0, T, X)$

$$x_k(t) \to 0 \text{ strongly in } X \text{ when } k \to +\infty \qquad \forall t \in [0, T], \tag{2.69}$$

where $x_k(t)$ is given by (2.52).

Let us return to our case $l \in L^\infty(0, T, X)$. Let $\varepsilon > 0$ and let us denote by $l_\varepsilon \in C^1([0, T], X)$ such that $\int_0^T \|l_\varepsilon(s) - l(s)\|_E \,\mathrm{d}s < \varepsilon$. We have

$$\|x_k(t)\|_E \leq \int_0^t \|l_\varepsilon(s) - l(s)\|_E \,\mathrm{d}s + \left\| \int_0^t U_k(t-s) l_\varepsilon(s) \,\mathrm{d}s \right\|_E$$

for all $t \in [0, T]$, and using (2.69) we obtain the statement of Theorem 2.3.

Remark 2.2 *Theorem* 2.3 *can also be proved in the particular case* $b, g, f \equiv 0$ *using Theorem* 2.1 *in the paper of Kurtz* (1973) *and the energy norm* (2.47).

3 An approach to perfect plasticity

In this section we use the same notations as in Section 1 and we consider the problems (1.1)–(1.4) in which (1.2) is replaced by the constitutive equation (1.8), (1.7) from Chapter 3. As in Section 3 of Chapter 3, we study the behaviour of the solution of this problem when the viscosity constant μ tends to zero.

3.1 A convergence result

Let K be a closed convex subset of $\mathscr{S}_N$ and $\tilde{G}_\mu$ the function defined by (3.2) of Chapter 3, where P_K is the projector map on K; it is easy to see that $\tilde{G}_\mu$ satisfies (1.6) with $\theta \equiv 0$; hence, if (1.5), (1.7), (1.8)–(1.12) hold, by Theorem 1.1 we get that for every $\mu > 0$ there exists a unique pair of functions $(u_\mu; \sigma_\mu)$ which satisfies (3.1) of Chapter 3, (1.1), (1.3), (1.4) and having the regularity (1.14) and (1.15). Let $\dot{u}_\mu = v_\mu$; we are interested here in the study of the behaviour of v_μ, σ_μ when $\mu \to 0$. For this reason, let us define $\mathscr{K}$ by (3.3) of Chapter 3, and let us consider the following assumptions:

$$f \in V'_{\Gamma_1} \tag{3.1}$$

$$\sigma_0 \in \mathscr{K}. \tag{3.2}$$

We also define the set $\mathscr{H}_{ad}$ by

$$\mathscr{H}_{ad} = \{\tau \in \mathscr{H}_2 \,|\, \tau\nu = f \text{ on } \Gamma_2\}. \tag{3.3}$$

Remark 3.1 *If* (3.1) *and* $(1.10)_3$ *hold it follows that* (1.12) *also holds with* $\tilde{\sigma} = \sigma_0$ *independent of time.*

We have the following result:

Theorem 3.1 *Under the hypotheses* (1.5), (1.7), (1.8)–(1.10), $(1.13)_1$ *and* (3.1)–(3.2) *there exists a pair of functions* $v \in W^{1,\infty}(0, T, \boldsymbol{H})$, $\sigma \in W^{1,\infty}(0, T, \mathscr{H}) \cap L^\infty(0, T, \mathscr{H}_2)$ *such that*

$$v_\mu \to v, \quad \dot{v}_\mu \to \dot{v} \qquad \text{weakly* in } L^\infty(0, T, \boldsymbol{H}) \tag{3.4}$$

$$\sigma_\mu \to \sigma, \quad \dot{\sigma}_\mu \to \dot{\sigma} \qquad \text{weakly* in } L^\infty(0, T, \mathscr{H}) \tag{3.5}$$

$$\sigma_\mu \to \sigma \qquad \text{weakly* in } L^\infty(0, T, \mathscr{H}_2) \tag{3.6}$$

when $\mu \to 0$. *Moreover,* v *and* σ *satisfy the following problem:*

$$\sigma(t) \in \mathscr{K} \cap \mathscr{H}_{ad} \qquad \forall t \in [0, T] \tag{3.7}$$

$$\rho\dot{v}(t) = \operatorname{Div}\sigma(t) + \rho b(t) \qquad \text{a.e. on } (0, T) \tag{3.8}$$

$$\langle \tilde{\mathscr{A}}\dot{\sigma}(t), \tau - \sigma(t)\rangle_{\mathscr{H}} + \langle v(t), \operatorname{Div}\tau - \operatorname{Div}\sigma(t)\rangle_{\boldsymbol{H}} \geq \langle \gamma_2\tau - \gamma_2\sigma(t), \dot{g}(t)\rangle_{\boldsymbol{H}'_\Gamma, \boldsymbol{H}_\Gamma}$$

for all $\tau \in \mathscr{K} \cap \mathscr{H}_{ad}$ *a.e. on* $(0, T)$. (3.9)

$$\sigma(0) = \sigma_0, \; v(0) = v_0 \qquad \text{in } \Omega. \tag{3.10}$$

PROOF. For every $\mu > 0$ we have

$$\left.\begin{aligned} \rho\dot{v}_\mu &= \operatorname{Div}\sigma_\mu + \rho b \\ \tilde{\mathscr{A}}\dot{\sigma}_\mu + \tilde{G}_\mu(\sigma_\mu) &= \varepsilon(v_\mu) \end{aligned}\right\} \quad \text{in } (0, T) \times \Omega \tag{3.11, 3.12}$$

$$v_\mu = \dot{g} \quad \text{on } (0, T) \times \Gamma_1, \qquad \sigma_\mu \nu = f \quad \text{on } (0, T) \times \Gamma_2 \tag{3.13}$$

$$v_\mu(0) = v_0, \qquad \sigma_\mu(0) = \sigma_0. \tag{3.14}$$

Let $\bar{v}_\mu$ and $\bar{\sigma}_\mu$ be the functions defined by

$$\bar{v}_\mu = v_\mu - \tilde{v}, \qquad \bar{\sigma}_\mu = \sigma_\mu - \sigma_0 \tag{3.15}$$

where $\tilde{v} = \dot{\tilde{u}}$ and $\tilde{u} = z, g$. It is easy to see that $\bar{v}_\mu$ and $\bar{\sigma}_\mu$ satisfy the following problem:

$$\left.\begin{aligned} \rho\dot{\bar{v}}_\mu &= \operatorname{Div}\bar{\sigma}_\mu + k \\ \tilde{\mathscr{A}}\dot{\bar{\sigma}}_\mu + \tilde{G}_\mu(\bar{\sigma}_\mu + \sigma_0) &= \varepsilon(\bar{v}_\mu) + h \end{aligned}\right\} \quad \text{in } (0, T) \times \Omega \tag{3.16, 3.17}$$

$$\bar{v}_\mu = 0 \quad \text{on } (0, T) \times \Gamma_1, \qquad \sigma_\mu \nu = 0 \quad \text{on } (0, T) \times \Gamma_2 \tag{3.18}$$

$$\bar{v}_\mu(0) = \bar{v}_0, \qquad \bar{\sigma}_\mu(0) = 0 \qquad \text{in } \Omega \tag{3.19}$$

where

$$k = \operatorname{Div}\tilde{\sigma} + \rho b - \rho\dot{\tilde{v}}, \qquad h = \varepsilon(\tilde{v}) - \mathscr{A}\dot{\tilde{\sigma}} \tag{3.20}$$

and

$$\bar{v}_0 = v_0 - \tilde{v}(0), \qquad \bar{\sigma}_0 = \sigma_0. \tag{3.21}$$

A priori estimates I. From (3.16)–(3.18) and using the Green-type formula (2.58) from Chapter 2, we get

$$\langle \rho\dot{\bar{v}}_\mu, \bar{v}_\mu\rangle_{\boldsymbol{H}} + \langle \tilde{\mathscr{A}}\dot{\bar{\sigma}}_\mu, \bar{\sigma}_\mu\rangle_{\mathscr{H}} + \langle \tilde{G}_\mu(\bar{\sigma}_\mu + \sigma_0), \bar{\sigma}_\mu\rangle_{\mathscr{H}}$$
$$= \langle k, \bar{v}_\mu\rangle_{\boldsymbol{H}} + \langle h, \bar{\sigma}_\mu\rangle_{\mathscr{H}} \qquad \text{a.e. on } (0, T). \tag{3.22}$$

Using (3.18) from Chapter 3 for $\tau = \sigma_0$ and $\sigma = \bar{\sigma}_\mu + \sigma_0$, by (3.2) we get $\langle \tilde{G}_\mu(\bar{\sigma} + \sigma_0), \bar{\sigma}_\mu\rangle_{\mathscr{H}} \geq \mathscr{G}_\mu(\bar{\sigma}_\mu + \sigma_0)$ for all $t \in [0, T]$. So, (3.22) becomes

$$\langle \rho\dot{\bar{v}}_\mu, \bar{v}_\mu\rangle_{\boldsymbol{H}} + \langle \tilde{\mathscr{A}}\dot{\bar{\sigma}}_\mu, \bar{\sigma}_\mu\rangle_{\mathscr{H}} + \mathscr{G}_\mu(\bar{\sigma}_\mu + \sigma_0)$$
$$\leq \langle k, \bar{v}_\mu\rangle_{\boldsymbol{H}} + \langle h, \bar{\sigma}_\mu\rangle_{\mathscr{H}} \qquad \text{a.e. on } (0, T). \tag{3.23}$$

It follows that

$$\langle \rho\dot{\bar{v}}_\mu, \bar{v}_\mu\rangle_{\boldsymbol{H}} + \langle \mathscr{A}\dot{\bar{\sigma}}_\mu, \bar{\sigma}_\mu\rangle_{\mathscr{H}} \leq \langle k, \bar{v}_\mu\rangle_{\boldsymbol{H}} + \langle h, \bar{\sigma}_\mu\rangle_{\mathscr{H}} \qquad \text{a.e. on } (0, T)$$

and using a standard technique we obtain

$$(\bar{v}_\mu)_\mu \text{ (and } (v_\mu)_\mu) \text{ is a bounded sequence in } L^\infty(0, T, \boldsymbol{H}), \tag{3.24}$$

$$(\bar{\sigma}_\mu)_\mu \text{ (and } (\sigma_\mu)_\mu) \text{ is a bounded sequence in } L^\infty(0, T, \mathscr{H}). \tag{3.25}$$

Moreover, from (3.23)–(3.25) we get

$$\int_0^T \mathcal{G}_\mu(\sigma_\mu) \leq C \tag{3.26}$$

where C is a strictly positive constant.

A priori estimates II. Setting $t = 0$ in (3.23) since $\mathcal{G}_\mu(\sigma_0) = 0$ (see (3.2)) we get

$$\left.\begin{array}{l} \dot{\bar{v}}_\mu(0) \text{ remains in a bounded set of } \boldsymbol{H} \\ \dot{\bar{\sigma}}_\mu(0) \text{ remains in a bounded set of } \mathcal{H}. \end{array}\right\} \tag{3.27}$$

Differentiating (3.16) and (3.17) with respect to t (this is legitimate if we approximate (3.16) and (3.17) by the Galerkin method) and multiplying these equations by $\dot{\bar{v}}_\mu$ and $\dot{\bar{\sigma}}_\mu$ respectively after using (2.58) from Chapter 2, we get

$$\langle \rho \ddot{\bar{v}}_\mu, \dot{\bar{v}}_\mu \rangle_{\boldsymbol{H}} + \langle \tilde{\mathcal{A}} \ddot{\bar{\sigma}}_\mu, \dot{\bar{\sigma}}_\mu \rangle_{\mathcal{H}} + \left\langle \frac{\mathrm{d}}{\mathrm{d}t} \tilde{G}_\mu(\bar{\sigma}_\mu + \sigma_0), \dot{\bar{\sigma}}_\mu \right\rangle_{\mathcal{H}}$$
$$= \langle \dot{k}, \dot{\bar{v}}_\mu \rangle_{\boldsymbol{H}} + \langle \dot{h}, \dot{\bar{\sigma}}_\mu \rangle_{\mathcal{H}} \qquad \text{a.e. on } (0, T). \tag{3.28}$$

Since $\tilde{G}_\mu$ is monotone we get

$$\left\langle \frac{\mathrm{d}}{\mathrm{d}t} \tilde{G}_\mu(\bar{\sigma}_\mu + \sigma_0), \dot{\bar{\sigma}}_\mu \right\rangle_{\mathcal{H}} \geq 0;$$

hence from (3.27)–(3.28) we obtain

$$(\dot{\bar{v}}_\mu)_\mu \text{ (and } (\dot{v}_\mu)_\mu) \text{ is a bounded sequence in } L^\infty(0, T, \boldsymbol{H}) \tag{3.29}$$

$$(\dot{\bar{\sigma}}_\mu)_\mu \text{ (and } (\dot{\sigma}_\mu)_\mu) \text{ is a bounded sequence in } L^\infty(0, T, \mathcal{H}). \tag{3.30}$$

The behaviour of v_μ, σ_μ when $\mu \to 0$. Using (3.24), (3.25), (3.29), and (3.30) we find that extracting subsequences again denoted by $(v_\mu)_\mu$ and $(\sigma_\mu)_\mu$, there exists $v \in W^{1,\infty}(0, T, \boldsymbol{H})$ and $\sigma \in W^{1,\infty}(0, T, \mathcal{H})$ such that (3.4) and (3.5) are satisfied. Let us prove the regularity $\sigma \in L^\infty(0, T, \mathcal{H}_2)$ and (3.6). From (3.11) and (3.29) we get that $(\mathrm{Div}\, \sigma_\mu)_\mu$ is a bounded sequence in $L^\infty(0, T, \boldsymbol{H})$, hence we can suppose

$$\mathrm{Div}\, \sigma_\mu \to \chi \qquad \text{weakly* in } L^\infty(0, T, \boldsymbol{H}), \tag{3.31}$$

where χ belongs to $L^\infty(0, T, \boldsymbol{H})$. From (3.31) and (3.5) we get $\mathrm{Div}\, \sigma = \chi$, hence $\sigma \in L^\infty(0, T, \mathcal{H}_2)$. Using again (3.31) and (3.5) we get (3.6).

In order to prove (3.7) note that from (3.18) of Chapter 3 we have

$$\mathcal{G}_\mu(\sigma) \leq \mathcal{G}_\mu(\sigma_\mu) - \langle \tilde{G}_\mu \sigma, \sigma_\mu - \sigma \rangle_{\mathcal{H}};$$

hence by (3.2) of Chapter 3 we get $\mu \mathcal{G}_\mu(\sigma) \leq \mu \mathcal{G}_\mu(\sigma_\mu) - \langle \sigma - P_K \sigma, \sigma_\mu - \sigma \rangle_{\mathcal{H}}$

for all $t \in [0, T]$. It follows that

$$\int_0^T \mu \mathscr{G}_\mu(\sigma)\,\mathrm{d}t \leq \mu \int_0^T \mathscr{G}_\mu(\sigma_\mu) - \int_0^T \langle \sigma - P_K\sigma, \sigma_\mu - \sigma \rangle_{\mathscr{H}}\,\mathrm{d}t$$

and using (3.17) of Chapter 3 and (3.26), (3.5), we get $\frac{1}{2}\int_0^T \|\sigma - P_K\sigma\|^2_{\mathscr{H}}\,\mathrm{d}t = 0$. So, $\sigma = P_K\sigma$ on $[0, T]$, which means

$$\sigma(t) \in \mathscr{K} \qquad \forall t \in [0, T]. \tag{3.32}$$

Let

$$S = \{\tau \in W^{1,\infty}(0, T, \mathscr{H}) \mid \tau(t) \in \mathscr{H}_{ad}\ \forall t \in [0, T]\}.$$

Using (1.3) and (3.3) we have $\sigma_\mu \in S$ and by (3.5), (3.6), we get $\sigma \in S$, i.e.

$$\sigma(t) \in \mathscr{H}_{ad} \qquad \forall t \in [0, T]. \tag{3.33}$$

Hence by (3.32) and (3.33) we get (3.7). In order to prove (3.8) let $t \in [0, T]$, $\varphi \in \boldsymbol{D}$, and let us denote by χ_t the characteristic function of $[0, t]$, i.e. $\chi_t\tilde{t} = 1$ if $\tilde{t} \in [0, t]$ and $\chi_t\tilde{t} = 0$ if $\tilde{t} \notin [0, t]$. For every $\mu > 0$, by (3.11) we get

$$\int_0^T \langle \rho\dot{v}_\mu, \chi_t\varphi \rangle_{\boldsymbol{H}}\,\mathrm{d}t = \int_0^T \langle \operatorname{Div} \sigma_\mu, \chi_t\varphi \rangle_{\boldsymbol{H}}\,\mathrm{d}t + \int_0^T \langle \rho b, \chi_t\varphi \rangle_{\boldsymbol{H}}\,\mathrm{d}t.$$

Since $\langle \operatorname{Div} \sigma_\mu, \chi_t\varphi \rangle_{\boldsymbol{H}} = -\langle \sigma_\mu, \chi_t\varepsilon(\varphi) \rangle_{\mathscr{H}}$ and $\chi_t\varphi \in L^1(0, T, \boldsymbol{H})$, $\chi_t\varepsilon(\varphi) \in L^1(0, T, \mathscr{H})$, using (3.4) and (3.5) we get

$$\int_0^T \langle \rho\dot{v}, \chi_t\varphi \rangle_{\boldsymbol{H}}\,\mathrm{d}t + \int_0^T \langle \sigma, \chi_t\varepsilon(\varphi) \rangle_{\mathscr{H}}\,\mathrm{d}t = \int_0^T \langle \rho b, \chi_t\varphi \rangle_{\boldsymbol{H}}\,\mathrm{d}t$$

or, equivalently,

$$\int_0^t \langle \rho\dot{v}, \varphi \rangle_{\boldsymbol{H}}\,\mathrm{d}t + \int_0^t \langle \sigma, \varepsilon(\varphi) \rangle_{\mathscr{H}}\,\mathrm{d}t = \int_0^t \langle \rho b, \varphi \rangle_{\boldsymbol{H}}\,\mathrm{d}t$$

and by the classical use of Lebesgue points on an L^1-function we get

$$\langle \rho\dot{v}, \varphi \rangle_{\boldsymbol{H}} + \langle \sigma, \varepsilon(\varphi) \rangle_{\mathscr{H}} = \langle \rho b, \varphi \rangle_{\boldsymbol{H}} \qquad \text{a.e. on } (0, T). \tag{3.34}$$

Since φ is an arbitrary element of $\boldsymbol{D}$, using (2.9) in Chapter 2, from (3.34) we get (3.8).

In order to obtain (3.9) we multiply (3.12) by $\tau - \sigma_\mu$, where τ is an arbitrary element of $\mathscr{K} \cap \mathscr{H}_{ad}$; after integrating on Ω we get

$$\langle \tilde{\mathscr{A}}\dot{\sigma}_\mu, \tau - \sigma_\mu \rangle_{\mathscr{H}} + \langle \tilde{G}_\mu\sigma_\mu, \tau - \sigma_\mu \rangle_{\mathscr{H}} = \langle \varepsilon(v_\mu), \tau - \sigma_\mu \rangle_{\mathscr{H}} \qquad \text{a.e. on } (0, T). \tag{3.35}$$

Using (2.44) from Chapter 2 and (3.13) we get

$$\langle \varepsilon(v_\mu), \tau - \sigma_\mu \rangle_{\mathscr{H}} = \langle \gamma_2\tau - \gamma_2\sigma, \dot{g} \rangle_{\boldsymbol{H}_\Gamma', \boldsymbol{H}_\Gamma} - \langle \operatorname{Div} \tau - \operatorname{Div} \sigma_\mu, v_\mu \rangle_{\boldsymbol{H}}.$$

Then (3.35) becomes

$$\langle \tilde{\mathscr{A}}\dot{\sigma}_\mu, \tau - \sigma_\mu \rangle_{\mathscr{H}} + \langle \tilde{G}_\mu \sigma_\mu, \tau - \sigma_\mu \rangle_{\mathscr{H}} + \langle \operatorname{Div} \tau - \operatorname{Div} \sigma_\mu, v_\mu \rangle_{\boldsymbol{H}}$$
$$= \langle \gamma_2 \tau - \gamma_2 \sigma_\mu, \dot{g} \rangle_{\boldsymbol{H}'_\Gamma, \boldsymbol{H}_\Gamma} \quad \text{a.e. on } (0, T). \quad (3.36)$$

Since $\tau \in \mathscr{K}$ we have $\mathscr{G}_\mu(\tau) = 0$ and by (3.18) of Chapter 3 we get $\langle \tilde{G}_\mu \sigma_\mu, \tau - \sigma_\mu \rangle_{\mathscr{H}} \leq 0$; from (3.11), we deduce

$$\langle \tilde{\mathscr{A}}\dot{\sigma}_\mu, \tau - \sigma_\mu \rangle_{\mathscr{H}} + \langle \operatorname{Div} \tau - \rho \dot{v}_\mu + \rho b, v_\mu \rangle_{\boldsymbol{H}}$$
$$\geq \langle \gamma_2 \tau - \gamma_2 \sigma_\mu, \dot{g} \rangle_{\boldsymbol{H}'_\Gamma, \boldsymbol{H}_\Gamma} \quad \text{a.e. on } (0, T).$$

By integration it follows that

$$\int_0^t \langle \tilde{\mathscr{A}}\dot{\sigma}_\mu, \tau \rangle_{\mathscr{H}} \, \mathrm{d}s + \int_0^t \langle \operatorname{Div} \tau + \rho b, v_\mu \rangle_{\boldsymbol{H}} \, \mathrm{d}s$$
$$\geq \int_0^t \langle \tilde{\mathscr{A}}\dot{\sigma}_\mu, \sigma_\mu \rangle_{\mathscr{H}} \, \mathrm{d}s + \int_0^t \langle \rho \dot{v}_\mu, v_\mu \rangle_{\boldsymbol{H}} \, \mathrm{d}s$$
$$+ \int_0^t \langle \gamma_2 \tau - \gamma_2 \sigma_\mu, \dot{g} \rangle_{\boldsymbol{H}'_\Gamma, \boldsymbol{H}_\Gamma} \, \mathrm{d}s \qquad \forall t \in [0, T]. \quad (3.37)$$

By a lower semicontinuity argument from (3.4) and (3.5) we deduce

$$\lim_{\mu \to 0} \int_0^t \langle \tilde{\mathscr{A}}\dot{\sigma}_\mu, \sigma_\mu \rangle_{\mathscr{H}} \, \mathrm{d}s \geq \int_0^t \langle \tilde{\mathscr{A}}\dot{\sigma}, \sigma \rangle_{\mathscr{H}} \, \mathrm{d}s$$

and

$$\lim_{\mu \to 0} \int_0^t \langle \rho \dot{v}_\mu, v_\mu \rangle_{\boldsymbol{H}} \, \mathrm{d}s \geq \int_0^t \langle \rho v, v \rangle_{\boldsymbol{H}} \, \mathrm{d}s.$$

Therefore, using (3.4) and (3.5), since $\gamma_2 \sigma_\mu \to \gamma_2 \sigma$ weakly* in $L^\infty(0, T, \boldsymbol{H}'_\Gamma)$ by (3.37), we obtain

$$\int_0^t \langle \tilde{\mathscr{A}}\dot{\sigma}, \tau \rangle_{\mathscr{H}} \, \mathrm{d}s + \int_0^t \langle \operatorname{Div} \tau + \rho b, v \rangle_{\boldsymbol{H}} \, \mathrm{d}s$$
$$\geq \int_0^t \langle \tilde{\mathscr{A}}\dot{\sigma}, \sigma \rangle_{\mathscr{H}} \, \mathrm{d}s + \int_0^t \langle \rho \dot{v}, v \rangle_{\boldsymbol{H}} \, \mathrm{d}s + \int_0^t \langle \gamma_2 \tau - \gamma_2 \sigma, \dot{g} \rangle_{\boldsymbol{H}'_\Gamma, \boldsymbol{H}_\Gamma} \, \mathrm{d}s$$

for all $t \in [0, T]$, and, after the classical use of Lebesgue points for an L^1 function we get

$$\langle \tilde{\mathscr{A}}\dot{\sigma}, \tau - \sigma \rangle_{\mathscr{H}} + \langle \operatorname{Div} \tau + \rho b - \rho \dot{v}, v \rangle_{\boldsymbol{H}} \geq \langle \gamma_2 \tau - \gamma_2 \sigma, g \rangle_{\boldsymbol{H}'_\Gamma, \boldsymbol{H}_\Gamma}$$

a.e. on $(0, T)$. Now using (3.8) we get (3.9).

Let us remark finally that (3.10) is a simple consequence of (3.14) and (3.4), (3.5).

In order to finish the proof of Theorem 3.1 note that since (3.7)–(3.10) has a unique solution $(v; \sigma)$ (see Remark 3.2) the sequences $(v_\mu)_\mu$, $(\dot{v}_\mu)_\mu$, $(\sigma_\mu)_\mu$, and $(\dot{\sigma}_\mu)_\mu$ satisfy (3.4)–(3.6); this means it is not necessary to extract some subsequences in order to have (3.4)–(3.6).

3.2 Dynamic processes in perfect plasticity

The problem (3.7)–(3.10) represents the weak formulation of a dynamic process in perfect plasticity considered by Duvaut and Lions (1972, Ch. V). In order to motivate (3.7)–(3.10) we start from the strong formulation of a dynamic problem for perfect plastic materials; this formulation consists of finding the displacement function u: $[0, T] \times \Omega \to \mathbb{R}^N$ and the stress function σ: $[0, T] \times \Omega \to \mathscr{S}_N$ such that (3.40)–(3.43) of Chapter 3 hold in $\Omega \times (0, T)$ and moreover

$$\rho\ddot{u} = \text{Div}\,\sigma + \rho b \qquad \text{in } (0, T) \times \Omega \tag{3.38}$$

$$u = g \quad \text{on } (0, T) \times \Gamma_1, \qquad \sigma\nu = f \quad \text{on } (0, T) \times \Gamma_2 \tag{3.39}$$

$$u(0) = u_0, \quad \dot{u}(0) = v_0, \quad \sigma(0) = \sigma_0 \text{ in } \Omega. \tag{3.40}$$

As usual, in (3.38)–(3.40) ρ represents the mass density, b is the body forces, g and f are the boundary data and u_0, v_0, σ_0 are the initial data. Supposing that (3.40)–(3.43) of Chapter 3 and (3.38)–(3.40) have a smooth solution $(u; \sigma)$, we denote $\dot{u} = v$ and from (3.40)–(3.43) of Chapter 3 we have

$$\langle \tilde{\mathscr{A}}\dot{\sigma}, \tau - \sigma\rangle_{\mathscr{H}} \geq \langle \varepsilon(v), \tau - \sigma\rangle_{\mathscr{H}} \qquad \forall \tau \in \mathscr{K}, \quad \text{a.e. on } (0, T). \tag{3.41}$$

From (3.39) we get $v = \dot{g}$ on $\Gamma_1 \times (0, T)$; defining $\mathscr{H}_{ad}$ by (3.3) if we take in (3.41) $\tau \in \mathscr{K} \cap \mathscr{H}_{ad}$ and we use (2.44) of Chapter 2, we obtain (3.9); by $(3.39)_2$ and (3.42) of Chapter 3 we get (3.7); (3.8) is a simple consequence of (3.38) and finally (3.10) results from (3.40). Hence, we arrive at the weak formulation of the problem (3.40)–(3.43) of Chapter 3, (3.38)–(3.40):

Find the velocity field v: $[0, T] \times \Omega \to \mathbb{R}^N$ and the stress tensor σ: $[0, T] \times \Omega \to \mathscr{S}_N$ such that (3.7)–(3.10) are satisfied.

From this point of view Theorem 3.1 represents an existence result for the problem (3.7)–(3.10) and at the same time shows that the solution $(v; \sigma)$ of this 'irregular' problem can be approximated in the sense given by (3.4)–(3.6) by the solution of the 'regular' problem (3.11)–(3.14).

Remark 3.2 *The uniqueness of the solution in* (3.7)–(3.10) *can be easily proved. Indeed, if* $(v_1; \sigma_1)$ *and* $(v_2; \sigma_2)$ *are two solutions of* (3.7)–(3.10) *having the regularity* $v_i \in W^{1,\infty}(0, T, \boldsymbol{H})$, $\sigma_i \in W^{1,\infty}(0, T, \mathscr{H}) \cap L^\infty(0, T, \mathscr{H}_2)$, $i = 1, 2$, *from* (3.8), (3.9) *we have*

$$\langle \tilde{\mathscr{A}}\dot{\sigma}_1 - \tilde{\mathscr{A}}\dot{\sigma}_2, \sigma_1 - \sigma_2\rangle_{\mathscr{H}} + \langle \rho\dot{v}_1 - \rho\dot{v}_2, v_1 - v_2\rangle_{\boldsymbol{H}} \leq 0 \qquad \text{a.e. on } (0, T). \tag{3.42}$$

By integration and using (1.5), (1.7), and (3.10) we get

$$\|\sigma_1(s) - \sigma_2(s)\|_{\mathscr{H}}^2 + \|v_1(s) - v_2(s)\|_H^2 \leq 0 \qquad \forall s \in [0, T].$$

Hence it follows that $\sigma_1 = \sigma_2$ and $v_1 = v_2$.

4 Dynamic processes for rate-type elastic–viscoplastic materials with internal state variables

In this section we consider constitutive equations of the form (5.1) of Chapter 3 involving an internal state variable denoted by κ. Mechanical interpretations and examples concerning this type of constitutive laws have already been presented in Section 5.1 of Chapter 3. As in that section we shall consider here, for simplicity only, the case of a single internal state variable (i.e. $m = 1$).

4.1 Problem statement and constitutive assumptions

Using the same notations as in Section 1 we consider the following mixed problem:

Find the displacement function u: $[0, T] \times \Omega \to \mathbb{R}^N$, the stress function σ: $[0, T] \times \Omega \to \mathscr{S}_N$ and the internal state variable κ: $[0, T] \times \Omega \to \mathbb{R}$ such that

$$\rho\ddot{u}(t) = \text{Div}\,\sigma(t) + \rho b(t) \tag{4.1}$$

$$\dot{\sigma}(t) = \mathscr{A}\varepsilon(\dot{u}(t)) + G(t, \sigma(t), \varepsilon(u(t)), \kappa(t)) \tag{4.2}$$

$$\dot{\kappa}(t) = \varphi(t, \sigma(t), \varepsilon(u(t)), \kappa(t)) \qquad \text{in } \Omega \tag{4.3}$$

$$u(t) = g(t) \quad \text{on } \Gamma_1, \quad \sigma(t)\nu = f(t) \quad \text{on } \Gamma_2 \quad \text{for } t \in (0, T) \tag{4.4}$$

$$u(0) = u_0, \quad \dot{u}(0) = v_0, \quad \sigma(0) = \sigma_0, \quad \kappa(0) = \kappa_0 \qquad \text{in } \Omega. \tag{4.5}$$

The problem (4.1)–(4.5) is similar to (1.1)–(1.4) in which (1.2) was replaced by a more general constitutive law (4.2), equation (4.3) was added for the unknown parameter κ and the initial data κ_0 was added in (4.5).

In order to study problem (4.1)–(4.5) we suppose that functions

$$G\colon \Omega \times [0, T] \times \mathscr{S}_N \times \mathscr{S}_N \times \mathbb{R} \to \mathscr{S}_N$$

and

$$\varphi\colon \Omega \times [0, T] \times \mathscr{S}_N \times \mathscr{S}_N \times \mathbb{R} \to \mathbb{R}$$

have the following properties:

G satisfies (5.16) of Chapter 3 on $[0, T]$ with $\theta \in W^{1,1}(0, T, H)$ (4.6)

φ satisfies (5.17) of Chapter 3) on $[0, T]$ with $\psi \in W^{1,1}(0, T, H)$ (4.7)

(see Remarks 1.3.2 and 5.1, both of Chapter 3, for some interpretations of the assumptions (4.6) and (4.7)).

We also assume that

$$\kappa_0 \in H. \tag{4.8}$$

4.2 Existence, uniqueness and continuous dependence of the solution

For the problem (4.1)–(4.5) we have the following existence and uniqueness result:

Theorem 4.1 *Let* (1.5), (4.6)–(4.8), (1.7)–(1.12) *hold. Then there exists a unique solution of the problem* (4.1)–(4.5) *which satisfies* (1.14), (1.15), *and* $\kappa \in W^{2,1}(0, T, H)$.

PROOF. The same technique as in the proof of Theorem 1.1 can be used. After the same homogenization of the boundary data as in Lemma 1.1 we can use the same arguments in order to prove that $x = (u^*; v^*; \sigma^*; \kappa)$ is the solution of the following Cauchy problem on $X = \boldsymbol{V}_1 \times \boldsymbol{H} \times \mathscr{H} \times H$:

$$\dot{x}(t) = Ax(t) + h(t, x(t)) \qquad \text{on } (0, T), x(0) = x_0 \tag{4.9}$$

where $D(A) = \boldsymbol{V}_1 \times \boldsymbol{V}_1 \times \mathscr{V}_1 \times \boldsymbol{H}$, $A: D(A) \subset X \to X$ is given by

$$A(u; v; \sigma; \kappa) = \left(v; \frac{1}{\rho} \operatorname{Div} \sigma; \mathscr{A}\varepsilon(v); 0\right) \tag{4.10}$$

and $h: [0, T] \times X \to X$ is given by

$$\begin{aligned} h(t, (u; v; \sigma; \kappa)) = (0; b(t) - \ddot{\tilde{u}}(t) + \rho^{-1} \operatorname{Div} \tilde{\sigma}(t); G(t, \tilde{\sigma}(t) + \sigma, \varepsilon(\tilde{u}(t)) + \varepsilon(u); \kappa) \\ + \mathscr{A}\varepsilon(\dot{\tilde{u}}(t) - \dot{\tilde{\sigma}}(t); \varphi(t, \tilde{\sigma}(t) + \sigma, \varepsilon(\tilde{u}(t)) + \varepsilon(u), \kappa)) \end{aligned} \tag{4.11}$$

for all $t \in [0, T]$ and $(u; v; \sigma; \kappa) \in X$. We can use the same arguments as in Lemma 1.3 in order to prove that A is the infinitesimal generator of a C_0-semigroup $(S(t))_{t \geq 0}$ given by

$$S(t)(u; v; \sigma; \kappa) = \left(u + \int_0^t T_1(s)(v; \sigma)\, ds; T(s)(v; \sigma); \kappa\right) \tag{4.12}$$

where $(T(t))_{t \geq 0}$ is the semigroup generated by B given by (1.24). Since (A3.39) also holds, we can use Theorem (A3.9) as well in order to obtain the statement of Theorem 4.1.

Using a similar technique to that in the proof of Theorem 1.2 we get the continuous dependence of the solution of (4.1)–(4.5) upon all input data, i.e.:

Theorem 4.2 *Let* (1.5), (4.6), (4.7), (1.7) *hold and let* $(u_i; \sigma_i; \kappa_i)$ *be the solutions of* (4.1)–(4.5) *for the data* b_i, g_i, f_i, u_{0i}, v_{0i}, σ_{0i}, κ_{0i}, $i = 1, 2$, *which satisfies* (1.8)–(1.10), (1.13), (4.8). *Then there exists* $C(T) > 0$ *which depends* *only on* Ω, Γ_1, $\mathscr{A}$, ρ, L, $\tilde{L}$, *and* T *such that*

$$\|u_1 - u_2\|_{0,T,H_1} + \|\dot{u}_1 - \dot{u}_2\|_{0,T,H} + \|\sigma_1 - \sigma_2\|_{0,T,\mathscr{H}} + \|\kappa_1 - \kappa_2\|_{1,T,H}$$
$$\leq C(T)[\|u_{01} - u_{02}\|_{H_1} + \|v_{01} - v_{02}\|_{H} + \|\sigma_{01} - \sigma_{02}\|_{\mathscr{H}} + \|\kappa_{01} - \kappa_{02}\|_{H}$$
$$+ \|b_1 - b_2\|_{0,T,H} + \|g_1 - g_2\|_{2,T,H_\Gamma} + \|f_1 - f_2\|_{1,T,V_{\Gamma 1}}]. \tag{4.13}$$

We can also easily deduce the following result:

Corollary 4.1 *Let* (1.5), (1.7), (1.8), (1.11), (1.12), (4.6), (4.7) *hold and let* $(u_i; \sigma_i; \kappa_i)$ *be the solutions of* (4.1)–(4.5) *for the initial data* $(u_{0i}; v_{0i}; \kappa_{0i})$, $i = 1, 2$ *which satisfies* (1.9), (1.10), (4.8). *Then there exists* $C(T) > 0$ *which depends only on* Ω, Γ_1, ρ, $\mathscr{A}$, L, $\tilde{L}$, *and* T *such that*

$$\|u_1 - u_2\|_{0,T,H_1} + \|\dot{u}_1 - \dot{u}_2\|_{0,T,H} + \|\sigma_1 - \sigma_2\|_{0,T,\mathscr{H}} + \|\kappa_1 - \kappa_2\|_{1,T,H}$$
$$\leq C(T)[\|u_{01} - u_{02}\|_{H_1} + \|v_{01} - v_{02}\|_{H} + \|\sigma_{01} - \sigma_{02}\|_{\mathscr{H}} + \|\kappa_{01} - \kappa_{02}\|_{H}]. \tag{4.14}$$

Remark 4.1 *The concepts of stability and finite time stability of a solution of* (4.1)–(4.5) *can be formulated in the same way as was made precise in the last part of Section* 1.3. *From* (4.14) *we find that every solution of* (4.1)–(4.5) *is finite time stable.*

4.3 A local existence result

In this section we shall consider an initial–boundary-value problem for semilinear rate-type elastic viscoplastic materials with a hardening parameter. The assumptions made on constitutive functions will be weak enough to include the work-hardening case. The one-dimensional mechanical problem is reduced to a semilinear hyperbolic equation in a real Hilbert space and the local existence and uniqueness of the solution is proved. If the solution is not a global one then it will blow up at the end of its maximal interval of existence.

Let $N = 1$, $\Omega = (0, 1) \subset \mathbb{R}$ and $T > 0$. We consider the following mixed problem:

Find the displacement, stress and hardening parameter function

$$u, \sigma, \kappa\colon [0, T] \times \Omega \to \mathbb{R}$$

such that

$$\rho\ddot{u}(t, x) = \sigma_x(t, x) + \rho b(t, x) \tag{4.15}$$

$$\dot{\sigma}(t, x) = E\dot{u}_x(t, x) + G(t, \sigma(t, x), u_x(t, x), \kappa(t, x)) \tag{4.16}$$

$$\dot{\kappa}(t, x) = \varphi(t, \sigma(t, x), u_x(t, x), \kappa(t, x)) \qquad \forall (t, x) \in (0, T] \times \Omega, \tag{4.17}$$

$$u(t, 0) = u(t, 1) = 0 \qquad \forall t \in (0, T] \tag{4.18}$$

$$u(0, x) = u_0(x), u(0, x) = v_0(x), \quad \sigma(0, x) = \sigma_0(x), \quad \kappa(0, x) = \kappa_0(x) \qquad \forall x \in \Omega. \tag{4.19}$$

For the sake of simplicity only homogenous materials are considered and only displacement boundary data are imposed.

Remark 4.2 *If we consider instead of* (4.18) *non-homogenous boundary data:*

$$u(t, 0) = g_0(t), u(t, 1) = g_1(t) \qquad \forall t \in (0, T], \tag{4.20}$$

then after a standard homogenization technique we can reduce the non-homogenous problem to a homogenous one.

Indeed, let

$$\bar{u}(t, x) = u(t, x) - xg_1(t) + (x - 1)g_0(t),$$

$$\bar{G}(t, \sigma, \varepsilon, \kappa) = G(t, \sigma, \varepsilon + g_1(t) - g_0(t), \kappa),$$

$$\bar{\varphi}(t, \sigma, \varepsilon, \kappa) = \varphi(t, \sigma, \varepsilon + g_1(t) - g_0(t), \kappa),$$

$$\bar{u}_0(x) = u_0(x) - xg_1(0) + (x - 1)g_0(0),$$

$$\bar{v}_0(x) = v_0(x) - x\dot{g}_1(0) + (x - 1)\dot{g}_0(0);$$

then the non-homogeneous problem (4.15)–(4.17), (4.19)–(4.20) *is equivalent to* (4.15)–(4.19) *for the pair* $(\bar{u}; \sigma; \kappa)$, *where* G, φ, u_0, v_0 *have been replaced by* $\bar{G}$, $\bar{\varphi}$, $\bar{u}_0$, $\bar{v}_0$.

In the study of the problem (4.15)–(4.19) the following assumptions will be used:

The functions $G, \varphi: [0, T] \times \mathbb{R}^3 \to \mathbb{R}$ are continuous and locally Lipschitz continuous on $\mathbb{R}^3$, uniformly with respect to the time variable, i.e. there exists a non-decreasing function $L: \mathbb{R}_+ \to \mathbb{R}_+$ such that

$$|G(t, \sigma_1, \varepsilon_1, \kappa_1) - G(t, \sigma_2, \varepsilon_2, \kappa_2)| \le L(r)(|\sigma_1 - \sigma_2| + |\varepsilon_1 - \varepsilon_2| + |\kappa_1 - \kappa_2|) \tag{4.21}$$

$$|\varphi(t, \sigma_1, \varepsilon_1, \kappa_1) - \varphi(t, \sigma_2, \varepsilon_2, \kappa_2)| \le L(r)(|\sigma_1 - \sigma_2| + |\varepsilon_1 - \varepsilon_2| + |\kappa_1 - \kappa_2|) \tag{4.22}$$

$$\forall t \in [0, T], (\sigma_i; \varepsilon_i; \kappa_i) \in \mathbb{R}^3 \quad \text{with} \quad \sigma_i^2 + \varepsilon_i^2 + \kappa_i^2 \le r^2, \quad i = 1,2.$$

Moreover, using the notations $H = L^2(\Omega)$, $H_1 = H^1(\Omega)$ we shall also suppose that

$$E, \rho > 0, G(t, 0, 0, 0) = \varphi(t, 0, 0, 0) = 0 \qquad \forall t \in [0, T] \tag{4.23}$$

$$b \in C^1([0, T], H) \tag{4.24}$$

$$u_0 \in H^2(\Omega), v_0, \sigma_0, \kappa_0 \in H_1 \tag{4.25}$$

$$u_0(0) = u_0(1) = v_0(0) = v_0(1) = 0. \tag{4.26}$$

Remark 4.3 (1) *If we put $\varphi(t, \sigma, \varepsilon, \kappa) = \sigma\tilde{G}(t, \sigma, \varepsilon, \kappa)$ with $\tilde{G}$ a locally Lipschitz continuous function with respect to $(\sigma; \varepsilon; \kappa)$, uniformly with respect to t, then φ is also a locally Lipschitz continuous function with respect to $(\sigma; \varepsilon; \kappa)$, uniformly with respect t. Hence* (4.22) *is satisfied in the work-hardening case.*

(2) *Assumption* (4.26) *represents the compatibility condition between the initial and boundary data which ensures that no discontinuities (shocks) should occur at $t = 0+$.*

(3) *If G does not depend on κ and $\varphi \equiv 0$ then from Theorem* 4.3 *below one can deduce the local existence and the uniqueness of the solution for elastic viscoplastic materials without hardening, for G a locally Lipschitz continuous function with respect to $(\sigma; \varepsilon)$, uniformly with respect to t. Though this assumption on the constitutive function G seems to be quite physically* reasonable *and weak enough (from the mathematical point of view) to include a large class of functions, it is not verified in some models which give a finite relaxation time (see Cristescu and Suliciu* (1982, *p.* 126) *and the references given there*).

The main result of this section is the following theorem:

Theorem 4.3 *Let* (4.21)–(4.26) *hold. Then there exist $T_0 \in (0, T]$ and a unique solution $(u; \sigma; \kappa)$ of the mixed problem* (4.15)–(4.19) *with the following regularity*:

$$u \in C^2([0, T_0), H) \cap C^1([0, T_0), H_1) \cap C^0([0, T_0), H^2(\Omega)) \tag{4.27}$$

$$\sigma, \kappa \in C^1([0, T_0), H) \cap C^0([0, T_0), H_1). \tag{4.28}$$

Moreover, if $T_0 < T$, then

$$\lim_{t \uparrow T_0} (\|\sigma(t)\|_{C(\bar{\Omega})} + \|u_x(t)\|_{C(\bar{\Omega})} + \|\kappa(t)\|_{C(\bar{\Omega})} + \|\dot{u}(t)\|_{C(\bar{\Omega})}) = +\infty. \tag{4.29}$$

Remark 4.4 (1) *From Theorem* 4.3 *we deduce that if the solution is not a global one, then it will blow up in the norm of $C(\bar{\Omega})$ of the end of its existence interval. If one can prove, in some particular problems, that the stress, deformation, velocity and hardening parameter functions are bounded in $C(\bar{\Omega})$ on the interval of existence, then we can conclude that the solution is a global one. Unfortunately, the energy estimates obtained by Suliciu and Sabac* (1988) *imply that $\sigma(t)$, $u_x(t)$, $\dot{u}(t)$ are bounded in $L^p(\Omega)$ and $\kappa(t)$ is bounded in $L^1(\Omega)$, hence these estimates are not so useful for our approach.*

(2) *One can suppose that* G, φ *are defined on a constitutive open set* $\mathscr{D} \subset \mathbb{R}^3$ *with* $(0; 0; 0) \in \mathscr{D}$ *and the solution must also satisfy the constitutive constraint*

$$(\sigma(t, x); u_x(t, x); \kappa(t, x)) \in \mathscr{D} \qquad \forall(t; x) \in [0, T] \times \Omega. \tag{4.30}$$

For instance, in order to have a finite density we can consider $\mathscr{D}$ to be included in the half-space

$$\{(\sigma; \varepsilon; \kappa) \in \mathbb{R}^3 | \varepsilon > -1\}. \tag{4.31}$$

If we suppose that

$$(\sigma_0(x); u_{0x}(x); \kappa_0(x)) \in \mathscr{D} \qquad \forall x \in \bar{\Omega}, \tag{4.32}$$

then we can use Theorem 4.3 in order to obtain the uniqueness and the local existence of the solution for the problem (4.15)–(4.19), (4.30). Indeed, since $\sigma_0, u_{0x}, \kappa_0 \in C(\bar{\Omega})$, we deduce that $K = \{(\sigma_0(x); u_{0x}(x); \kappa_0(x)) | x \in \bar{\Omega}\}$ is a compact subset of $\mathscr{D}$. Let $r_0 > 0$ such that $K_0 = \overline{K + B(0, r_0)} \subset \mathscr{D}$. If we suppose that G, φ are locally Lipschitz continuous functions on $\mathscr{D}$, then we may construct $\bar{G}, \bar{\varphi}: [0, T] \times \mathbb{R}^3 \to \mathbb{R}$ such that $\bar{G} = G$, $\bar{\varphi} = \varphi$ on K_0 and $\bar{G}, \bar{\varphi}$ satisfy (4.21), (4.22). We can now use Theorem 4.3 to deduce that there exists a unique local solution for the problem (4.15)–(4.19), where G and φ were replaced by $\bar{G}$ and $\bar{\varphi}$. From (4.27)–(4.28) we deduce that σ, u_x, $\kappa \in C^0([0, T_0]), C(\bar{\Omega}))$; hence there exists $T_1 \in (0, T_0)$ small enough such that $(\sigma(t, x); u_x(t, x), \kappa(t, x)) \in K_0$ for all $(t; x) \in [0, T_1] \times \Omega$. Hence $(u; \sigma; \kappa)$ is the unique solution of (4.15)–(4.19), (4.30) on $[0, T_1]$.

PROOF OF THEOREM 4.3. Let us denote by $B: D(B) \subset H \times H \to H \times H$, $D(B) = H_0^1(\Omega) \times H_1$ the operator given by

$$B(v; \sigma) = (\rho^{-1}\sigma_x; Ev_x) \qquad \forall(v; \sigma) \in D(B). \tag{4.33}$$

From Lemma 1.3 we deduce that $B^* = -B$ (in a suitable inner product on $H \times H$) hence B is the infinitesimal generator of a C^0-semigroup denoted by $T(t) \in B(H \times H)$. Let $T_1(t), T_2(t) \in B(H \times H, H)$ be such that $T(t)\,(v; \sigma) = (T_1(t)(v; \sigma); T_2(t)(v; \sigma))$, and let $X = H_0^1(\Omega) \times H \times H \times H$. We consider the C^0-semigroup $(S(t))_{t \geq 0} \subset B(X)$ given by

$$S(t)(u; v; \sigma; \kappa) = \left(u + \int_0^t T_1(s)(v; \sigma)\, \mathrm{d}s; T(s)(v; \sigma); \kappa\right) \tag{4.34}$$

for all $t \geq 0$ and $(u; v; \sigma; \kappa) \in X$. As it follows from Lemma 1.2, the infinitesimal generator of $(S(t))_{t \geq 0}$ is $A: D(A) \subset X \to X$ given by

$$A(u; v; \sigma; \kappa) = (v; \rho^{-1}\sigma_x; Ev_x; 0) \tag{4.35}$$

for all $(u; v; \sigma; \kappa) \in D(A) = H_0^1(\Omega) \times H_0^1(\Omega) \times H_1 \times H_1$.

Let $Y = H^2(\Omega) \cap H_0^1(\Omega) \times H_0^1(\Omega) \times H_1 \times H_1$ and let us prove that $S(t)(Y) \subset Y$ for all $t \geq 0$ and $(\tilde{S}(t))_{t \geq 0}$ is a C^0-semigroup acting on Y

where $\tilde{S}(t)$ is the restriction of $S(t)$ to Y, for all $t \geq 0$. Indeed, let $\tilde{T}(t)$ be the restriction of $T(t)$ on $[D(B)] = H_0^1(\Omega) \times H_1$ for all $t \geq 0$. From Lemma A3.3 we deduce that $(\tilde{T}(t))_{t \geq 0}$ is a C^0-semigroup acting on $[D(B)]$, hence the map $(v; \sigma) \to \tilde{T}(t)(v; \sigma)$ belongs to $B(H_0^1(\Omega) \times H_1)$. If we bear in mind that the infinitesimal generator of $(\tilde{T}(t))_{t \geq 0}$ is the operator $\tilde{B}: D(B^2) \subset [D(B)] \to [D(B)]$ defined as the restriction of B to $D(B^2)$, then we obtain that $(v; \sigma) \to \int_0^t \tilde{T}(s)(v; \sigma)\, ds$ belongs to $B([D(B)], [D(B^2)])$. Hence for the first component we have $(v; \sigma) \to \int_0^t T_1(s)(v; \sigma)\, ds$ belongs to $B(H_0^1(\Omega) \times H_1, H^2(\Omega) \cap H_0^1(\Omega))$ for all $t \geq 0$. From (4.34) one can easily deduce now that $(\tilde{S}(t))_{t \geq 0}$ is a C^0-semigroup acting on Y. Let us denote by $p: Y \to \mathbb{R}_+$ the following norm:

$$p(u; v; \sigma; \kappa) = \|u_x\|_{C(\bar{\Omega})} + \|v\|_{C(\bar{\Omega})} + \|\sigma\|_{C(\bar{\Omega})} + \|\kappa\|_{C(\bar{\Omega})} \qquad \forall (u; v; \sigma; \kappa) \in Y. \tag{4.36}$$

Let us prove subsequently that there exists $P > 0$ such that

$$p(\tilde{S}(t)y) \leq Pp(y) \qquad \forall (t; y) \in [0, T] \times Y. \tag{4.37}$$

In order to prove (4.37) let us denote by $q: H_0^1(\Omega) \times H_1 \to \mathbb{R}$ the functional defined by

$$q(v; \sigma) = \|v\|_{C(\bar{\Omega})} + \|\sigma\|_{C(\bar{\Omega})} \tag{4.38}$$

and let us prove first that there exists $Q > 0$ such that

$$q(\tilde{T}(t)(v_0; \sigma_0)) \leq Qq(v_0; \sigma_0) \qquad \forall (v_0; \sigma_0) \in H_0^1(\Omega) \times H_1, \quad t \in [0, T]. \tag{4.39}$$

If $(v_0; \sigma_0) \in H_0^1(\Omega) \times H_1$, then we can put the hyperbolic system $(\dot{v}(t); \dot{\sigma}(t)) = B(v(t), \sigma(t))$, $(v(0); \sigma(0)) = (v_0; \sigma_0)$ in the characteristic form to get

$$v(t, x) = \tfrac{1}{2}[v_0(x - ct) + v_0(x + ct)] + \frac{1}{2\sqrt{(\rho E)}}[\sigma_0(x + ct) - \sigma_0(x - ct)] \tag{4.40}$$

$$\sigma(t, x) = \tfrac{1}{2}[\sigma_0(x - ct) + \sigma_0(x + ct)] + \frac{\sqrt{(\rho E)}}{2}[v_0(x + ct) - v_0(x - ct)] \tag{4.41}$$

for all $(t; x)$ from the triangle $\{(t; x) \mid 0 \leq t \leq c^{-1} \min(x; 1 - x)\}$, where $c = \sqrt{(E/\rho)}$. A similar formula can be obtained in the triangles

$$\{(t; x) | c^{-1}x \leq t \leq c^{-1}/2 \quad \text{or} \quad c^{-1}(1 - x) \leq t \leq c^{-1}/2\}.$$

Hence we deduce that there exists $Q_0 > 0$ such that $q(\tilde{T}(t)(v_0; \sigma_0)) \leq Q_0 q((v_0; \sigma_0))$ for all $t \in [0, c^{-1}/2]$. Let $Q = Q_0^n$, where $n = [2Tc] + 1$; then (4.39) follows.

Let $u_0 \in H^2(\Omega) \cap H_0^1(\Omega)$ and $u(t) = u_0 + \int_0^t \tilde{T}_1(s)(v_0; \sigma_0) ds$. If we bear in mind that $u_x(t) = u_{0x} + E^{-1}(\sigma(t) - \sigma_0)$, then from (4.39) we obtain

$$\|u_x(t)\|_{C(\bar\Omega)} \le \|u_{0x}\|_{C(\bar\Omega)} + E^{-1}(Q+1)\cdot(\|v_0\|_{C(\bar\Omega)} + \|\sigma_0\|_{C(\bar\Omega)}),$$

and (4.37) easily follows.

Let us consider f: $[0, T] \to X$ and F: $[0, T] \times Y \to X$ given by

$$f(t) = (0; b(t); 0; 0) \tag{4.42}$$

$$F(t, (u; v; \sigma; \kappa)) = (0; 0; G(t, \sigma, u_x, \kappa); \varphi(t, \sigma, u_x, \kappa)) \tag{4.43}$$

for all $t \in [0, T]$, $(u; v; \sigma; \kappa) \in Y$. Let $\dot u = v$, $y_0 = (u_0; v_0; \sigma_0; \kappa_0)$, $y(t) = (u(t); v(t); \sigma(t); \kappa(t))$; then we note that (4.15)–(4.19) is equivalent to the following Cauchy problem:

$$\dot y(t) = Ay(t) + F(t, y(t)) + f(t) \tag{4.44}$$

$$y(0) = y_0. \tag{4.45}$$

In order to prove a local existence result for (4.44), (4.45) the following lemmas will be useful:

Lemma 4.1 *Let F be given by* (4.43). *Then for all $t \in [0, T]$, $y \in Y$ we have $F(t, y) \in Y$ and the function $t \to F(t, y)$ is continuous from $[0, T]$ to X. Moreover, there exists $\bar C > 0$ such that for all $t \in [0, T]$, $r \ge 0$, $y, \bar y \in Y$ with $p(y), p(\bar y) \le r$, the following inequalities hold*:

$$\|F(t, y)\|_Y \le \bar C L(r) \|y\|_Y \tag{4.46}$$

$$p(F(t, y)) \le \bar C L(r) \tag{4.47}$$

$$\|F(t, y) - F(t, \bar y)\|_X \le \bar C L(r) \|y - \bar y\|_X \tag{4.48}$$

PROOF. Let $y = (u; v; \sigma; \kappa)$, $\bar y = (\bar u; \bar v; \bar\sigma; \bar\kappa) \in Y$ such that $p(y)$, $p(\bar y) \le r$. From (4.21) we deduce that $G \in W^{1,\infty}_{\mathrm{loc}}(\mathbb{R}^3)$ which implies

$$\frac{\mathrm{d}}{\mathrm{d}x} G(t, \sigma, u_x, \kappa) = G_\sigma(t, \sigma, u_x, \kappa)\sigma_x + G_\varepsilon(t, \sigma, u_x, \kappa)u_{xx} + G_\kappa(t, \sigma, u_x, \kappa)\kappa_x$$

belongs to H and

$$\left\| \frac{\mathrm{d}}{\mathrm{d}x} G(t, \sigma, u_x, \kappa) \right\|_H^2 \le 3L^2(r)(\|\sigma_x\|_H^2 + \|u_{xx}\|_H^2 + \|\kappa_x\|_H^2). \tag{4.49}$$

Using (4.21) and (4.23) we can easily deduce that

$$\|G(t, \sigma, u_x, \kappa)\|_H^2 \le 3L^2(r)(\|\sigma\|_H^2 + \|u_x\|_H^2 + \|\kappa\|_H^2) \tag{4.50}$$

$$\|G(t, \sigma, u_x, \kappa)\|_{C(\bar\Omega)} \le L(r)(\|\sigma\|_{C(\bar\Omega)} + \|u_x\|_{C(\bar\Omega)} + \|\kappa\|_{C(\bar\Omega)}). \tag{4.51}$$

Inequalities similar to (4.49)–(4.51) for φ can be obtained and (4.46)–(4.47) will follow. Inequality (4.48) is a direct consequence of (4.21)–(4.22). Since G and φ are uniformly continuous on the compact set

$$\{(t; \sigma; \varepsilon; \kappa) \in \mathbb{R}^4 \,|\, t \in [0, T], \sigma^2 + \varepsilon^2 + \kappa^2 \leq r^2\},$$

we deduce that $t \to F(t, y)$ is continuous from $[0, T]$ to X.

If we denote by

$$h(t) = \int_0^t S(t-s) f(s)\, \mathrm{d}s \qquad \forall t \in [0, T] \tag{4.52}$$

with an arbitrary $f \in C^1([0, T], X)$, then we have $h \in C^0([0, T], [D(A)])$ and since $Y \subsetneqq [D(A)]$, h does not belong to $C^0([0, T], Y)$. But if f has the particular form given by (4.42) we have the following result:

Lemma 4.2 *If* (4.24) *holds then h given by* (4.52), (4.42) *belongs to*

$$C^1([0, T], X) \cap C^0([0, T], Y).$$

PROOF. If we compute $h(t)$ from (4.34) we obtain

$$h(t) = \left(\int_0^t \int_0^{t-s} T_1(\tau)(b(s); 0)\, \mathrm{d}\tau\, \mathrm{d}s; \int_0^t T(t-s)(b(s); 0)\, \mathrm{d}s; 0 \right) \qquad \forall t\, [0, T]. \tag{4.53}$$

Since $(b; 0) \in C^1([0, T], H \times H)$ we deduce that (see Lemma A3.5)

$$B\left(\int_0^t T(t-s)(b(s); 0)\, \mathrm{d}s \right) = T(t)(b(0); 0) - (b(t); 0) + \int_0^t T(t-s)(\dot{b}(s); 0)\, \mathrm{d}s \qquad \forall t\, [0, T] \tag{4.54}$$

and $t \to \int_0^t T(t-s)(b(s); 0)\, \mathrm{d}s$ belongs to $C^0([0, T], [D(B)])$. In order to prove that

$$t \to \int_0^t \int_0^{t-s} T_1(\tau)(b(s); 0)\, \mathrm{d}\tau\, \mathrm{d}s$$

belongs to $C^0([0, T], H^2(\Omega) \cap H_0^1(\Omega))$ note that $(t; s) \to \int_0^{t-s} T(\tau)(b(s); 0)\, \mathrm{d}\tau$ belongs to $C^0([0, T] \times [0, T], [D(B)])$ and

$$B\left(\int_0^{t-s} T(\tau)(b(s); 0)\, \mathrm{d}\tau \right) = T(t-s)(b(s); 0) - (b(s); 0).$$

Hence

$$(t; s) \to \int_0^{t-s} T_1(\tau)(b(s); 0)\, \mathrm{d}\tau \qquad \text{belongs to } C^0([0, T] \times [0, T], H_0^1(\Omega))$$

and we have

$$E\frac{\mathrm{d}}{\mathrm{d}x}\left(\int_0^{t-s} T_1(\tau)(b(s);0)\,\mathrm{d}\tau\right) = T_2(t-s)(b(s);0). \tag{4.55}$$

From (4.54) we obtain

$$\rho^{-1}\frac{\mathrm{d}}{\mathrm{d}x}\int_0^t T_2(t-s)(b(s);0)\,\mathrm{d}s = T_1(t)(b(0);0) - b(t) + \int_0^t T_1(t-s)(\dot{b}(s);0)\,\mathrm{d}s \tag{4.56}$$

If we use now (4.55), (4.56) we can easily deduce that

$$E\rho^{-1}\frac{\mathrm{d}^2}{\mathrm{d}x^2}\left(\int_0^t\int_0^{t-s} T_1(\tau)(b(s);0\,\mathrm{d}\tau\,\mathrm{d}s\right)$$

$$= T_1(t)(b(0);0) - b(t) + \int_0^t T_1(t-s)(\dot{b}(s);0)\,\mathrm{d}s.$$

Hence

$$t \to \int_0^t\int_0^{t-s} T_1(\tau)(b(s);0)\,\mathrm{d}\tau\,\mathrm{d}s \qquad \text{belongs to } C([0,T], H^2(\Omega)\cap H_0^1(\Omega)).$$

We can use now Lemmas 4.1–4.2 and Theorem A3.10 in order to obtain the statement of Theorem 4.3.

5 Other functional methods in the study of dynamic problems

In this section we deal with dynamic problems for rate-type materials in the case when the plastic rate of deformation does not depend on the deformation tensor. In this particular case some of the existence and uniqueness results proved in Sections 1 and 4 can also be obtained by monotony methods or by fixed-point methods. So, we investigate the results stated in Theorems 1.1, 1.2, 4.1, and 4.2 with a technique which avoids any semigroup-theory arguments.

5.1 Monotony methods

Under the same notations as in Section 1, we consider the following mixed problem:

Find the displacement function u: $[0, T] \times \Omega \to \mathbb{R}^N$ and the stress function σ: $[0, T] \times \Omega \to \mathcal{S}_N$ such that

$$\rho\ddot{u}(t) = \operatorname{Div}\sigma(t) + \rho b(t) \tag{5.1}$$

$$\tilde{\mathscr{A}}\dot{\sigma}(t) + \tilde{G}(t,\sigma(t)) = \varepsilon(\dot{u}(t)) \qquad \text{in } \Omega \tag{5.2}$$

$$u(t) = g(t) \quad \text{on } \Gamma_1, \quad \sigma(t)\nu = f(t) \quad \text{on } \Gamma_2, \qquad t \in (0, T] \tag{5.3}$$

$$u(0) = u_0, \qquad \dot{u}(0) = v_0, \qquad \sigma(0) = \sigma_0. \tag{5.4}$$

The problem (5.1)–(5.4) is a particular form of (1.1)–(1.4) in which (1.2) has been replaced by the constitutive law (5.2) in which $\tilde{\mathscr{A}}$ is the inverse of a tensor $\mathscr{A}$. In order to study problem (5.1)–(5.4) we consider the following assumptions on the constitutive function $\tilde{G}: \Omega \times [0, T] \times \mathscr{S}_N \to \mathscr{S}_N$:

(a) $(\tilde{G}(x, t, \sigma_1) - \tilde{G}(x, t, \sigma_2))\cdot(\sigma_1 - \sigma_2) \geq 0$ for all $\sigma_1, \sigma_2 \in \mathscr{S}_N$, $t \in [0, T]$, a.e. in Ω;
(b) there exists $L > 0$ such that

$$|\tilde{G}(x, t_1, \sigma_1) - \tilde{G}(x, t_2, \sigma_2)| \leq L(|t_1 - t_2| + |\sigma_1 - \sigma_2|)$$

for all $\sigma_1, \sigma_2 \in \mathscr{S}_N$, $t_1, t_2 \in [0, T]$, a.e. in Ω;
(c) $x \to \tilde{G}(x, t, \sigma)$ is a measurable function with respect to the Lebesgue measure on Ω, for all $t \in [0, T]$ and $\sigma \in \mathscr{S}_N$;
(d) $x \to \tilde{G}(x, t, 0_N) \in \mathscr{H}$ for all $t \in [0, T]$. (5.5)

The main results of this section are the following:

Theorem 5.1 *Let* (1.5), (1.7)–(1.12), (5.5) *hold. Then there exists a unique solution of the problem* (5.1)–(5.4) *having the regularity* (1.14), (1.15).

Theorem 5.2 *Let* (1.5), (1.7), (5.5) *hold and let* $(u_i; \sigma_i)$ be the solutions of (5.1)–(5.4) for the data $b_i, g_i, f_i, u_{0i}, v_{0i}, \sigma_{0i}, i = 1, 2$ *which satisfy* (1.8)–(1.10), (1.13). *Then there exists* $C(T) > 0$ *which depends only on* Ω, Γ_1, $\mathscr{A}$, ρ, G, *and* T *such that* (1.31) *holds.*

In order to prove Theorem 5.1 we need some preliminary results. Let us homogenize the boundary conditions (5.3) with the notation

$$u^* = u - \tilde{u}, \qquad v^* = \dot{u} - \dot{\tilde{u}}, \qquad \sigma^* = \sigma - \tilde{\sigma} \tag{5.6}$$

and

$$u_0^* = u_0 - \tilde{u}(0), \qquad v_0^* = v_0 - \dot{\tilde{u}}(0), \qquad \sigma_0^* = \sigma_0 - \tilde{\sigma}(0). \tag{5.7}$$

From (1.10) we notice that $u_0^*, v_0^* \in \boldsymbol{V}_1$, $\sigma_0^* \in \mathscr{V}_1$, hence we can easily deduce the following lemma:

Lemma 5.1 *The pair* $(u; \sigma)$ *is a solution of* (5.1)–(5.4) *such that* (1.14), (1.15) *hold iff* $u^* \in W^{1,\infty}(0, T, \boldsymbol{V}_1)$, $v^* \in W^{1,\infty}(0, T, \boldsymbol{H}) \cap L^\infty(0, T, \boldsymbol{V}_1)$,

$$\sigma^* \in W^{1,\infty}(0, T, \mathscr{H}) \cap L^\infty(0, T, \mathscr{V}_1)$$

and

$$\dot{u}^*(t) = v^*(t) \tag{5.8}$$

$$\dot{v}^*(t) = \rho^{-1} \operatorname{Div} \sigma^*(t) + a(t) \tag{5.9}$$

$$\dot{\sigma}^*(t) + \mathscr{A}\tilde{G}(t, \sigma^*(t) + \tilde{\sigma}(t)) = \mathscr{A}\varepsilon(v^*(t)) + H(t) \quad \text{a.e. on } (0, T) \quad (5.10)$$

$$u^*(0) = u_0^*, \qquad v^*(0) = v_0^*, \qquad \sigma^*(0) = \sigma_0^* \quad (5.11)$$

where a: $[0, T] \to \boldsymbol{H}$ and H: $[0, T] \to \mathscr{H}$ are given by

$$a(t) = b(t) - \ddot{\tilde{u}}(t) + \rho^{-1} \operatorname{Div} \tilde{\sigma}(t) \quad (5.12)$$

$$H(t) = \mathscr{A}\varepsilon(\dot{\tilde{u}}(t)) - \dot{\tilde{\sigma}}(t) \quad (5.13)$$

for all $t \in [0, T]$.

Let us consider now the Hilbert space $X = \boldsymbol{H} \times \mathscr{H}$ with the inner product

$$\langle (v_1; \tau_1), (v_2; \tau_2) \rangle_X = \langle \rho v_1, v_2 \rangle_{\boldsymbol{H}} + \langle \tilde{\mathscr{A}} \tau_1, \tau_2 \rangle_{\mathscr{H}} \quad (5.14)$$

which generates an equivalent norm on X (see (1.5), (1.7)). Let $D(B) = \boldsymbol{V}_1 \times \mathscr{V}_1 \subset X$ and B: $D(B) \subset X \to X$ be given by

$$B(v; \sigma) = (\rho^{-1} \operatorname{Div} \sigma; \mathscr{A}\varepsilon(v)) \qquad \forall v \in \boldsymbol{V}_1, \sigma \in \mathscr{V}_1. \quad (5.15)$$

We also consider the operator $\mathscr{G}$: $[0, T] \times X \to X$ and the function h: $[0, T] \to X$ defined by

$$\mathscr{G}(t, (v; \tau)) = (0; \mathscr{A}\tilde{G}(t, \tau + \tilde{\sigma}(t))) \quad (5.16)$$

$$h(t) = (a(t); H(t)) \quad (5.17)$$

for all $t \in [0, T]$ and $(v; \tau) \in X$.

The following lemma can easily be proved:

Lemma 5.2 *The functions* $(u^*; v^* \sigma^*)$ *have the regularity defined in Lemma* 5.1 *and satisfy* (5.8)–(5.13) *iff* $x = (v^*; \sigma^*) \in W^{1,\infty}(0, T, X) \cap L^\infty(0, T, [D(B)])$ *and*

$$u^*(t) = \int_0^t v(s)\, \mathrm{d}s + u_0^* \qquad \forall t \in [0, T] \quad (5.18)$$

$$\dot{x}(t) + \mathscr{G}(t, x(t)) = Bx(t) + h(t) \quad \textit{a.e. on } (0, T) \quad (5.19)$$

$$x(0) = x_0 \quad (5.20)$$

where $x_0 = (v_0^*; \sigma_0^*)$.

PROOF OF THEOREM 5.1. The operators $\mathscr{G}$ and B defined by (5.16) and (5.15) satisfy (A3.68)–(A3.70); indeed (5.5a) implies (A3.68), (1.5), (5.5b) imply (A3.69) and (A3.70) was proved in Lemma 1.3. Moreover, from (1.7), (1.8), (1.11) and (1.12) we get $a \in W^{1,1}(0, T, \boldsymbol{H})$ and by (1.5) we get $H \in W^{1,1}(0, T, \mathscr{H})$. Hence, by (5.17) we get $h \in W^{1,1}(0, T, X)$. Since (1.9),

(1.10), and (5.7) imply $x_0^* = (v_0^*; \sigma_0^*) \in V_1 \times \mathscr{V}_1 = D(B)$, we can apply Lemma A3.9. Hence follow the existence and the uniqueness of the solution $x = (v^*; \sigma^*)$ for the problem (5.19), (5.20) such that $x(t) \in D(B)$ for all $t \in [0, T]$ and $x \in W^{1,\infty}(0, T, X)$; since from (5.19) we get $Bx \in L^\infty(0, T, X)$, using (5.15), (1.5), and (1.7) we deduce $\operatorname{Div} \sigma \in L^\infty(0, T, \boldsymbol{H})$ and $\varepsilon(v) \in L^\infty(0, T, \mathscr{H})$. It follows that $v^* \in L^\infty(0, T, V_1)$ and $\sigma^* \in L^\infty(0, T, \mathscr{V}_1)$. This means that $x^* \in L^\infty(0, T, [D(B)])$. Theorem 5.1 follows now from Lemmas 5.1 and 5.2.

PROOF OF THEOREM 5.2. From the proof of Theorem 5.1 we have

$$u_i = u_i^* + \tilde{u}_i, \quad \dot{u}_i = v_i^* + \dot{\tilde{u}}_i, \quad \sigma_i = \sigma_i^* + \tilde{\sigma}_i, \qquad i = 1, 2 \tag{5.21}$$

where $\tilde{u}_i$, $\tilde{\sigma}_i$ are given in Remark 1.1 for the data g_i and f_i and $(u_i; v_i; \sigma_i)$ satisfies

$$u_i^*(t) = \int_0^t v_i^*(s)\,\mathrm{d}s + u_{0i}^* \qquad \forall t \in [0, T] \tag{5.22}$$

$$\dot{x}_i(t) + \mathscr{G}_i(t, x_i(t)) = Bx_i(t) + h_i(t) \qquad \text{a.e. on } (0, T) \tag{5.23}$$

$$x_i(0) = x_{0i} \tag{5.24}$$

where

$$x_i = (v_i^*; \sigma_i^*), \qquad x_{0i} = (v_{0i}^*; \sigma_{0i}^*) \tag{5.25}$$

$$u_{0i}^* = u_{0i} - \tilde{u}_i(0), \qquad v_{0i}^* = v_{0i} - \dot{\tilde{u}}_i(0), \qquad \sigma_{0i}^* = \sigma_{0i} - \tilde{\sigma}_i(0) \tag{5.26}$$

and $\mathscr{G}_i$, h_i are defined by (5.16), (5.17), (5.12), (5.13) in which b, $\tilde{u}$, and $\tilde{\sigma}$ are replaced by b_i, $\tilde{u}_i$, and $\tilde{\sigma}_i$, $i = 1, 2$.

Using (5.19) and (A3.73) we get

$$\langle \dot{x}_i(t) - \dot{x}_2(t), x_1(t) - x_2(t)\rangle_X + \langle \mathscr{G}_1(t, x_1(t)) - \mathscr{G}_2(t, x_2(t)), x_1(t) - x_2(t)\rangle_X$$
$$= \langle h_1(t) - h_2(t), x_1(t) - x_2(t)\rangle_X \qquad \text{a.e. on } (0, T).$$

By integration and using (5.24), (5.16), (1.5), and (5.5b) we have

$$\begin{aligned}
&\tfrac{1}{2}\|x_1(s) - x_2(s)\|_X^2 \\
&\quad = \tfrac{1}{2}\|x_{01} - x_{02}\|_X^2 - \int_0^s \langle \mathscr{G}_1(t, x_1(t)) - \mathscr{G}_2(t, x_2(t)), x_1(t) - x_2(t)\rangle_X\,\mathrm{d}t \\
&\qquad + \int_0^s \langle h_1(t) - h_2(t), x_1(t) - x_2(t)\rangle_X\,\mathrm{d}t \\
&\quad \le \tfrac{1}{2}\|x_{01} - x_{02}\|_X^2 \\
&\qquad + C\bigg(\int_0^s (\|\tilde{\sigma}_1(t) - \tilde{\sigma}_2(t)\|_{\mathscr{H}} + \|h_1(t) - h_2(t)\|_X)\|x_1(t) - x_2(t)\|_X\,\mathrm{d}t \\
&\qquad + \int_0^s \|x_1(t) - x_2(t)\|_X^2\,\mathrm{d}t\bigg)
\end{aligned}$$

for all s [0, T]. From Lemma A4.13 we get

$$\|x_1(s) - x_2(s)\|_X \leq C\left(\|x_{01} - x_{02}\|_X + \int_0^s \|\tilde{\sigma}_1(t) - \tilde{\sigma}_2(t)\|_{\mathscr{H}}\, dt + \int_0^s \|h_1(t) - h_2(t)\|_X\, dt \right) \qquad \forall s \in [0, T].$$

It follows that

$$\|x_1 - x_2\|_{0,T,X} \leq C(\|x_{01} - x_{02}\|_X + \|\tilde{\sigma}_1 - \tilde{\sigma}_2\|_{0,T,\mathscr{H}} + \|h_1 - h_2\|_{0,T,X}). \tag{5.27}$$

From (5.17), (5.12), (5.13), (1.5), and (1.7) we deduce

$$\|h_1 - h_2\|_{0,T,X} \leq C(\|b_1 - b_2\|_{0,T,\boldsymbol{H}} + \|\ddot{\tilde{u}}_1 - \ddot{\tilde{u}}_2\|_{0,T,\boldsymbol{H}} + \|\varepsilon(\dot{\tilde{u}}_1) - \varepsilon(\dot{\tilde{u}}_2)\|_{0,T,\mathscr{H}} + \|\mathrm{Div}\, \tilde{\sigma}_1 - \mathrm{Div}\, \tilde{\sigma}_2\|_{0,T,\boldsymbol{H}} + \|\dot{\tilde{\sigma}}_1 - \dot{\tilde{\sigma}}_2\|_{0,T,\mathscr{H}}$$

and using (1.32), (1.34), (1.35), (1.37) we get

$$\|h_1 - h_2\|_{0,T,X} \leq C(\|b_1 - b_2\|_{0,T,\boldsymbol{H}} + \|g_1 - g_2\|_{2,T,\boldsymbol{H}_\Gamma} + \|f_1 - f_2\|_{1,T,V'_{\Gamma 1}}). \tag{5.28}$$

From (5.25), (5.27), and (5.28) after some algebra we get

$$\|v_1^* - v_2^*\|_{0,T,\boldsymbol{H}} + \|\sigma_1^* - \sigma_2^*\|_{0,T,\mathscr{H}} \leq C(\|v_{01} - v_{02}\|_{\boldsymbol{H}} + \|\sigma_{01} - \sigma_{02}\|_{\mathscr{H}} + \|b_1 - b_2\|_{0,T,\boldsymbol{H}} + \|g_1 - g_2\|_{2,T,\boldsymbol{H}_\Gamma} + \|f_1 - f_2\|_{1,T,V'_{\Gamma 1}})$$

and using (5.21), (5.22), and (5.26) we get (1.31).

5.2 A fixed point method

Under the same notations as in Section 4 we consider the following dynamic problem for rate-type materials involving a strain hardening parameter:

Find the displacement function u: $[0, T] \times \Omega \to \mathbb{R}^N$, the stress function σ: $[0, T] \times \Omega \to \mathscr{S}_N$ and the hardening parameter κ: $[0, T] \times \Omega \to \mathbb{R}$ such that

$$\rho\ddot{u}(t) = \mathrm{Div}\, \sigma(t) + \rho b(t) \tag{5.29}$$

$$\tilde{\mathscr{A}}\dot{\sigma}(t) + \tilde{G}(t, \sigma(t), \kappa(t)) = \varepsilon(\dot{u}(t)) \tag{5.30}$$

$$\dot{\kappa}(t) = \varphi(t, \sigma(t), \kappa(t)) \qquad \text{in } \Omega \tag{5.31}$$

$$u(t) = g(t) \quad \text{on } \Gamma_1, \quad \sigma(t)\nu = f(t) \quad \text{on } \Gamma_2 \qquad \text{for } t \in (0, T] \tag{5.32}$$

$$u(0) = u_0, \quad \dot{u}(0) = v_0, \quad \sigma(0) = \sigma_0, \quad \kappa(0) = \kappa_0 \qquad \text{in } \Omega. \tag{5.33}$$

The problem (5.29)–(5.33) is a particular form of (4.1)–(4.5) in which (4.2) was replaced by the constitutive law (5.30) in which $\tilde{\mathscr{A}}$ is the inverse of a tensor $\mathscr{A}$. For simplicity we supposed that φ in (5.31) does not depend on the deformation tensor.

In order to study problem (5.29)–(5.33) we consider the following assumptions on the constitutive functions $\tilde{G}: \Omega \times [0, T] \times \mathscr{S}_N \times \mathbb{R} \to \mathscr{S}_N$ and $\varphi: \Omega \times [0, T] \times \mathscr{S}_N \times \mathbb{R} \to \mathbb{R}_N$:

$$\left.\begin{array}{ll}
\text{(a)} & (\tilde{G}(x, t, \sigma_1, \kappa) - \tilde{G}(x, t, \sigma_2, \kappa))\cdot(\sigma_1 - \sigma_2) \geq 0 \text{ for all } t \in [0, T], \\
& \sigma_1, \sigma_2 \in \mathscr{S}_N,\ \kappa \in \mathbb{R}, \text{ a.e. in } \Omega. \\
\text{(b)} & \text{there exists } L > 0 \text{ such that } |\tilde{G}(x, t_1, \sigma_1, \kappa_1) - \tilde{G}(x, t_2, \sigma_2, \kappa_2)| \\
& \leq L(|t_1 - t_2| + |\sigma_1 - \sigma_2| + |\kappa_1 - \kappa_2|) \text{ for all } t_1, t_2 \in [0, T], \\
& \sigma_1, \sigma_2 \in \mathscr{S}_N,\ \kappa_1, \kappa_2 \in \mathbb{R}, \text{ a.e. in } \Omega. \\
\text{(c)} & x \to \tilde{G}(x, t, \sigma, \kappa) \text{ is a measurable function with respect to the} \\
& \text{Lebesgue measure on } \Omega, \text{ for all } t \in [0, T],\ \sigma \in \mathscr{S}_N \text{ and } \kappa \in \mathbb{R}. \\
\text{(d)} & x \to \tilde{G}(x, t, 0_N, 0) \in \mathscr{H} \text{ for all } t \in [0, T].
\end{array}\right\} \quad (5.34)$$

$$\left.\begin{array}{ll}
\text{(a)} & \text{there exists } \tilde{L} > 0 \text{ such that } |\varphi(x_1, t_1, \sigma_1, \kappa_1) - (x, t_2, \sigma_2, \kappa_2)| \\
& \leq \tilde{L}(|t_1 - t_2| + |\sigma_1 - \sigma_2| + |\kappa_1 - \kappa_2| \text{ for all } t_1, t_2 \in [0, T], \\
& \sigma_1, \sigma_2 \in \mathscr{S}_N,\ \kappa_1, \kappa_2 \in \mathbb{R}, \text{ a.e. in } \Omega. \\
\text{(b)} & x \to \varphi(x, t, \sigma, \kappa) \text{ is a measurable function with respect to the} \\
& \text{Lebesgue measures on } \Omega, \text{ for all } t \in [0, T],\ \sigma \in \mathscr{S}_N \text{ and } \kappa \in \mathbb{R}. \\
\text{(c)} & x \to \varphi(x, t, 0_N, 0) \in H \text{ for all } t \in [0, T].
\end{array}\right\} \quad (5.35)$$

In this section we use monotony arguments and a fixed-point technique in order to prove the following result:

Theorem 5.3 *Let* (1.5), (5.34), (5.35), (1.7)–(1.12), (4.8) *hold. Then there exists a unique solution of* (5.29)–(5.33) *which satisfies* (1.14), (1.15), *and* $\kappa \in W^{2,\infty}(0, T, H)$.

PROOF. *The uniqueness part.* Let $(u_1; \sigma_1; \kappa_1)$ and $(u_2; \sigma_2; \kappa_2)$ be two solutions of the problem (5.29)–(5.33). Let $u = u_1 - u_2$, $v_1 = \dot{u}_1$, $v_2 = \dot{u}_2$, $v = v_1 - v_2$, $\sigma = \sigma_1 - \sigma_2$, $\kappa = \kappa_1 - \kappa_2$; we get

$$\left.\begin{array}{l}
\dot{u}(t) = v(t) \\
\rho\dot{v}(t) = \operatorname{Div} \sigma(t) \\
\tilde{\mathscr{A}}\dot{\sigma}(t) + \tilde{G}(t, \sigma_1(t), \kappa_1(t)) - \tilde{G}(t, \sigma_2, \kappa_2(t)) = \varepsilon(v(t)) \\
\dot{\kappa}(t) = \varphi(t, \sigma_1(t), \kappa_1(t)) - \varphi(t, \sigma_2(t), \kappa_2(t)) \qquad \text{in } \Omega \\
u(t) = 0 \quad \text{on } \Gamma_1, \quad \sigma(t)\nu = 0 \quad \text{on } \Gamma_2 \qquad \text{for } t \in (0, T] \\
u(0) = 0, \quad v(0) = 0, \quad \sigma(0) = 0, \quad \kappa(0) = 0 \qquad \text{in } \Omega
\end{array}\right\} \quad (5.36)$$

Using (5.36) and (2.58) in Chapter 2 we get

$$\begin{aligned}
&\langle \rho\dot{v}(t), v(t)\rangle_H + \langle \tilde{\mathscr{A}}\dot{\sigma}(t), \sigma(t)\rangle_{\mathscr{H}} + \langle \dot{\kappa}(t), \kappa(t)\rangle_H \\
&\qquad + \langle \tilde{G}(t, \sigma_1(t), \kappa_1(t)) - \tilde{G}(t, \sigma_2(t), \kappa_2(t)), \sigma(t)\rangle_{\mathscr{H}} \\
&\quad = \langle \varphi(t, \sigma_1(t), \kappa_1(t)) - \varphi(t, \sigma_2(t), \kappa_2(t)), \kappa(t)\rangle_H \qquad \text{a.e. on } (0, T).
\end{aligned}$$

Using (5.34b) and (5.35a) we get

$$\langle \rho\dot{v}(t), v(t)\rangle_H + \langle \tilde{\mathscr{A}}\dot{\sigma}(t), \sigma(t)\rangle_{\mathscr{H}} + \langle \dot{\kappa}(t), \kappa(t)\rangle_H$$
$$\leq C(\|\sigma(t)\|_{\mathscr{H}} + \|\kappa(t)\|_H)^2$$
$$\leq C(\|\sigma(t)\|^2_{\mathscr{H}} + \|\kappa(t)\|^2_H) \qquad \text{a.e. on } (0, T).$$

By integration and using the initial conditions in (5.36) and (1.5), (1.7) we obtain

$$\|v(s)\|^2_H + \|\sigma(s)\|^2_{\mathscr{H}} + \|\kappa(s)\|^2_H \leq C \int_0^s (\|\sigma(t)\|^2_{\mathscr{H}} + \|\kappa(t)\|^2_H)\,dt$$

for all $s \in [0, T]$. Applying now Corollary A4.1 we get $v = 0$, $\sigma = 0$, $\kappa = 0$ and from (5.36) it follows that $u = 0$.

The existence part. Let $\eta \in L^\infty(0, T, H)$ and κ_η be the function defined by

$$\kappa_\eta(t) = \int_0^t \eta(s)\,ds + \kappa_0 \tag{5.37}$$

for all $t \in [0, T]$. From (4.8) we get

$$\kappa_\eta \in W^{1,\infty}(0, T, H). \tag{5.38}$$

Let $\tilde{G}\colon \Omega \times [0, T] \times \mathscr{S}_N \to \mathscr{S}_N$ be the function defined by

$$G(x, t, \sigma) = \tilde{G}(x, t, \sigma, \kappa_\eta(t, x)) \qquad \forall t \in [0, T], \quad \sigma \in \mathscr{S}_N, \quad \text{a.e. in } \Omega. \tag{5.39}$$

(5.34) and (5.38) imply (5.5) for $\tilde{G}$; hence by Theorem 5.1 we get the existence and uniqueness of a pair $(u_\eta; \sigma_\eta)$ such that

$$u_\eta \in W^{2,\infty}(0, T, \boldsymbol{H}) \cap W^{1,\infty}(0, T, \boldsymbol{H}_1) \tag{5.40}$$

$$\sigma_\eta \in W^{1,\infty}(0, T, \mathscr{H}) \cap L^\infty(0, T, \mathscr{H}_2). \tag{5.41}$$

$$\left.\begin{aligned} &\rho\ddot{u}_\eta(t) = \operatorname{Div}\sigma_\eta(t) + \rho b(t) \\ &\tilde{\mathscr{A}}\dot{\sigma}_\eta(t) + \tilde{G}(t, \sigma_\eta(t), \kappa_\eta(t)) = \varepsilon(\dot{u}_\eta(t)) \qquad \text{in } \Omega \\ &u_\eta(t) = g(t) \quad \text{on } \Gamma_1, \quad \sigma_\eta(t)\lambda = f(t) \quad \text{on } \Gamma_2, \qquad t \in (0, T] \\ &u_\eta(0) = u_0, \quad \dot{u}_\eta(0) = v_0, \quad \sigma_\eta(0) = \sigma_0 \qquad \text{in } \Omega. \end{aligned}\right\} \tag{5.42}$$

Using (5.35), (5.38), and (5.41) we can define the operator $\Lambda\colon L^\infty(0, T, H) \to W^{1,\infty}(0, T, H)$ as follows

$$\Lambda\eta(t) = \varphi(t, \sigma_\eta(t), \kappa_\eta(t)) \tag{5.43}$$

for all $t \in [0, T]$ and $\eta \in L^\infty(0, T, H)$. We shall prove that Λ has a unique fixed point. Indeed, let $\eta_1, \eta_2 \in L^\infty(0, T, H)$, $\kappa_1 = \kappa_{\eta_1}$, $\kappa_2 = \kappa_{\eta_2}$, $u_1 = u_{\eta_1}$, $u_2 = u_{\eta_2}$, $\sigma_1 = \sigma_{\eta_1}$, $\sigma_2 = \sigma_{\eta_2}$, the functions defined by (5.37), (5.40)–(5.42). For simplicity we introduce the following notation: $\eta_1 - \eta_2 = \eta$, $u_1 - u_2 = u$, $\dot{u}_1 = v_1$, $\dot{u}_2 = v_2$, $v_1 - v_2 = v$, $\sigma_1 - \sigma_2 = \sigma$, $\kappa_1 - \kappa_2 = \kappa$. From (5.42) we have that $(u; v; \sigma)$ satisfies the following problem:

$$\left.\begin{aligned}
&\dot{u}(t) = v(t)\\
&\rho\dot{v}(t) = \operatorname{Div}\sigma(t)\\
&\tilde{A}\dot{\sigma}(t) + \tilde{G}(t, \sigma_1(t), \kappa_1(t)) - \tilde{G}(t, \sigma_2(t), \kappa_2(t)) = \varepsilon(v(t)) \qquad \text{in } \Omega\\
&u(t) = 0 \quad \text{on } \Gamma_1, \quad \sigma(t)\nu = 0 \quad \text{on } \Gamma_2, \qquad t \in (0, T]\\
&u(0) = 0, \quad v(0) = 0, \quad \sigma(0) = 0 \qquad \text{in } \Omega.
\end{aligned}\right\} \tag{5.44}$$

Using (5.43) we have

$$\Lambda\eta_1(t) - \Lambda\eta_2(t) = \varphi(t, \sigma_1(t), \kappa_1(t)) - \varphi(t, \sigma_2(t), \kappa_2(t))$$

for all $t \in [0, T]$ and from (5.35a) we get

$$\|\Lambda\eta_1(t) - \Lambda\eta_2(t)\|_H \le C(\|\sigma(t)\|_{\mathscr{H}} + \|\kappa(t)\|_H) \qquad \forall t \in [0, T]. \tag{5.45}$$

In the same way, from (5.44) and (2.58) from Chapter 2 we get

$$\begin{aligned}
&\langle \rho\dot{v}(t), v(t)\rangle_H + \langle \tilde{\mathscr{A}}\dot{\sigma}(t), \sigma(t)\rangle_{\mathscr{H}}\\
&\qquad + \langle \tilde{G}(t, \sigma_1(t), \kappa_1(t) - \tilde{G}(t, \sigma_2(t), \kappa_2(t)), \sigma(t)\rangle_{\mathscr{H}} = 0 \quad \text{a.e. on } [0, T]
\end{aligned}$$

and using (5.34b) we get

$$\langle \rho\dot{v}(t), v(t)\rangle_H + \langle \tilde{\mathscr{A}}\dot{\sigma}(t), \sigma(t)\rangle_{\mathscr{H}} \le C(\|\sigma(t)\|_{\mathscr{H}} + \|\kappa(t)\|_H)\|\sigma(t)\|_{\mathscr{H}}$$

a.e. on $(0, T)$. By integration and using (1.5) it follows that

$$\tfrac{1}{2}\|\sigma(s)\|^2_{\mathscr{H}} \le C\left(\int_0^s \|\sigma(t)\|^2_{\mathscr{H}}\, dt + \int_0^s \|\kappa(t)\|_H \|\sigma(t)\|_{\mathscr{H}}\, dt\right)$$

for all $s \in [0, T]$ and from Lemma A4.13 we deduce

$$\|\sigma(s)\|_{\mathscr{H}} \le C \int_0^s \|\kappa(t)\|_H\, dt \qquad \forall s \in [0, T]. \tag{5.46}$$

Finally, let us notice that from (5.37) we get

$$\|\kappa(t)\|_H \le \int_0^t \|\eta(s)\|_H\, ds \qquad \forall t \in [0, T]. \qquad (5.47$$

Using now (5.45), (5.46), and (5.47) we deduce

$$\|\Lambda\eta_1(t) - \Lambda\eta_2(t)\|_H \le C\left\{\int_0^t \int_0^s \|\eta(r)\|_H\, dr\, ds + \int_0^t \|\eta(s)\|_H\, ds\right\} \qquad \forall t \in [0, T]. \tag{5.48}$$

For every $p \in \mathbb{N}$ we define by recurrence the operators Λ^p: $L^\infty(0, T, H) \to L^\infty(0, T, H)$ and I^p: $C([0, T], \mathbb{R}) \to C([0, T], \mathbb{R})$ as follows: $\Lambda^1 = \Lambda$, $\Lambda^p\psi = \Lambda(\Lambda^{p-1}\psi)$, $(I^1 z)(t) = \int_0^t z(s)\,\mathrm{d}s$, $I^p z = I(I^{p-1}z)$ for all $\psi \in L^\infty(0, T, H)$ and $z \in C([0, T], \mathbb{R})$. With these notations, (5.48) becomes

$$\|\Lambda\eta_1(t) - \Lambda\eta_2(t)\|_H \leq C\{(I^2\|\eta\|_H)(t) + (I^1\|\eta\|_H)(t) \qquad \forall t \in [0, T]$$

and since $I^p \cdot I^q = I^{p+q}$, using again (5.48) we deduce

$$\|\Lambda^p\eta_1(t) - \Lambda^p\eta_2(t)\|_H \leq C^p\left\{\sum_{k=0}^{p} C_p^k(I^{2p-k}\|\eta\|_H)(t)\right\} \tag{5.49}$$

for all $t \in [0, T]$ and $p \in \mathbb{N}$. But

$$\begin{aligned}(I^{2p-k}\|\eta\|_H)(t) &\leq (I^{2p-k}\|\eta\|_{0,T,H})(t)\\ &\leq (I^{2p-k}\|\eta\|_{0,T,H})(T)\\ &= \frac{T^{2p-k}}{(2p-k)!}\|\eta\|_{0,T,H}\end{aligned}$$

hence from (5.49) we get

$$\|\Lambda^p\eta_1 - \Lambda^p\eta_2\|_{0,T,H} \leq C^p\left\{\sum_{k=0}^{p} C_p^k\frac{T^{2p-k}}{(2p-k)!}\right\}\|\eta\|_{0,T,H} \tag{5.50}$$

for all $p \in \mathbb{N}$.

Since

$$\sum_{k=0}^{p} C_p^k\frac{T^{2p-k}}{(2p-k)!} = \sum_{k=0}^{p} C_p^k\frac{T^p}{p!}\frac{p!}{(2p-k)!}T^{p-k} \leq \frac{T^p}{p!}\sum_{k=0}^{p} C_p^k T^{p-k} = \frac{T^p(1+T)^p}{p!}$$

from (5.50) we obtain

$$\|\Lambda^p\eta_1 - \Lambda^p\eta_2\|_{0,T,H} \leq \frac{C^p}{p!}\|\eta\|_{0,T,H} \qquad \forall p \in \mathbb{N}. \tag{5.51}$$

But $\lim_p C^p/p! = 0$, hence (5.51) implies that for p large enough the operator Λ^p is a contraction in $L^\infty(0, T, H)$. Then, there exists a unique $\eta^* \in L^\infty(0, T, H)$ such that $\Lambda^p\eta^* = \eta^*$; it follows that $\Lambda(\Lambda^p\eta^*) = \Lambda\eta^*$ hence $\Lambda^p(\Lambda\eta^*) = \Lambda\eta^*$ which implies $\Lambda\eta^* = \eta^*$. Moreover, η^* is the unique fixed point of Λ.

Using now (5.37) and (5.43) we get

$$\kappa_{\eta^*}(t) = \varphi(t, \sigma_{\theta^*}(t), \kappa_{\eta^*}(t)) \qquad \text{a.e. on } (0, T). \tag{5.52}$$

From (5.42), (5.52), and (5.37) we get that $(u_{\eta^*}; \sigma_{\eta^*}; \kappa_{\eta^*})$ satisfies (5.29)–(5.33); the regularity (1.14), (1.15) results from (5.40) and (5.41) and finally $\kappa_{\eta^*} \in W^{2,\infty}(0, T, H)$ by (5.52) and (5.35a).

Remark 5.1 *The uniqueness part of Theorem* 5.3 *can be also proved by using the uniqueness of any fixed point for the operator* Λ.

Remark 5.2 *The same technique as in the proof of Theorem* 5.2 *can be used in order to obtain* (4.13) *for every two solutions of* (5.29)–(5.33) *for two sets of input data. This technique is based on a direct estimation for a problem with homogeneous boundary conditions similar to* (5.8)–(5.13) *and avoid formulas of the form* (1.43), (1.44) *based on arguments of semigroups theory.*

6 Perturbations of homogeneous simple shear and strain localization

Shear bands, which are narrow regions of intense plastic shearing, are commonly observed during high strain-rate deformation processes as high-speed machining, forming and penetration. They have been extensively studied, beginning with the early work of Zenner and Holloman (1944). Usually it is assumed that a homogeneous solution exists and the thermo-plastic instability of this homogeneous solution is regarded as the mechanism for the formation of these shear bands.

The aim of this paragraph is to prove the existence, uniqueness, and continuous dependence of the solution for a simple shearing problem and to analyse the evolution of the perturbation of the homogeneous solution in order to point out the mechanism of strain localization. This study was performed for a class of viscoplastic materials with thermal softening.

Our intention is to give a general analysis of the perturbation which can be applied to different particular forms of constitutive equations used in the literature. The link between the initial perturbation and the final one is also studied.

In Section 6.1 the problem is stated and in Section 6.2 the local existence and uniqueness of smooth solutions are proved. Moreover it is also proved that if a solution is not a global one then it will blow up at the end of its interval of existence. Theorem 6.2 of this section shows that the length of the interval of existence of a solution and the solution at a finite time depend continuously upon the initial data (i.e. finite time stability).

In Section 6.3 the homogeneous solution is introduced and some mathematical and physical arguments against a classical (linear) stability analysis of the homogeneous solution are given. In order to obtain some information on the value and the shape of the perturbation, without any calculation performed on the non-homogeneous perturbation, a parameter, which can be calculated using only the homogeneous solution, is introduced. The analysis of the evolution of the perturbation is deduced from the behaviour of this parameter. The link between the initial and final perturbation is pointed out. In this way it is shown, in some cases, that for an arbitrary initial perturbation the final perturbation has a particular form which can lead to strain localization in some special zones.

In Section 6.4 some numerical results are presented in order to illustrate the conclusions of Section 6.3. Three different cases for the same homogeneous solution are considered. Two types of initial perturbation (in temperature and velocity) are presented. For the initial velocity perturbations, the 'blocking' phenomenon of the elastic waves in the plastic domain is pointed out.

6.1 Problem statement

Let us consider the one-dimensional shearing of an infinite slab bounded by the planes $\bar{x} = 0$ and $\bar{x} = h$ which extends to infinity in the directions $\bar{y}$ and $\bar{z}$. Assume that displacements are zero with respect to the direction $\bar{x}$ and $\bar{z}$ and the velocity $\bar{v}$ in the $\bar{y}$ direction is a function of $\bar{x}$ and the time $\bar{t}$ and it will be subjected to the boundary conditions

$$\bar{v}(\bar{t}, 0) = 0, \qquad \bar{v}(\bar{t}, h) = V \qquad \text{for } \bar{t} > 0$$

where V is the constant velocity of the upper edge of the layer. We assume that the strain rate $\dot{\bar{\gamma}} = \bar{v}_{\bar{x}}$ can be written as the sum of the elastic and inelastic strain rates

$$\dot{\bar{\gamma}} = \dot{\bar{\gamma}}^E + \dot{\bar{\gamma}}^I \tag{6.1}$$

which are given by

$$\dot{\bar{\gamma}}^E = \frac{1}{G}\dot{\bar{\tau}} \tag{6.2}$$

$$\dot{\bar{\gamma}}^I = \bar{H}(\bar{\tau}, \bar{\theta}) \tag{6.3}$$

where $\bar{\tau}$ denotes the shear stress, $\bar{\theta}$ is the temperature, and G is the elastic coefficient. Equation (6.3) represents a viscoplastic constitutive law with thermal softening and without hardening. For different and more concrete forms of function H, see for instance Fressengeas (1988), Fressengeas and Molinari (1987), Chen *et al.* (1989), Molinari and Leroy (1990), Wright and Walter (1987). The momentum balance law and the equation of energy reads as follows:

$$\rho \bar{v}_{\bar{t}} = \bar{\tau}_{\bar{x}} \tag{6.4}$$

$$C\bar{\theta}_{\bar{t}} = K\bar{\theta}_{\bar{x}\bar{x}} + \beta\bar{\tau}\cdot\dot{\bar{\gamma}}^I, \tag{6.5}$$

where ρ is the density, C is the specific heat, K is the thermal conductivity, β is the Taylor–Quiney coefficient which expresses the fraction of plastic work that is converted into heat and the subscripts $\bar{t}, \bar{x}$ denote partial differentiation with respect to $\bar{t}$ and $\bar{x}$ respectively. Let us introduce the dimensionless variables:

$$v = \bar{v}/V, \qquad \tau = \bar{\tau}/\Sigma, \qquad \theta = \bar{\theta}/\Theta \tag{6.6}$$

$$x = \bar{x}/h, \qquad t = \bar{t}V/h \tag{6.7}$$

where Σ, Θ are the specific stress and temperature constants. If we put

$$H(\tau, \theta) = h/V\bar{H}(\tau\Sigma, \theta\Theta)$$

and we use the above equations we can formulate the following initial and boundary-value problem:

Find the functions v, τ, θ: $\mathbb{R}_+ \times (0, 1) \to \mathbb{R}$ such that

$$\dot{v}(t, x) = a\tau_x(t, x) \tag{6.8}$$

$$\dot{\tau}(t, x) = b[v_x(t, x) - H(\tau(t, x), \theta(t, x))] \tag{6.9}$$

$$\dot{\theta}(t, x) = c\theta_{xx}(t, x) + f\tau(t, x)H(\tau(t, x), \theta(t, x)) \tag{6.10}$$

$$v(t, 0) = 0, \quad v(t, 1) = 1, \quad \theta_x(t, 0) = \theta_x(t, 1) = 0, \quad t > 0 \tag{6.11}$$

$$v(0, x) = v_0(x), \quad \tau(0, x) = \tau_0(x), \quad \theta(0, x) = \theta_0(x), \qquad x \in (0, 1), \tag{6.12}$$

where $a = \Sigma/(\rho V^2)$, $b = G/\Sigma$, $c = K/(CVh)$, $f = \beta\Sigma/(C\Theta)$, and the dots above symbols represent the partial derivation with respect to the time variable t.

Remark 6.1 *One can easily prove that if* $H(0, \theta) = 0$, *which is a natural assumption and holds for all examples we have in mind, then the problem* (6.8)–(6.12) *has no stationary solution for* $V \neq 0$. *Indeed if* $\dot{v} = \dot{\tau} = \dot{\theta} = 0$ *from* (6.8) *we get* $\tau(x) = \tau$ *and if we integrate* (6.10) *and* (6.9) *from* 0 *to* 1 *we obtain* $f\tau \int_0^1 H(\tau, \theta(x))\,dx = 0$ *and* $1 = \int_0^1 H(\tau, \theta(x))\,dx$; *hence* $f\tau = 0$, *and since* $f \neq 0$ *we get* $\tau = 0$, *a contradiction.*

6.2 Existence and uniqueness of smooth solutions

In this section we prove the local existence and uniqueness of smooth solutions for the problem (6.8)–(6.12). Moreover it is proved that if the solution is not a global one then it will blow up in the $H^1(\Omega)$ norm at the end of its interval of existence. If we can prove, in some particular problems, that the velocity, stress, and temperature are bounded in $H^1(\Omega)$ on the interval of existence, then we can conclude that the solution is global.

The following assumptions will be used:

$$H \in W^{2,\infty}_{\text{loc}}(\mathbb{R}^2) \tag{6.13}$$

$$v_0, \quad \tau_0, \quad \theta_0 \in H^2(\Omega) \tag{6.14}$$

$$v_0(0) = 0, \quad v_0(1) = 1, \quad \tau_{0x}(0) = \tau_{0x}(1) = 0, \quad \theta_{0x}(0) = \theta_{0x}(1) = 0, \tag{6.15}$$

where $\Omega = (0, 1)$.

The main result of this section is the following:

Theorem 6.1 *Let us suppose that* (6.13)–(6.15) *hold. Then there exist* $T_{\max} = T_m(v_0, \tau_0, \theta_0) \in (0, +\infty]$ *and a unique solution* v, τ, θ *of* (6.8)–(6.12) *with the* *following regularity*:

$$v, \tau \in C^1([0, T_{\max}), H^1(\Omega)) \cap C^0([0, T_{\max}), H^2(\Omega)) \tag{6.16}$$

$$\theta \in C^1([0, T_{\max}), L^2(\Omega)) \cap C^0([0, T_{\max}), H^2(\Omega)). \tag{6.17}$$

Moreover if $T_{\max} < +\infty$, *then*

$$\lim_{t \uparrow T_{\max}} (\|v(t)\|_{H^1(\Omega)} + \|\tau(t)\|_{H^1(\Omega)} + \|\theta(t)\|_{H^1(\Omega)}) = +\infty. \tag{6.18}$$

Note that if v, τ, θ *is the solution of* (6.8)–(6.12) *having the regularity* (6.16), (6.17), *then* (6.14), (6.15) *hold.*

PROOF. Consider the spaces $X = H_0^1(\Omega) \times H^1(\Omega) \times L^2(\Omega)$,

$$V = \{\tau \in H^2(\Omega)/\tau_x(0) = \tau_x(1) = 0\},$$

$W_0 = H_0^1(\Omega) \cap H^2(\Omega)$ and $D(A) = W \times V \times V$. Let A: $D(A) \subset X \to X$ be given by

$$A(u; \sigma; q) = (a\sigma_x; bu_x; cq_{xx}) \qquad \forall(u; \sigma; q) \in D(A), \tag{6.19}$$

which generates a C^0-semigroup of contractions denoted by $(S(t))_{t \geq 0}$ acting on X. Let $Y = [D(A)]$, i.e. Y represents the space $D(A)$ endowed with the graph norm of A, and let p: $[D(A)] \to \mathbb{R}$ be the norm

$$p(u, \sigma, q) = \|u\|_{H^1(\Omega)} + \|\sigma\|_{H^1(\Omega)} + \|q\|_{H^1(\Omega)} \tag{6.20}$$

for all $(u, \sigma, q) \in [D(A)]$. One can easily check that p is a continuous norm and (A3.46) holds. Let F: $[D(A)] \to [D(A)]$ be given by

$$F((u; \sigma; q)) = (0; b(1 - H(\sigma, q)); f\sigma H(\sigma, q)) \tag{6.21}$$

for all $(u; \sigma; q) \in D(A)$. From (6.13) we can deduce that there exists $L(r)$ and $C > 0$ such that (A3.47)–(A3.49) hold. If we put $v^*(t, x) = v(t, x) - x$, $v_0^*(x) = v_0(x) - x$, $y(t) = (v^*(t), \tau(t), \theta(t))$, $y_0 = (v_0^*, \tau_0, \theta_0)$, then we get that v, τ, θ is the solution of (6.8)–(6.12) iff y is the solution of (A3.53)–(A3.54). Since from (6.14), (6.15) we get that $y_0 \in [D(A)]$ we can use now Theorem A3.10, to obtain the statement of Theorem 6.1.

The following theorem gives the continuous dependence of the blow-up time $T_m(v_0, \tau_0, \theta_0)$ and of the solution v, τ, θ upon the initial data v_0, τ_0, θ_0, which assures the finite time stability of smooth solutions.

Theorem 6.2 *Let us suppose that the assumptions of Theorem* 6.1 *hold. Then for all* $T \in (0, T_m(v_0, \tau_0, \theta_0))$ *there exists* $\delta = \delta(T, v_0, \tau_0, \theta_0)$ *and* $C = C(T, v_0, \tau_0, \theta_0)$ *such that for all* $v_0^1, \tau_0^1, \theta_0^1$ *which satisfy* (6.14), (6.15) *and*

$$\|v_0^1 - v_0\|_{H^1(\Omega)} + \|\tau_0^1 - \tau_0\|_{H^1(\Omega)} + \|\theta_0^1 - \theta_0\|_{H^1(\Omega)} \leq \delta \tag{6.22}$$

we have $T_m(v_0^1, \tau_0^1, \theta_0^1) > T$. *Moreover if we denote by* v^1, τ^1, θ^1 *the solution of* (6.8)–(6.12) *for the initial data* $v_0^1, \tau_0^1, \theta_0^1$, *then we have*

$$\begin{aligned}\|v(t) - v^1(t)\|_{H^1(\Omega)} + \|\tau(t) - \tau^1(t)\|_{H^1(\Omega)} + \|\theta(t) - \theta^1(t)\|_{H^1(\Omega)} \\ \leq C[\|v_0 - v_0^1\|_{H^1(\Omega)} + \|\tau_0 - \tau_0^1\|_{H^1(\Omega)} + \|\theta_0 - \theta_0^1\|_{H^1(\Omega)}] \qquad \forall t \in [0, T].\end{aligned} \tag{6.23}$$

PROOF. We shall use the same notations as in the proof of Theorem 6.1. One can easily prove that

$$p(F(x) - F(y)) \leq L(r)p(x - y) \qquad \forall x, y \in D(A) \quad \text{with } p(x), p(y) \leq r. \tag{6.24}$$

Let us suppose that there exists $T \in (0, T_m(v_0, \tau_0, \theta_0))$ such that for all $n \geq 1$ there exist $v_0^n, \tau_0^n, \theta_0^n$ such that

$$\|v_0 - v_0^n\|_{H^1(\Omega)} + \|\tau_0 - \tau_0^n\|_{H^1(\Omega)} + \|\theta_0 - \theta_0^n\|_{H^1(\Omega)} \leq \frac{1}{n}$$

and $T_n = T_m(v_0^n, \tau_0^n, \theta_0^n) < T$. Let us denote by v_n, τ_n, θ_n the solution of (6.8)–(6.12) for the initial data $(v_0^n, \tau_0^n, \theta_0^n)$ and let $v_0^{*n}(x) = v_0^n(x) - x$, $v_n^*(t, x) = v_n(t, x) - x$. If we put $y_0^n = (v_0^{*n}, \tau_0^n, \theta_0^n) \in D(A)$, $y_n(t) = (v_n^*(t), \tau_n(t), \theta_n(t))$, then from (A3.53) and (A3.54) we get

$$y(t) = S(t)y_0 + \int_0^t S(t - s)F(y(s))\,ds \qquad \forall t \in [0, T_{\max}) \tag{6.25}$$

$$y_n(t) = S(t)y_0^n + \int_0^t S(t - s)F(y_n(s))\,ds \qquad \forall t \in [0, T_n). \tag{6.26}$$

Let $r > 0$ such that $p(y(t)) \leq r/2$ for all $t \in [0, T]$. Since from the statement of Theorem 6.1, we have that $\lim_{t \uparrow T_n} p(y_n(t)) = +\infty$, we can deduce that there exists $\bar{T}_n < T_n$ such that $p(y_n(\bar{T}_n)) = r$ and $p(y_n(t)) \leq r$ for all $t \in [0, \bar{T}_n]$. If we subtract (6.26) from (6.25) and we make use of (6.24) we get

$$p(y(t) - y_n(t)) \leq p(y_0 - y_0^n) + L(r)\int_0^t p(y(s) - y_n(s))\,ds \qquad \forall t \in [0, \bar{T}_n].$$

From this inequality and the Gronwal lemma we get

$$p(y(t) - y_n(t)) \leq p(y_0 - y_0^n)\,e^{L(r)T} \qquad \forall t \in [0, \bar{T}_n]. \tag{6.27}$$

If we bear in mind that $p(y_0 - y_0^n) < 1/n$, we get

$$r = p(y_n(\bar{T}_n)) \leq r/2 + \frac{1}{n}\,e^{L(r)T} \qquad \forall n \in \mathbb{N},$$

a contradiction. We have just proved the first statement of the theorem. The second statement easily follows from (6.27) which is now valid for all $t \in [0, T]$.

6.3 Perturbations of the homogeneous solutions

In this section the behaviour of the perturbation of the homogeneous solution is studied. In order to do this, a parameter $P(t)$, called non-homogeneous stability parameter, which depend only on the homogeneous solution, is introduced. The behaviour of the perturbation is deduced from this parameter.

If the initial data are homogeneous, i.e.

$$\dot{\gamma}_0(x) = v_{0x}(x) = 1, \qquad \tau_0(x) = \tau_0^0, \qquad \theta_0(x) = \theta_0^0 \tag{6.28}$$

(which imply $v_0(x) = v_0^0(x) = x$) then the problem (6.8)–(6.12) has a homogeneous solution

$$v^0(t, x) = x, \qquad \tau^0(t, x) = \tau^0(t), \qquad \theta^0(t, x) = \theta^0(t)$$

(which imply $\dot{\gamma}^0(t, x) = v_x^0(t, x) = 1$) where τ^0, θ^0 is the solution of the following Cauchy problem:

$$\dot{\tau}^0(t) = b[1 - H(\tau^0(t), \theta^0(t))] \tag{6.29}$$

$$\dot{\theta}^0(t) = f\tau^0(t)H(\tau^0(t), \theta^0(t)) \tag{6.30}$$

$$\tau^0(0) = \tau_0^0, \qquad \theta^0(0) = \theta_0^0. \tag{6.31}$$

Let $T_0 = T_m(v_0^0, \tau_0^0, \theta_0^0) \in (0, +\infty]$ and $[0, T_0)$ be the interval of existence of the homogeneous solution v^0, τ^0, θ^0 (see Theorem 6.1).

Remark 6.2 *If one supposes (for physical reasons) that $\tau^0(t)H(\tau^0(t), \theta^0(t)) \geq 0$ for all t from $[0, T_0)$, then $T_0 = +\infty$, i.e. the homogeneous solution is global. Indeed from* (6.29) *and* (6.30) *we get*

$$\frac{\mathrm{d}}{\mathrm{d}t} D^2(t) = 2fb\tau^0(t) \quad \text{where } D^2(t) = f(\tau^0(t))^2 + 2b(\theta^0(\tau) - \theta^0).$$

If we have in mind that $\theta^0(t) \geq \theta^0$ then we get $2bf\tau^0(t) \leq gD(t)$ with $g = 2bf^{1/2}$, hence $(\mathrm{d}/\mathrm{d}t)D^2(t) \leq gD(t)$. If we use now this last inequality we get $D(t) \leq D(0) + gt/2$ for all t from $\{0, T_0)$ and from Theorem 6.1 *we deduce $T_0 = +\infty$.*

A classical (linear) stability analysis of the homogeneous solution will give the behaviour of the perturbations for $t \to +\infty$. Our opinion is that such an analysis cannot be performed in our case for different reasons. A mathematical reason is that generally $T_0 < +\infty$, i.e. the homogeneous solution blows up in finite time. If this mathematical difficulty is removed (i.e. one can prove that in some special cases the homogeneous solution is global (see Remark 6.2) and there exists a neighbourhood of the initial data $v_0^0, \tau_0^0, \theta_0^0$ where global existence of the solution is assured) then from the physical point of view it must be checked whether the constitutive equations moreover remain valid for very large time (for instance, the growth of the temperature predicted by the constitutive relations used in some papers could generate phase change or rupture).

On the other hand the finite time stability result obtained in Theorem 6.2 does not give any information upon the value of the perturbation at a given time.

In order to study the evolution of the perturbation $(\hat{v}, \hat{\tau}, \hat{\theta})$ let

$$\hat{v}(t, x) = v(t, x) - x, \qquad \hat{\tau}(t, x) = \tau(t, x) - \tau^0(t), \qquad \hat{\theta}(t, x) = \theta(t, x) - \theta^0(t) \tag{6.32}$$

where v, τ, θ is the solution of (6.8)–(6.12). If we denote the initial perturbation by $\hat{v}_0, \hat{\tau}_0, \hat{\theta}_0$, i.e.

$$\hat{v}_0(x) = v_0(x) - x, \qquad \hat{\tau}_0(x) = \tau_0(x) - \tau_0^0, \qquad \hat{\theta}_0(x) = \theta_0(x) - \theta_0^0,$$

then we can easily deduce that $\hat{v}, \hat{\tau}, \hat{\theta}$ is the solution of the following problem:

$$\dot{\hat{v}}(t, x) = a\hat{\tau}_x(t, x) \tag{6.33}$$

$$\dot{\hat{\tau}}(t, x) = b[\hat{v}_x(t, x) - (H(\tau^0(t) + \hat{\tau}(t, x), \theta^0(t) + \hat{\theta}(t, x)) - H(\tau^0(t), \theta^0(t)))] \tag{6.34}$$

$$\begin{aligned}\dot{\hat{\theta}}(t, x) = c\hat{\theta}_{xx}(t, x) + f[(\tau^0(t) + \hat{\tau}(t, x))H(\tau^0(t) + \hat{\tau}(t, x), \theta^0(t) + \hat{\theta}(t, x)) \\ - \tau^0(t)H(\tau^0(t), \theta^0(t))]\end{aligned} \tag{6.35}$$

$$\hat{v}(t, 0) = \hat{v}(t, 1) = 0, \qquad \hat{\theta}_x(t, 0) = \hat{\theta}_x(t, 1) = 0 \tag{6.36}$$

$$\hat{v}(0, x) = \hat{v}_0(x), \qquad \hat{\tau}(0, x) = \hat{\tau}(x), \qquad \hat{\theta}(0, x) = \hat{\theta}_0(x). \tag{6.37}$$

Let $T < T_0$ and note that from Theorem 6.2 we find that if the initial perturbations $\hat{v}_0, \hat{\tau}_0, \hat{\theta}_0$ are small enough, then the solution $\hat{v}, \hat{\tau}, \hat{\theta}$ exists on $[0, T]$. Suppose now that $\hat{\tau}$ and $\hat{\theta}$ are small on $[0, T]$ and consider the following approximation:

$$H(\tau^0(t) + \hat{\tau}, \theta^0(t) + \hat{\theta}) - H(\tau^0(t), \theta^0(t)) \simeq H_\tau(\tau^0(t), \theta^0(t))\hat{\tau} + H_\theta(\tau^0(t), \theta^0(t))\hat{\theta} \tag{6.38}$$

$$\begin{aligned}(\tau^0(t) + \hat{\tau})H(\tau^0(t) + \hat{\tau}, \theta^0(t) + \hat{\theta}) - \tau^0(t)H(\tau^0(t), \theta^0(t)) \\ \simeq [H(\tau^0(t), \theta^0(t)) + \tau^0(t)H_\tau(\tau^0(t), \theta^0(t))]\hat{\tau} + \tau^0(t)H_\theta(\tau^0(t), \theta^0(t))\hat{\theta}.\end{aligned} \tag{6.39}$$

Let us denote by $\tilde{v}, \tilde{\tau}, \tilde{\theta}$ the linear perturbation which is the solution of the following linearized version of (6.33)–(6.37):

$$\dot{\tilde{v}}(t, x) = a\tilde{\tau}_x(t, x) \tag{6.40}$$

$$\dot{\tilde{\tau}}(t, x) = b[\tilde{v}_x(t, x) - H_\tau(\tau^0(t), \theta^0(t))\tilde{\tau}(t, x) - H_\theta(\tau^0(t), \theta^0(t))\tilde{\theta}(t, x)] \tag{6.41}$$

$$\begin{aligned}\dot{\tilde{\theta}}(t, x) = c\tilde{\theta}_{xx}(t, x) + f[(H(\tau^0(t), \theta^0(t)) + \tau^0(t)H_\tau(\tau^0(t), \theta^0(t))\tilde{\tau}(t, x) \\ + \tau^0(t)H_\theta(\tau^0(t), \theta^0(t))\tilde{\theta}(t, x)]\end{aligned} \tag{6.42}$$

$$\tilde{v}(t, 0) = \tilde{v}(t, 1) = 0, \qquad \tilde{\theta}_x(t, 0) = \tilde{\theta}_x(t, 1) = 0 \tag{6.43}$$

$$\tilde{v}(0, x) = \hat{v}_0(x), \qquad \tilde{\tau}(0, x) = \hat{\tau}_0(x), \qquad \tilde{\theta}(0, x) = \hat{\theta}_0(x). \tag{6.44}$$

Remark 6.3 *It is still an open problem to prove that the conclusions of the analysis performed on the linear problem (6.40)–(6.44) are valid for the non-linear system (6.33)–(6.37). However, numerical experiments (see Section 6.4) show that there are no important qualitative and quantitative differences between the solutions (i.e.* $\hat{v} \simeq \tilde{v}$, $\hat{\tau} \simeq \tilde{\tau}$, $\hat{\theta} \simeq \tilde{\theta}$*) if the initial perturbation of the homogeneous solution* $\hat{v}_0$, $\hat{\tau}_0$, $\hat{\theta}_0$ *are smaller than 10% of the homogeneous initial data* v_0^0, τ_0^0, θ_0^0.

If we have in mind the boundary conditions (6.43) and equation (6.40) which imply (for smooth solutions) that $\tilde{\tau}_x(t, 0) = \tilde{\tau}_x(t, 1) = 0$ then we can develop the solution $\tilde{v}$, $\tilde{\tau}$, $\tilde{\theta}$ in Fourier series as follows:

$$\tilde{v}(t, x) = \sum_{n=1}^{\infty} p_n(t) \sin(n\pi x) \tag{6.45}$$

$$\tilde{\tau}(t, x) = \sum_{n=0}^{\infty} q_n(t) \cos(n\pi x) \tag{6.46}$$

$$\tilde{\theta}(t, x) = \sum_{n=0}^{\infty} r_n(t) \cos(n\pi x). \tag{6.47}$$

Since in the study of strain localization the evolution of the homogeneous perturbations (i.e. for $n = 0$) plays no role, we shall consider only $n \geq 1$.

Let $x_n(t) = (p_n(t); q_n(t); r_n(t))$ and define the matrix $A_n(t)$ by

$$A_n(t) =$$

$$\left\| \begin{matrix} 0 & -n\pi a & 0 \\ n\pi b & -bH_\tau(\tau^0(t), \theta^0(t)) & -bH_\theta(\tau^0(t), \theta^0(t)) \\ 0 & f[H(\tau^0(t), \theta^0(t)) + \tau^0(t)H_\tau(\tau^0(t), \theta^0(t))] & f(\tau^0(t) \cdot H_\theta(\tau^0(t), \theta^0(t)) - n^2\pi^2 c \end{matrix} \right\| \tag{6.48}$$

then from (6.40)–(6.44) we get the following sequence of linear Cauchy problems:

$$\dot{x}_n(t) = A_n(t)x_n(t) \tag{6.49}$$

$$x_n(0) = x_0^n, \tag{6.50}$$

where $x_0^n = (p_n^0; q_n^0; r_n^0)$ are the Fourier coefficients of the initial data

$$\hat{v}_0(x) = \sum_{n=1}^{\infty} p_n^0 \sin(n\pi x), \tag{6.51}$$

$$\hat{\tau}_0(x) = \sum_{n=0}^{\infty} q_n^0 \cos(n\pi x), \tag{6.52}$$

$$\hat{\theta}_0(x) = \sum_{n=0}^{\infty} r_n^0 \cos(n\pi x). \tag{6.53}$$

Let us suppose in the following that:

$$H(\tau^0(t), \theta^0(t)) > 0; \qquad H_\tau(\tau^0(t), \theta^0(t)) > 0; \qquad H_\theta(\tau^0(t), \theta^0(t)) > 0 \quad (6.54)$$

for $t \in (0, T]$ which is a natural assumption and holds for all concrete examples we have in mind.

Let $E_n(t)$ be given by

$$E_n(t) = bp_n^2(t) + aq_n^2(t) + abr_n^2(t)/(4f). \tag{6.55}$$

From (6.49) we obtain

$$(E_n(t))_t = 2ab[C_1(t)q_n^2(t) + C_2(t)r_n(t)q_n(t) + C_3(t)q_n^2(t)],$$

where $C_1(t) = -H_\tau(\tau^0(t), \theta^0(t))$,

$$C_2(t) = [H(\tau^0(t), \theta^0(t)) + \tau^0(t)H_\tau(\tau^0(t), \theta^0(t))]/4 - H_\theta(\tau^0(t), \theta^0(t))$$

and $C_3(t) = [\tau^0(t)H_\theta(\tau^0(t), \theta^0(t)) - n^2\pi^2 c/f]/4$. Let $\Delta_n(t) = C_2^2(t) - 4C_1(t)C_3(t)$ and let us denote by

$$d = c\pi^2/f = \pi^2 K\Theta/(\Sigma V h\beta) \tag{6.56}$$

and by $P(t)$ the following function:

$$P(t) = \tau^0(t)H_\theta(\tau^0(t), \theta^0(t)) + [H(\tau^0(t), \theta^0(t)) + \tau^0(t)H_\tau(\tau^0(t), \theta^0(t)) - 4H_\theta(\tau^0(t), \theta^0(t))]^2/[16H_\tau(\tau^0(t), \theta^0(t))] \tag{6.57}$$

which will be called in the following the (non-homogeneous) stability parameter. After some work we obtain that $P(t) \le n^2 d$ iff $\Delta_n(t) \le 0$ which imply $(E_n(t))_t \le 0$. In this way we have deduced that if $P(t) \le n^2 d$ for all t from $[T_1, T_2]$ then $E_n(t) \le E_n(T_1)$ for all t from $[T_1, T_2]$.

From the above analysis we can deduce the following conclusions regarding the behaviour of the perturbation on $[T_1, T_2]$:

1. If $P(t) \le d$ for all t from $[T_1, T_2]$ then $E_k(t) \le E_k(T_1)$ for all t from $[T_1, T_2]$ and $k \ge 1$, i.e. the non-homogeneous perturbations of the homogeneous solution will not increase on $[T_1, T_2]$ and the strain localization does not occur during the time interval $[T_1, T_2]$.

2. If there exists $n \ge 1$ such that $n^2 d < P(t) \le (n+1)^2 d$ for all t from $[T_1, T_2]$ then $E_k(t)$ will not increase on $[T_1, T_2]$ for $k > n$ but an increasing of $E_k(t)$ for $k \le n$ is possible. In this case we can expect that the strain will be localized in the points where $|\cos k\pi x|$ takes its maximum value, i.e. in the points $x_{i,k} = i/k$ with $0 \le i \le k$ and $1 \le k \le n$. If n is small then we can expect that the strain will be localized in some special zones. For example, for $n = 1$ the strain can be localized at the boundary and for $n = 2$ at the boundary or in the middle. If n is large then there exist no special zones where the strain will be localized and the localization of the strain will depend on the shape of the perturbation at $t = T_1$.

The above analysis performed on the time interval $[T_1, T_2]$ is significant for the behaviour of the solution on $[0, T]$ if the interval $[T_1, T_2]$ is sufficiently large.

To obtain some information about the behaviour of the perturbations and possible localization one can compute the homogeneous solution τ^0, θ^0 and the parameter P and compare it with $n^2 d$. The time interval $[0, T]$ is partitioned into intervals $[T_K, T_{K+1}]$, $[0, T] = \bigcup_{K=1}^{P} [T_K, T_{K+1}]$, where different conclusions of the above analysis hold. If these conclusions on $[T_K, T_{K+1}]$ are added then one has an idea of the behaviour of the perturbation on $[0, T]$ without any computations performed on the non-linear partial differential system (6.33)–(6.37).

Remark 6.4 *If we denote* $Q(t) = \tau^0(t) H_\theta(\tau^0(t), \theta^0(t))$ *then, after some work, we deduce that if* $Q(t) < n^2 d$ *then the real parts of the eigenvalues of* $A_n(t)$ *are negative and if* $Q(t) > n^2 d$ *then* $A_n(t)$ *has at least a positive eigenvalue. No important qualitative and quantitative differences between the parameters P and Q were remarked during the numerical experiments presented in Section* 6.4.

Remark 6.5 *Since d depends on the thermal conductivity K and the stability parameter P(t) do not depend on it, we get from the above analysis that for large values of K (i.e. for large d) the strain localization does not occur.*

Remark 6.6 *The above analysis agrees with the following, usually accepted, mechanism of strain localization. Non-uniform initial data induce non-uniform heating, which in turn increases plastic flow in hotter regions and reduces it in colder regions. But plastic flow will induce still greater heating in the hotter regions and this feedback mechanism creates destabilization, which tends to create plastic shear bands. This destabilizing mechanism is opposed by the thermal conductivity, which tends to make the temperature field homogeneous and to break the above feedback. Whether localization will occur depends on the result of the competition between thermal softening (present in P(t)) and thermal conductivity (present in d). Finally, if strain localization occurs then the position of the shear bends will depend on n.*

6.4 Numerical results

In this section we present some numerical experiences in order to verify the conclusions of Section 6.3.

Following Fressengeas and Molinari (1987) or Fressengeas (1988) let us consider the constitutive equation for the plastic stain rate (6.3) of the form

$$\bar{\tau} = \mu(\bar{\theta})^{\nu}(\dot{\bar{\gamma}}^{I})^{m},$$

where μ is a viscosity parameter, which implies

$$\bar{H}(\bar{\tau}, \bar{\theta}) = \left(\frac{\tau}{\mu}\right)^{1/m} (\bar{\theta})^{-\nu/m}. \tag{6.58}$$

Let us consider a process with homogeneous strain rate $\dot{\Gamma}$ and let $V = \dot{\Gamma}h$. If we put $\Sigma = \mu\Theta^{\nu}(\dot{\Gamma})^m$, then we deduce that

$$H(\tau, \theta) = \tau^{1/m}\theta^{-\nu/m},$$
$$a = \Sigma/(\rho\dot{\Gamma}^2h^2) \quad \text{and} \quad c = K/(C\dot{\Gamma}h^2). \tag{6.59}$$

Note that the coefficients b and f do not depend on h, hence the homogeneous solution τ^0, θ^0 will be the same for all h, which implies that the stability parameter P will not depend on h. Since $d = \pi^2 K\Theta/(\Sigma\dot{\Gamma}h^2\beta)$ we deduce that for different values of h we shall obtain different values of d, hence different conclusions on the linear stability of the homogeneous solution v^0, τ^0, θ^0. Let us denote by $\gamma(t, x)$ the strain given by

$$\gamma(t, x) = \int_0^t v_x(s, x)\, \mathrm{d}s$$

and by $\gamma^0(t) = t$ the homogeneous strain. Hence the time and strain are the same in the dimensionless variables. The initial homogeneous data that will be considered in what follows are $\tau_0^0 = 0$, $\theta_0^0 = 1$.

As follows from Fressengeas and Molinari (1987) we have considered for a low-carbon cold-rolled steel (CRS)AISI 1018 the following constants: $\mu = 4.36 \times 10^8$ SI, $\nu = -0.38$, $m = 0.019$, $\dot{\Gamma} = 10^4\ \mathrm{s}^{-1}$, $\rho = 7800\ \mathrm{kg/m^3}$, $K = 54\ \mathrm{W/(m\ K)}$, $C = 500\ \mathrm{J/(kg\ K)}$, $\beta = 0.9$, $G = 8.4 \times 10^{10}$ Pa, $\Theta = 300$ K, which imply that $\Sigma = 0.5945 \times 10^8$ Pa, $b = 199.2$, $f = 356.7$.

Figure 6.1 shows the homogeneous solution τ^0, θ^0.

Since the homogeneous strain $\gamma^0(t) = t$ we remark in Figure 6.1 that the material is initially elastic, and in the plastic zone we have a softening behaviour. The temperature is constant as long as the material is elastic but increases when plastic strain appears.

In order to verify the conclusions from the end of the previous Section (especially (2)) let us consider three cases for three different values of h which imply different values of V, c and d:

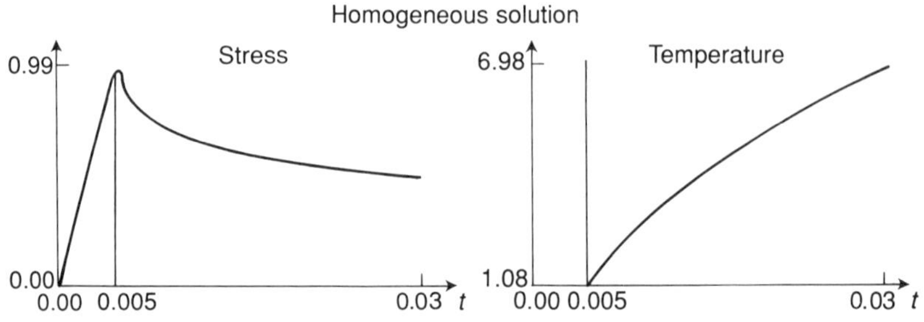

Fig. 6.1 Evolution of the homogeneous solution.

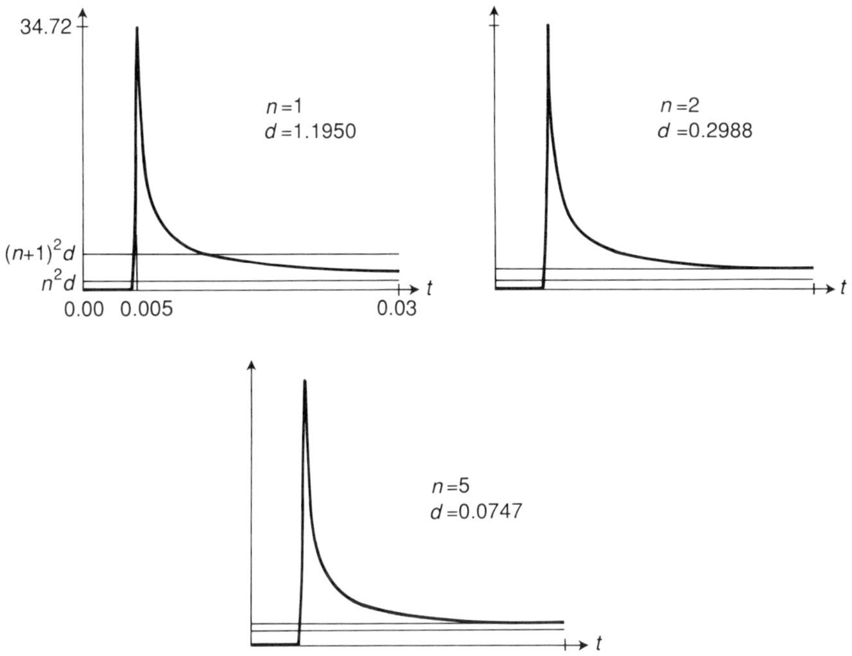

Fig. 6.2 The evolution of the non-homogeneous stability parameter $P(t)$ compared with n^2d for different values of d and n.

Case 1. $h = 5 \times 10^{-4}$ m, $V = 5$ m/s, $a = 304.9$, $c = 43.2$, $d = 1.1195$.
Case 2. $h = 10^{-3}$ m, $V = 10$ m/s, $a = 76.22$, $c = 10.8$, $d = 0.2988$.
Case 3. $h = 2 \times 10^{-3}$ m, $V = 20$ m/s, $a = 19.06$, $c = 2.7$, $d = 0.0747$.

Figure 6.2 shows the evolution of the non-homogeneous stability parameter P compared with n^2d and $(n + 1)^2$ for the three cases considered above. We remark that $P(t) < d$ in the elastic zone; hence the material is stable as far as it is elastic. When the plastic effects become important P quickly increases and after that it decreases up to a positive value P_∞. If we now compare P_∞ with n^2d we get (6.66) for $n = 1$ in Case 1, for $n = 2$ in Case 2 and for $n = 5$ in Case 3.

In Figures 6.3–9 show the evolution of the perturbation for two types of initial perturbations in the three cases. The numerical results were obtained by using a Galerkin-type numerical method for the problem (6.33)–(6.37). In order to do the space discretization, a finite-dimensional subspace which consists of $2 \times 41 + 40$ Fourier functions (i.e. in (6.45)–(6.47) we take only the first 41 coefficients) was considered. In order to integrate the non-linear ordinary differential system which results after the space discretization, a Runge–Kutta method was used. The same method was used for the numerical integration of the problem (6.40)–(6.44), and no significant

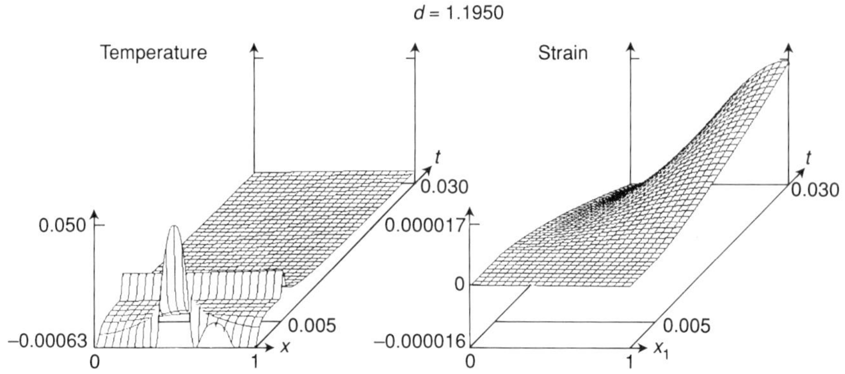

Fig. 6.3 The evolution of the perturbation for a temperature perturbation in Case 1.

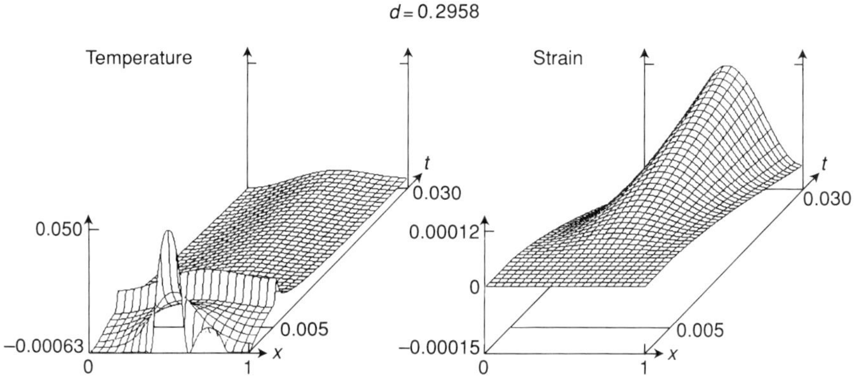

Fig. 6.4 The evolution of the perturbation for an initial temperature perturbation in Case 2.

differences in the numerical evaluations of the solutions of the two problems were remarked. This is a consequence of the fact that in the plastic zone, when the non-linear terms are important, we have $\hat{\tau}(t, x) \ll \tau^0(t)$ and $\hat{\theta}(t, x) \ll \theta^0(t)$, hence (6.38), (6.39) are quite good approximations.

Figures 6.3–5 show the evolutions of the perturbations for the same initial temperature perturbation $\hat{\theta}_0$ ($\hat{v}_0 = \hat{\tau}_0 = 0$) in all cases.

One can see in Figure 6.3 that in the first case the central perturbation which is greater than the other one (i.e. $r_1 \ll r_2$ in (6.47)) will not lead to a strain concentration in the middle. Since $n = 1$ in (6.66) in this case the strain concentration at the boundary agrees with conclusion (2) of Section 6.3. In Figures 6.4 and 6.5 we can see that in the other two cases we get a strain concentration in the middle which is more accentuated in the third case. In all three cases we note that in the neighbourhood of the time when the body 'becomes inelastic' the temperature and stress perturbations quickly increase but after a while they decrease. No final stress concentration is obtained.

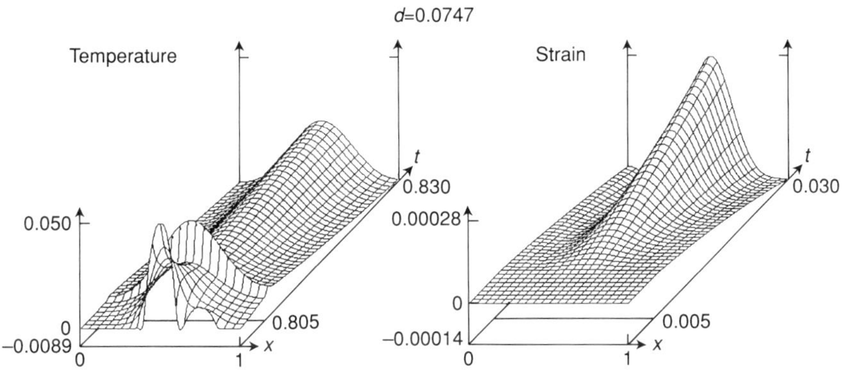

Fig. 6.5 The evolution of the perturbation for an initial temperature perturbation in Case 3.

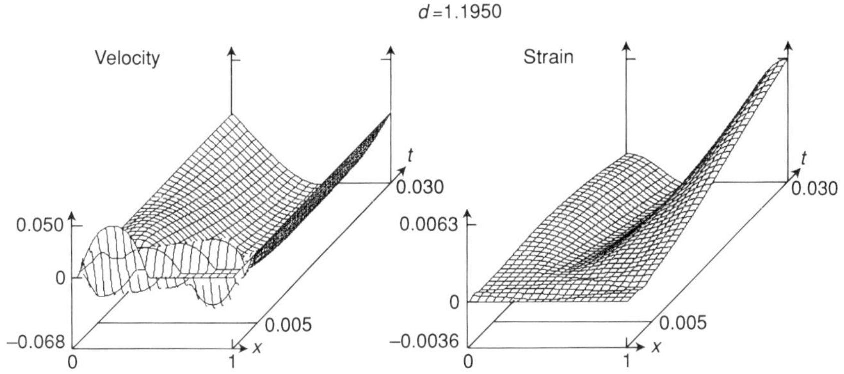

Fig. 6.6 The evolution of the perturbation for an initial velocity perturbation in Case 1.

In the first time period when the body is still elastic, the temperature perturbation rapidly decreases, hence it will be very small when the process becomes unstable (i.e. in the plastic zone). This can explain the small values of final strain perturbation (less than 1% of the final homogeneous strain). Hence in this particular example we did not obtain a strain localization. If instead of an initial temperature perturbation we consider an initial velocity perturbation, then this perturbation will travel as an elastic wave during the time when the body is elastic and its value will be conserved when the body becomes plastic, i.e. when the process becomes unstable.

In Figures 6.6–8 the evolutions of the perturbations for the same initial velocity perturbation $\hat{v}_0$ ($\hat{\tau}_0 = \hat{\theta}_0 = 0$) in all cases are plotted. As long as the body is elastic the waves are travelling and reflect at the boundary. Since the speed of the elastic waves is equal to $\sqrt{(ab)}$ and decrease when h increases, it will be the greatest in the first case and smallest in the third case. When the body becomes inelastic the elastic waves are blocked (or they travel with a very small speed) and the instability process begins.

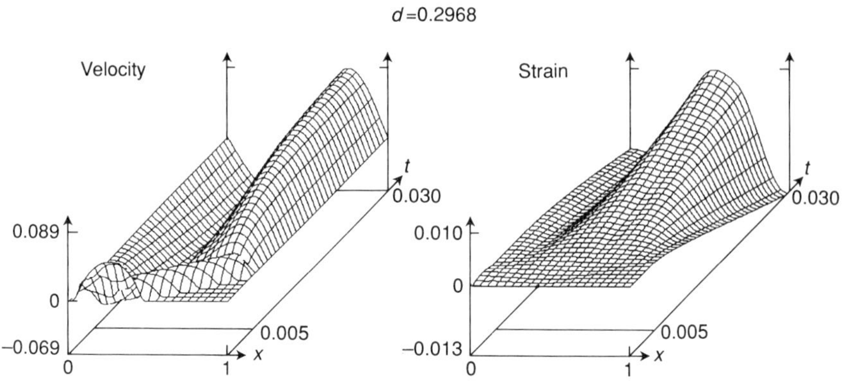

Fig. 6.7 The evolution of the perturbation for an initial velocity perturbation in Case 2.

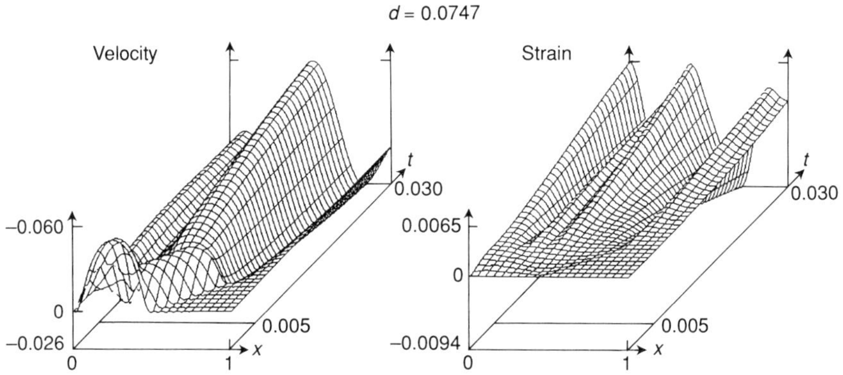

Fig. 6.8 The evolution of the perturbation for an initial velocity perturbation in Case 3.

As we can see from Figures 6.6 and 6.7 respectively, we obtain a strain concentration at the boundary in the first case and in the middle in the second case. In the third case we find the strain concentrated in three or four zones (see Figure 6.8). All these results, which agree with the conclusions of Section 6.3, show that the final concentration of the strain depends on the shape of the perturbation at the moment when the body becomes plastic (i.e. the moment when the elastic waves are 'blocked') and on n. As in the case of initial temperature perturbation, no final stress concentration is obtained.

Since the final strain perturbation is between 21% and 36% of the final homogeneous strain we have obtained in this case a strain localization. The final temperature perturbation is less than 7% of the final homogeneous temperature, but the material is hotter in the regions where the strain is localized.

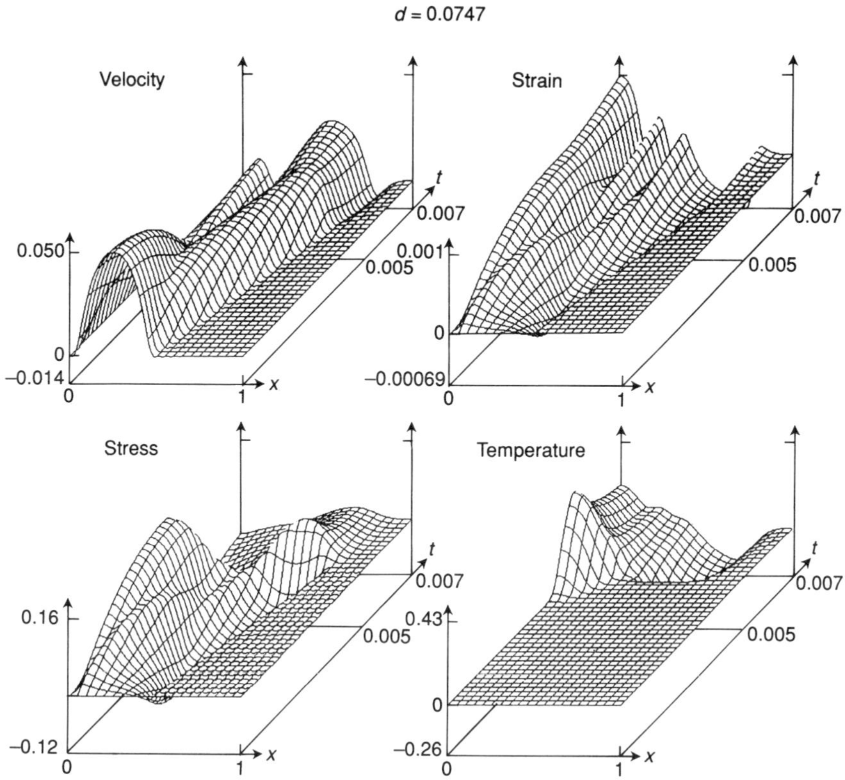

Fig. 6.9 The 'blocking' of the elastic waves in the plastic zone in Case 3.

In Figure 6.9 we can see the blocking phenomenon of the elastic waves. This can be 'explained' by the fact that when the elastic behaviour is more important (i.e. $\dot{\gamma}^E \gg \dot{\gamma}^I$) we have a hyperbolic problem, but when the plastic behaviour becomes dominant (i.e. $\dot{\gamma}^I \ll \dot{\gamma}^E$) we can neglect the elastic strain rate $\dot{\gamma}^E$ in (6.2) which will lead to a parabolic problem.

Bibliographical notes

Dynamic problems of the form (1.1)–(1.4) have already been studied by many authors, with different methods. So, in the case when f does not depend on time and G is given by (1.7) in Chapter 3, an existence and uniqueness result was obtained by Duvaut and Lions (1972, Ch. 5) using the constructive theory of partial differential equations including Galerkin approximations, regularization and monotony methods. In the case when f, and also $\mathscr{A}$ and G in (1.2), depend on time but G does not depend on ε, an existence and uniqueness result was given by Suquet (1981a) using a finite-difference discretization in time and also monotony arguments. Since in (1.2) G depends both on σ and ε, the monotony arguments used in the above-mentioned papers do not work here. For this reason a different technique was used in Section 1.2 in order to obtain an existence and uniqueness result. This technique based on the equivalence between the problem (1.1)–(1.4) and a semilinear evolution equation in a Hilbert space follows the paper of Ionescu (1988b). Section 1.4, concerning weak solutions, was written following Dascalu and Ionescu (1991). The existence of the energy function for hypoelastic materials was proved by Mihailescu–Suliciu and Suliciu (1979). The results presented in Sections 2.1 and 2.2 belong to Suliciu (1984). Some of these results were extended in the one-dimensional case for a non-monotone function F by Faciu and Mihailescu-Suliciu (1987), Suliciu and Sabac (1988), Faciu (1989). The results given in Section 2.3 were obtained by Ionescu (1988b, 1990b).

For more details as well as for comments on the results presented in Section 3 concerning dynamic processes in perfect plasticity we refer to Duvaut and Lions (1972), Ch. 5.

Theorem 4.1 was proved in the papers of Ionescu (1988b, 1990), and the local existence and uniqueness result presented in Theorem 4.3 was obtained by Ionescu (1990b, 1991).

Section 5.1 was written following the ideas of Sofonea (1987, 1988) and Theorem 5.3 was obtained in the papers of Sofonea (1989a, 1990a). A dynamical problem concerning rate-type equations of the form (5.30) was also considered by Laborde (1979) (see also Remark 5.1 in Chapter 4); in this paper an existence and uniqueness result is given using an approximation method based on the discretization of the evolution of the convex set and a similar fixed-point method.

Different approaches for strain localization for different viscoplastic thermal softening solids can be found, for instance, in Bai (1982), Burns and Trucano (1982), Douglas and Chen (1985), Douglas *et al.* (1987), Fressengeas and Molinari (1987), Fressengeas (1988), Chen *et al.* (1989), Molinari and Leroy (1990), Wright and Walter (1987) (see also the review paper of Bai, 1990 and the references given there).

Paragraph 6 was written following the ideas of Ionescu and Predeleanu (1991).

5

THE FLOW OF THE BINGHAM FLUID WITH FRICTION

This chapter deals with the study of some boundary-value problems describing the flow of the Bingham fluid with friction through a bounded domain Ω in $\mathbb{R}^3$. Existence and uniqueness results are given using elliptic variational inequality arguments. The blocking phenomenon is also studied and the link is given between the yield limit, the friction coefficient and the external forces for which the flow of the Bingham fluid is blocked. An iterative finite-element method for the approximation of the solution is presented and applied to wire-drawing processes.

1 Boundary-value problems for the Bingham fluid with friction

We start this section by presenting the constitutive equations of the Bingham fluid and the friction laws on the boundary. Afterwards we give the variational formulations for the problems studied and we present existence and uniqueness results of the solutions.

1.1 The constitutive equations of the Bingham fluid

The Bingham model is characterized by the following property: the material starts to flow only if the applied forces exceed a certain limit, called the *yield limit.*

In order to describe this model let us denote by u the material velocity; besides the rate deformation tensor

$$D = \tfrac{1}{2}(\nabla u + \nabla^T u) \tag{1.1}$$

we shall also consider its deviator

$$D' = D - \tfrac{1}{3}(\operatorname{tr} D)I_3. \tag{1.2}$$

We denote the Cauchy stress tensor by T and its deviator by

$$T' = T + pI_3. \tag{1.3}$$

In (1.3) the scalar $-p$ represents the spherical part of the stress tensor; one can identify p with the *pressure.*

Besides the deviators D' and T' another deviator tensor S will be introduced as the part of the stress which corresponds to the plastic properties of the material.

By a *process* we shall mean a collection of smooth functions $t \to D'(t)$, $T'(t)$, $S(t)$ for $t \in [0, \tau]$ where $\tau > 0$ is the duration of the process. The Bingham model for rigid–viscoplastic bodies assumes that for any process we have

$$T' = S + 2\eta D' \tag{1.4}$$

$$f(S) = |S|^2 - g^2 \leq 0 \tag{1.5}$$

$$D' = 2\lambda S \tag{1.6}$$

where $\eta > 0$ is the *viscosity coefficient*, $g/\sqrt{2}$ is the *yield stress in pure shear* and λ is a scalar function such that

$$\begin{aligned} \lambda(t) &= 0 \qquad \text{if } f(S) < 0 \text{ or } f(S) = 0 \text{ and } \dot{f}(S) < 0 \\ \lambda(t) &> 0 \qquad \text{if } f(S) = 0 \text{ and } \dot{f}(S) = 0. \end{aligned} \tag{1.7}$$

In (1.7) the derivative $\dot{f}(S)$ is the derivative with respect to time of the function $t \to f(S(t))$, for every smooth tensor valued function $t \to S(t)$.

The yield condition (1.5) is the *Mises condition.* According to it the invariant $S_{\mathrm{II}} = \frac{1}{2}S \cdot S = \frac{1}{2}|S|^2$ cannot exceed the square of the yield stress in pure shear $g/\sqrt{2}$. We shall refer in the next pages to g as the *yield limit* of the Bingham fluid. From (1.6) and (1.7) it follows that the deviator of the rate of deformation can vary only if S stays on the surface $f(S) = 0$, moving along it. For all other processes, D' is zero. This is the reason why the equality $|S| = g$ is called the *yield* (or flow) *condition.*

Usually for the Bingham model the incompressibility of the volume is also assumed, i.e.

$$\operatorname{tr} D = 0 \tag{1.8}$$

for any process of any duration $\tau > 0$.

Further on, the Bingham model will be considered with the tensors D and T' only. For this reason note that from (1.4) and (1.6) we get

$$T' = (1 + 4\eta\lambda)S \tag{1.9}$$

$$|T'| = (1 + 4\eta\lambda)|S|. \tag{1.10}$$

If $|T'| > g$ then (1.10), (1.5) implies $\lambda > 0$ and from (1.7) we get $|S| = g > 0$. From (1.10) is follows that

$$\lambda = \frac{1}{4\eta}\left(\frac{|T'|}{g} - 1\right)$$

and hence by (1.6), (1.9) we get

$$D' = \frac{2}{1 + 4\eta\lambda} T' = \frac{1}{2\eta}\left(1 - \frac{g}{|T'|}\right)T',$$

and using (1.2), (1.8) we obtain

$$D = \frac{1}{2\eta}\left(1 - \frac{g}{|T'|}\right)T'.$$

Let us suppose now $|T'| \leq g$; if $|S| = g$, from (1.10) we get $\lambda = 0$; if $|S| < g$, then from (1.7) we also get $\lambda = 0$. Hence, in this case by (1.6) and (1.8) we have $D = 0$.

In consequence we get the constitutive equations of the Bingham fluid

$$D = \begin{cases} \dfrac{1}{2\eta}\left(1 - \dfrac{g}{|T'|}\right)T' & \text{if } |T'| > g \\ 0 & \text{if } |T'| \leq g. \end{cases} \tag{1.11}$$

We can invert the constitutive equations (1.11). Indeed, let $|D| = 0$; from (1.11) we get $|T'| \leq g$; let $|D| \neq 0$; from (1.11) it follows that $|T'| > g$ and $|T'| = 2\eta|D| + g$; using this result and (1.8), the equation

$$D = \frac{1}{2\eta}\left(1 - \frac{g}{|T'|}\right)T'$$

leads to

$$T' = 2\eta D + g\frac{D}{|D'|}.$$

Hence, from (1.11) we obtain

$$\begin{cases} T' = 2\eta D + g\dfrac{D}{|D|} & \text{if } |D| \neq 0 \\ |T'| \leq g & \text{if } |D| = 0 \end{cases} \tag{1.12}$$

It is easy to observe that conversely, (1.12) implies (1.11) hence, (1.11) and (1.12) are equivalent. Subsequently we shall refer to (1.12) as the constitutive equations of the Bingham fluid.

Remark 1.1 *If in* (1.12) $g = 0$ *one recovers the constitutive law for a classical viscous incompressible fluid* (*Newtonian fluid*). *Thus, for small* g, *we can consider the Bingham fluid as a model close to the classical viscous fluid.*

The type of behaviour described by the equations (1.11) or (1.12) can be observed in the case of certain oils or certain sediments which are used in the process of oil drilling. If g is strictly positive, one can observe rigid zones in the interior of the flow. As g increases, these rigid zones become larger and may completely block the flow when g is sufficiently large. This property, called the *blocking property* for the Bingham fluid, will be more precisely stated in Section 2.

Though the Bingham model is a model of fluid body, it was also called the Bingham solid (see for instance Oldroyd, 1947a–c) and it was used to describe the deformation and flow of many solid bodies. This model was therefore often used in metal-forming processes; it was first introduced for wire drawing (Cristescu, 1975) and intensively used hereafter (Cristescu, 1976, 1977, 1980).

In what will follow we give an equivalent form of the constitutive equations (1.12) by means of the subdifferential operator. Let $\varphi: \mathscr{S}_3 \to \mathbb{R}$ be the function defined by

$$\varphi(D) = \eta|D|^2 + g|D| \qquad \forall D \in \mathscr{S}_3 \tag{1.13}$$

and let $\partial\varphi$ be the subdifferential of φ (see A2.6). We have the following result, used in Section 1.3:

Lemma 1.1 *The constitutive equations* (1.12) *are equivalent with the multi-valued equation* $T' \in \partial\varphi(D)$.

PROOF. Let $D, T' \in \mathscr{S}_3$ such that (1.12) holds and $S \in \mathscr{S}_3$. If $|D| = 0$ by (1.12) and (1.13) we have $T'\cdot(S - D) = T'\cdot S \le |T'|\cdot|S| \le g|S| \le \varphi(S) - \varphi(D)$ hence $T' \in \partial\varphi(D)$. If $|D| \ne 0$, then using (1.12), the Schwarz inequality, and (1.13) we get

$$\begin{aligned}
T'\cdot(S - D) &= \left(2\eta D + g\frac{D}{|D|}\right)\cdot(S - D) \\
&= 2\eta D\cdot S - 2\eta|D|^2 + \frac{g}{|D|}D\cdot S - g|D| \\
&\le 2\eta D\cdot S - 2\eta|D|^2 + g|S| - g|D| \\
&\le \eta|D|^2 + \eta|S|^2 - 2\eta|D|^2 + g|S| - g|D| \\
&= \varphi(S) - \varphi(D);
\end{aligned}$$

hence $T' \in \partial\varphi(D)$. So, we proved that (1.12) implies $T' \in \partial\varphi(D)$.

Conversely, let $D, T' \in \mathscr{S}_3$ such that $T' \in \partial\varphi(D)$. From (A2.6) we have

$$T'\cdot(S - D) \le \varphi(S) - \varphi(D) \qquad \forall S \in \mathscr{S}_3. \tag{1.14}$$

Let us suppose $|D| = 0$; if $|T'| = 0$ we have obviously $|T'| \le g$; if $T' \ne 0$, taking $S = \varepsilon T'/|T'|$ in (1.14) where $\varepsilon > 0$; we have $\varepsilon|T'| \le \eta\varepsilon^2 + g\varepsilon$. Dividing by ε and passing to the limit when $\varepsilon \to 0$ we obtain $|T'| \le g$. Hence, we have proved that

$$|D| = 0 \Rightarrow |\mathrm{T}'| \le g. \tag{1.15}$$

Let us suppose now that $|D| \ne 0$; taking $S = D + \varepsilon W$ in (1.14) where $\varepsilon > 0$

and $W \in \mathscr{S}_3$ we have $\varepsilon T' \cdot W \leq \varphi(D + \varepsilon W) - \varphi(D)$; hence

$$\begin{aligned}\varepsilon T' \cdot W &\leq \eta|D + \varepsilon W|^2 + g|D + \varepsilon W| - \eta|D|^2 - g|D| \\ &= \eta\varepsilon^2|W|^2 + 2\eta\varepsilon D \cdot W + g|D + \varepsilon W| - g|D|.\end{aligned}$$

Dividing by ε and passing to the limit when $\varepsilon \to 0$ we have

$$T' \cdot W \leq 2\eta D \cdot W + \lim_{\varepsilon \to 0} \frac{g|D + \varepsilon W| - g|D|}{\varepsilon}.$$

But

$$\lim_{\varepsilon \to 0} \frac{g|D + \varepsilon W| - g|D|}{\varepsilon} = g \frac{\mathrm{d}}{\mathrm{d}\varepsilon} (|D|^2 + \varepsilon^2|W|^2 + 2\varepsilon D \cdot W)^{1/2}_{\varepsilon = 0} = g \frac{D \cdot W}{|D|}$$

and hence we obtained the inequality

$$T' \cdot W \leq 2\eta D \cdot W + g \frac{D \cdot W}{|D|}$$

for all $W \in \mathscr{S}_3$. It follows that

$$T' \cdot W = 2\eta D \cdot W + g \frac{D}{|D|} \cdot W \qquad \forall W \in \mathscr{S}_3$$

and hence $T' = 2\eta D + gD/|D|$. Hence we have proved that

$$|D| \neq 0 \Rightarrow T' = 2\eta D + g \frac{D}{|D|}. \tag{1.16}$$

By (1.15) and (1.16) we have (1.12) hence the proof of Lemma 1.1 is complete.

1.2 Statement of the problems and friction laws

In this section we consider two boundary-value problems for the Bingham fluid with friction. For this let us consider a bounded domain $\Omega \subset \mathbb{R}^3$ with a smooth boundary $\partial\Omega = \Gamma$ and the following problem in Ω:

Find the velocity function $u: \Omega \to \mathbb{R}^3$ and the stress function $T': \Omega \to \mathscr{S}_3$ such that

$$\left.\begin{aligned} T' &= 2\eta D + g \frac{D}{|D|} && \text{if } |D| \neq 0 \\ |T'| &\leq g && \text{if } |D| = 0 \end{aligned}\right\} \tag{1.17}$$

$$D = D(u) = \tfrac{1}{2}(\nabla u + \nabla^T u) \tag{1.18}$$

$$\operatorname{Div} T + b = 0 \tag{1.19}$$

$$\operatorname{tr} D = \operatorname{div} u = 0. \tag{1.20}$$

Problem (1.17)–(1.20) describes the stationary flow of the Bingham fluid in Ω. Here Ω represents the region fulfilled by the fluid in the Eulerian frame, i.e. the Eulerian coordinate is taken as a spatial variable. Equations (1.17) represent the constitutive equations of the Bingham model and have already been discussed in the previous section. The kinematic equations (1.18), the incompressibility condition (1.20) and the equilibrium equations (1.19) in which b are the given body forces are discussed in Section 2 of Chapter 1. Subsequently we use the notation $D(u)$ for the rate deformation tensor defined by (1.18), in order to underline the dependence of this tensor on the velocity function u.

In order to complete (1.17)–(1.20) with the boundary conditions, let us consider that Γ is divided into five disjoint parts $\Gamma = \Gamma_1 \cup \Gamma_2 \cup \Gamma_3 \cup \Gamma_4 \cup \Gamma_5$. The following boundary conditions are imposed:

$$u = U \qquad \text{on } \Gamma_1 \tag{1.21}$$

$$t = F \qquad \text{on } \Gamma_2 \tag{1.22}$$

$$u_\tau = 0, \quad t \cdot v = f_v \qquad \text{on } \Gamma_3 \tag{1.23}$$

$$u \cdot v = 0, \qquad t_\tau = 0 \qquad \text{on } \Gamma_4. \tag{1.24}$$

In (1.21)–(1.24) v is the exterior unit normal of Γ, $t = T \cdot v$ is the stress vector and for all v we used the notation $v_\tau = v - (v \cdot v)v$ for the tangential component of v.

On Γ_5 we shall consider some friction laws. One of them is the following local friction law:

$$\left.\begin{array}{l} \qquad\qquad u \cdot v = 0 \\ |t_\tau| \leq \mu S \quad \begin{cases} \text{if } |t_\tau| < \mu S \text{ then } u_\tau = 0 \\ \text{if } |t_\tau| = \mu S \text{ then there exists } \lambda \geqslant 0 \\ \text{such that } u_\tau = -\lambda t_\tau \end{cases} \end{array}\right\} \text{ on } \Gamma_5 \tag{1.25}$$

in which μ is the *friction coefficient*.

If S is the modulus of the normal stress $t \cdot v$ then (1.25) is the classical Coulomb friction law. For some metal-forming problems it is more realistic to use a viscoplastic friction law (introduced in Cristescu, 1975) supposing that $S = |T'|$. Everywhere when (1.25) will be used we shall consider S as a given datum.

If we compute $|T'|$ from (1.17) we obtain $|T'| = g + 2\eta|D(u)|$. Hence if $S = |T'| = g + 2\eta|D(u)|$ then the non-linear boundary condition (1.25) depends on the solution itself. The functions of the set the solution must belong to are not smooth enough for this local friction law to make sense. We shall remove this difficulty by replacing the local friction law by a non-local one. Some heuristic arguments in support of a non-local friction law can be found in Oden and Pires (1981) and some mathematical arguments will be presented here.

Let us consider $w, w_h \in \mathscr{D}(\mathbb{R}^N)$, $h \in (0, 1]$ a regularization function given for instance by (1.8), (1.9) of Chapter 2. We shall denote by $r_h \in C^0(\Gamma)$ the function

$$r_h(x) = \int_\Omega w_h(x - y)\, dy \qquad \forall x \in \Gamma. \tag{1.26}$$

For all $v \in L^1_{\text{loc}}(\mathbb{R}^N)$ we also denote by $L_h v \in C^0(\Gamma)$ the function given by

$$(L_h v)(x) = r_h^{-1}(x) \int_\Omega w_h(x - y) v(y)\, dy. \tag{1.27}$$

The following theorem shows that $L_h v$ is a good approximation of the restriction of v on the boundary Γ if v is smooth enough for this restriction to make sense.

Theorem 1.1 *Let $\Omega \subset \mathbb{R}^N$ be a bounded domain.*

(i) If Ω has the cone property then there exists $d > 0$ such that

$$d \le r_h(x) \le 1 \qquad \forall x \in \Gamma, \quad h \in (0, 1). \tag{1.28}$$

Moreover for all $v \in C^0(\bar{\Omega})$ we have

$$L_h v \to v \qquad \text{in } C^0(\Gamma) \quad \text{when } h \to 0. \tag{1.29}$$

(ii) If Ω has the strong local Lipschitz property, then there exists $h_0 > 0$ and $C_p > 0$ such that

$$\| L_h v \|_{L^p(\Gamma)} \le C_p \| v \|_{W^{1,p}(\Omega)} \qquad \forall h \in (0, h_0), \quad v \in W^{1,p}(\Omega). \tag{1.30}$$

Moreover for all $v \in W^{1,p}(\Omega)$ we have

$$L_h v \to \gamma_1(v) \text{ strongly in } L^p(\Gamma) \text{ when } h \to 0. \tag{1.31}$$

Let $(\mathscr{L}_h \tau)_{ij} = L_h \tau_{ij}$ for all $\tau \in [L^1_{\text{loc}}(\mathbb{R}^N)]^{N \times N}$. The non-local friction law on Γ_5 can be formulated now as follows:

$$\left.\begin{array}{l} u \cdot v = 0 \\ |t_\tau| \le \mu(g + 2\eta |\mathscr{L}_h D(u)|) \text{ and} \\ \qquad \text{if } |t_\tau| < \mu(g + 2\eta |\mathscr{L}_h D(u)|) \text{ then } u_\tau = 0 \\ \qquad \text{if } |t_\tau| = \mu(g + 2\eta |\mathscr{L}_h D(u)|) \text{ then there exists } \lambda \ge 0 \\ \qquad\qquad \text{such that } u_\tau = -\lambda t_\tau. \end{array}\right\} \text{ on } \Gamma_5. \tag{1.32}$$

Remark 1.2 *One can consider in (1.32) $L_h |D(u)|$ instead of $|\mathscr{L}_h D(u)|$. All the results of this section still hold in this case.*

PROOF OF THEOREM 1.1. (i) Let $x \in \Gamma$ and C_x be a finite cone (which is congruent with a fixed cone C of vertex 0) of vertex x such that $C_x \subset \Omega$.

From (1.26) we have

$$r_h(x) \geq \frac{1}{h^N} \int_{C_x} w\left(\frac{x-y}{h}\right) \mathrm{d}y = \int_{D_h} w(z)\,\mathrm{d}z,$$

where $D_h = (1/h)(x - C_x)$ is a finite cone of vertex 0 which is congruent with $(1/h)C \supset C$. If we denote by $d = \int_C w(z)\,\mathrm{d}z$ then we get (1.28).

For all $v \in C^0(\bar{\Omega})$ and $x \in \Gamma$ we have

$$v(x) - (L_h v)(x) = r_h^{-1}(x) \int_\Omega w_h(x-y)(v(y) - v(x))\,\mathrm{d}y.$$

Since v is uniformly continuous on $\bar{\Omega}$, for all $\varepsilon > 0$ there exists $\delta_\varepsilon > 0$ such that $|v(x) - v(y)| < \varepsilon$ for all $x, y \in \bar{\Omega}$ with $|x - y| < \delta_\varepsilon$. If $h < \delta_\varepsilon$, then we have

$$|v(x) - (L_h v)(x)| \leq \varepsilon/d \int_\Omega w_h(x-y)\,\mathrm{d}y \leq \varepsilon/d.$$

Hence $\|v - L_h v\|_{C^0(\Gamma)} \leq \varepsilon/d$ for all $h < \delta_\varepsilon$.

(ii) Let us remark that for all $v \in L^1(\Omega)$, $x \in \Gamma$ we have $|L_h(v)(x)| \leq C/(h^N d)\|v\|_{L^1(\Omega)}$; hence

$$L_h \in B(L^1(\Omega), C^0(\Gamma)) \qquad \forall h \in (0, 1]. \tag{1.33}$$

From (1.33) we deduce $L_h \in B(W^{1,p}(\Omega), L^p(\Gamma))$ and since $C^1(\bar{\Omega})$ is a dense subset of $W^{1,p}(\Omega)$ we deduce that it is enough to prove (1.30) for all $v \in C^1(\bar{\Omega})$, $h \in (0, h_0)$.

Let $(U_i)_{i=\overline{1,M}}$ be a collection of open bounded sets such that $\Gamma \subset \bigcup_{i=1}^M U_i$, let $\tau_i \in \mathbb{R}^N$, $|\tau_i| = 1$, and $t_i > 0$, $i = \overline{1, M}$ such that $\overline{\Omega \cap U_i} + t\tau_i \subset \Omega$ for all $t \in (0, t_i]$. We consider r_i such that $\Delta_i = \overline{\Omega \cap U_i + t_i\tau_i + B(0, r_i/2)} \subset \Omega$. Let $(U_i')_{i=\overline{1,M}}$ such that $\Gamma \subset \bigcup_{i=1}^M U_i'$ and $U_i' \subset\subset U_i$ for all $i = \overline{1, M}$ and let $\Gamma_i' = \Gamma \cap U_i'$. We choose $h_0 > 0$ such that $U_i' + B(0, h_0) \subset\subset U_i$ and $h_0 < \frac{1}{2}\min_{i=\overline{1,M}}(t_i, r_i)$ and we denote by $\Gamma_i = \Gamma \cap U_i$. Let i and $v \in C^1(\bar{\Omega})$ be fixed and let us denote by $\Omega_i = \Omega \cap U_i$ and $\tilde{v}_i \in L^p(\mathbb{R}^N)$ such that $\tilde{v}_i(x) = v(x + t_i\tau_i)$ if $x + t_i\tau_i \in \Omega$ and $\tilde{v}_i(x) = 0$ if $x + t_i\tau_i \notin \Omega$. For all $u \in L^p(\Omega)$ we shall denote by $\tilde{u} \in L^p(\mathbb{R}^N)$ the function which is equal to u on Ω and vanishes outside Ω. For all $y \in \Gamma_i' + B(0, h_0)$ we have

$$|\tilde{v}|(y) \leq |\tilde{v}_i|(y) + \int_0^{t_i} |\widetilde{\nabla v}\cdot\tau_i|(y + \xi\tau_i)\,\mathrm{d}\xi;$$

hence for all $x \in \Gamma_i'$ we obtain

$$(w_h * \tilde{v})(x) \leq (w_h * |\tilde{v}_i|)(x) + \int_0^{t_i} (w_h * |\widetilde{\nabla v}\cdot\tau_i|)(x + \xi\tau_i)\,\mathrm{d}\xi.$$

But

$$\int_0^{t_i} (w_h * |\widetilde{\nabla v}\cdot\tau_i|)(x + \xi\tau_i)\,\mathrm{d}\xi \leq c \int_0^{t_i} \int_{B(0,1)} |\widetilde{\nabla v}|(x + hz + \xi\tau_i)\,\mathrm{d}z\,\mathrm{d}\xi;$$

hence there exists $C_p^1 > 0$ such that for all $x \in \Gamma_i'$ we have

$$|(w_h * \tilde{v})(x)|^p \le C_p^1 \left[|w_h * |\tilde{v}_i||^p(x) + \int_0^{t_i} \int_{B(0,1)} |\widetilde{\nabla v}|^p(x + hz + \xi\tau_i)\,\mathrm{d}z\,\mathrm{d}\xi \right]. \tag{1.34}$$

Since Γ_i' is the graph of a Lipschitz function there exists $C_\gamma > 0$ such that

$$\int_{\Gamma_i'} |w_h * |\tilde{v}_i||^p(x)\,\mathrm{d}x \le C_\gamma \left[\int_{\Omega_i} |w_h * |\tilde{v}_i||^p(x)\,\mathrm{d}x + \int_{\Omega_i} |\nabla(w_h * |\tilde{v}_i|)|^p(x)\,\mathrm{d}x \right]. \tag{1.35}$$

But

$$\begin{aligned} \int_{\Omega_i} |w_h * |\tilde{v}_i||^p(x)\,\mathrm{d}x &= \int_{\Omega_i + t_i\tau_i} |w_h * |\tilde{v}||^p(x)\,\mathrm{d}x \\ &\le c^p \int_{B(0,1)} \left[\int_{\Omega_i + t_i\tau_i} |v|^p(x + hz)\,\mathrm{d}x \right] \mathrm{d}z \\ &\le c^p \left(\int_{\Omega_i + t_i\tau_i + B(0,h_0)} |v|^p(x)\,\mathrm{d}x \right) \operatorname{vol}(B(0,1)); \end{aligned}$$

hence we have just obtained

$$\int_{\Omega_i} |w_h * |\tilde{v}_i||^p(x)\,\mathrm{d}x \le c^p \operatorname{vol}(B(0,1)) \|v\|^p_{L^p(\Omega)}. \tag{1.36}$$

Since $|\tilde{v}_i| \in W^{1,p}(\Omega_i + B(0, h_0))$ we have

$$\nabla(w_h * |\tilde{v}_i|)(x) = (w_h * \nabla|\tilde{v}_i|)(x) = (w_h * \nabla|\tilde{v}|)(x + t_i\tau_i)$$

for all $x \in \Omega_i$. Hence we can deduce

$$\int_{\Omega_i} |\nabla(w_h * |\tilde{v}_i|)|^p(x)\,\mathrm{d}x \le c^p \operatorname{vol}(B(0,1)) \|\nabla v\|^p_{L^p(\Omega)}. \tag{1.37}$$

From (1.35), (1.36) and (1.37) we deduce that there exists $C_p^2 > 0$ such that

$$\int_{\Gamma_i'} |w_h * |\tilde{v}_i||^p(x)\,\mathrm{d}x \le C_p^2 \|v\|_{W^{1,p}(\Omega)}; \tag{1.38}$$

If we have in mind that

$$\begin{aligned} &\int_{\Gamma'} \int_0^{t_i} \int_{B(0,1)} |\widetilde{\nabla v}|^p(x + hz + \xi\tau)\,\mathrm{d}z\,\mathrm{d}\xi\,\mathrm{d}x \\ &\qquad = \int_{B(0,1)} \int_{\Gamma'} \int_0^{t_i} |\widetilde{\nabla v}|^p(x + hz + \xi\tau)\,\mathrm{d}\xi\,\mathrm{d}x\,\mathrm{d}z \\ &\qquad \le \operatorname{vol}(B(0,1)) \|\nabla v\|^p_{L^p(\Omega_i)} \end{aligned}$$

from the above inequality and (1.34), (1.38) we deduce

$$\|w_h * \tilde{v}\|^p_{L^p(\Gamma)} \le C_p^3 \|v\|^p_{W^{1,p}(\Omega)}, \tag{1.39}$$

and since

$$|(L_h v)(x)|^p = |r_h^{-1}(x)(w_h * \tilde{v})(x)|^p \le \frac{1}{d^p} |w_h * \tilde{v}|^p(x)$$

from (1.39) we deduce (1.30).

The convergence (1.31) easily follows from (1.29)–(1.30) and the density of $C^1(\bar{\Omega})$ in $W^{1,p}(\Omega)$.

1.3 An existence and uniqueness result in the local friction law case

In this section we study the boundary-value problem (1.17)–(1.25). For this, let us consider the set of kinematic admissible fields defined by

$$K = \{v \in \boldsymbol{H}_1 \,|\, \operatorname{div} v = 0 \text{ in } \Omega,\ v|_{\Gamma_1} = U,\ v_\tau|_{\Gamma_3} = 0,\ v \cdot \nu|_{\Gamma_4 \cup \Gamma_5} = 0\}. \tag{1.40}$$

In connection with the data which appear in (1.21)–(1.25), (1.19) we suppose the following:

$$K \neq \varnothing \tag{1.41}$$

$$F \in [L^2(\Gamma_2)]^3 \tag{1.42}$$

$$f_\nu \in L^2(\Gamma_3) \tag{1.43}$$

$$b \in \boldsymbol{H} \tag{1.44}$$

$$\mu \in L^\infty(\Gamma_5), \qquad \mu \ge 0 \quad \text{a.e. on } \Gamma_5 \tag{1.45}$$

$$S \in L^\infty(\Gamma_5), \qquad S \ge 0 \quad \text{a.e. on } \Gamma_5. \tag{1.46}$$

Under the above assumptions we give the variational formulation for the problem (1.17)–(1.25) and prove an existence and uniqueness result. For this, let us introduce the following notations:

$$a\colon \boldsymbol{H}_1 \times \boldsymbol{H}_1 \to \mathbb{R}, \qquad a(u, v) = \eta \int_\Omega D(u) \cdot D(v)\, \mathrm{d}x \tag{1.47}$$

$$\tilde{j}\colon \boldsymbol{H}_1 \to \mathbb{R}, \qquad \tilde{j}(v) = g \int_\Omega |D(v)|\, \mathrm{d}x + \int_{\Gamma_5} \mu |S| |v_\tau|\, \mathrm{d}\Gamma \tag{1.48}$$

$$\tilde{f}\colon \boldsymbol{H}_1 \to \mathbb{R}, \qquad \tilde{f}\langle v \rangle = \int_\Omega b \cdot v\, \mathrm{d}x + \int_{\Gamma_2} F \cdot v\, \mathrm{d}\Gamma + \int_{\Gamma_3} f_\nu v \cdot \nu\, \mathrm{d}\Gamma \tag{1.49}$$

$$J\colon \boldsymbol{H}_1 \to \mathbb{R}, \qquad J(v) = \tfrac{1}{2} a(v, v) + \tilde{j}(v) - \tilde{f}\langle v \rangle. \tag{1.50}$$

We have the following result:

Theorem 1.2 *In hypothesis* (1.41)–(1.46) *if* $(u; T)$ *is a smooth solution for the problem* (1.17)–(1.25) *then* u *is a minimum point of* J *on* K, *i.e.*

$$u \in K, \quad J(u) \leq J(v) \qquad \forall v \in K. \tag{1.51}$$

PROOF. Let u and T be smooth functions satisfying (1.17)–(1.25). For all $v \in K$ using (1.19) and (1.20) we have

$$\int_{\Omega} b(v-u)\,dx = -\int_{\Omega} \operatorname{Div} T\cdot(v-u)\,dx = -\int_{\Gamma} t(v-u)\,d\Gamma + \int_{\Omega} T'\cdot D(v-u)\,dx.$$

Hence, from (1.21)–(1.25) it follows that

$$\begin{aligned}\int_{\Omega} b(v-u)\,dx &= -\int_{\Gamma_2} F(v-u)\,d\Gamma - \int_{\Gamma_3} f_{\nu}(v\cdot\nu - u\cdot\nu)\,d\Gamma - \int_{\Gamma_5} t_{\tau}(v_{\tau}-u_{\tau})\,d\Gamma \\ &\quad + \int_{\Omega} T'\cdot D(v-u)\,dx \\ &\leq -\int_{\Gamma_2} F(v-u)\,d\Gamma - \int_{\Gamma_3} f_{\lambda}(v\cdot\nu - u\cdot\nu)\,d\Gamma \\ &\quad + \int_{\Gamma_5} \mu S(|v_{\tau}| - |u_{\tau}|) + \int_{\Omega} T'\cdot D(v-u)\,dx.\end{aligned}$$

Using now Lemma 1.1 and (1.47)–(1.50) we get (1.51).

Remark 1.3 *Applying Lemma A2.6(i) with* $\varphi(v) = \tilde{j}(v) + \psi_K(v)$ *for all* $v \in \boldsymbol{H}_1$ *we get that the minimum problem* (1.51) *is equivalent to the following variational inequality on* K:

$$u \in K,\ a(u, v-u) + \tilde{j}(v) - \tilde{j}(u) \geq \langle \tilde{f}, v-u\rangle \qquad \forall v \in K. \tag{1.52}$$

Next, by means of a translation, we shall obtain homogeneous boundary conditions. Let $u_0 \in K$ (see (1.41)) and define

$$V = K - u_0 = \{v \in \boldsymbol{H}_1 \,|\, \operatorname{div} v = 0 \text{ in } \Omega,\ v|_{\Gamma_1} = 0,\ v_3|_{\Gamma_3} = 0,\ v\cdot\nu|_{\Gamma_4\cup\Gamma_5} = 0\}. \tag{1.53}$$

V is closed subspace of $\boldsymbol{H}_1$; we consider the functional $j\colon \boldsymbol{H}_1 \to \boldsymbol{R}$ and the element $f \in V'$ defined by

$$j(v) = \tilde{j}(v + u_0) \tag{1.54}$$

$$\langle f, v\rangle = \langle \tilde{f}, v\rangle + a(u_0, v) \tag{1.55}$$

for all $v \in V$. With these notations it is easy to verify the following result:

Remark 1.4 *The element $u \in K$ is a solution for the variational inequality* (1.52) *iff the element $\bar{u} = u - u_0 \in V$ is a solution for the variational inequality*

$$\bar{u} \in V, \quad a(\bar{u}, v - \bar{u}) + j(v) - j(\bar{u}) \geq \langle f, v - \bar{u} \rangle \qquad \forall v \in V. \tag{1.56}$$

As follows from Theorem 1.2 and Remarks 1.3, 1.4, if (u, T) is a smooth solution of the problem (1.17)–(1.25), then the function $u - u_0$ is a solution for the variational inequality (1.56). This result allows us to consider (1.56) as a *variational formulation* of the mechanical problem (1.17)–(1.25) and, except for a translation, we shall consider every solution of (1.56) as a solution of the problem (1.17)–(1.25).

Moreover, we have the following existence and uniqueness result:

Theorem 1.3 *Let* (1.41)–(1.46) *hold. Then, if* meas $\Gamma_1 > 0$, *there exists a unique solution $\bar{u} = u - u_0$ of the variational inequality* (1.56).

PROOF. It can be easily proved that the functional j defined by (1.54), (1.48) is convex and continuous on V and the form a defined by (1.47) is bilinear, symmetric and bounded; since meas $\Gamma_1 > 0$ we can apply Korn's inequality (2.32) from Chapter 2, and we have

$$\text{there exists } \rho > 0 \text{ so that } a(v, v) \geq 2\eta\rho \|v\|_{\boldsymbol{H}_1}^2 \tag{1.57}$$

for all $v \in V$; it results that a is a coercive form on V. The statement of Theorem 3.2 results now from Corollary A2.4.

1.4 An existence result in the non-local friction law case

In this section we shall give the variational formulation of the problem (1.17)–(1.24), (1.32) and we shall prove an existence result. The uniqueness is also studied in connection with the friction coefficient μ.

For the sake of simplicity let us consider $\mu(x) = \mu > 0$ constant on Γ_5. Let

$$d\colon \boldsymbol{H}_1 \times \boldsymbol{H}_1 \to \mathbb{R} \qquad d(u, v) = \int_{\Gamma_5} (g + 2\eta|\mathscr{L}_h D(u)|) \cdot |v| \, \mathrm{d}\Gamma \tag{1.58}$$

$$m\colon \boldsymbol{H}_1 \to \mathbb{R} \qquad m(u) = \int_{\Omega} g|D(u)| \, \mathrm{d}x. \tag{1.59}$$

If a is given by (1.47), $\tilde{f}$ by (1.49), and K by (1.40), then we have the following variational formulation of the problem (1.17)–(1.24), (1.32):

Theorem 1.4 *Let* (1.41)–(1.45) *hold. If $(u; T)$ is a smooth solution of the problem* (1.17)–(1.24), (1.32), *then u is the solution of the following variational inequality*:

$$u \in K, \quad a(u, u - v) + m(u) - m(v) + \mu d(u, u) - \mu d(u, v) \leq \langle \tilde{f}, v - u \rangle \quad \forall v \in K \tag{1.60}$$

The proof of this statement is almost the same as the proof of Theorem 1.2, so we can omit it here.

Concerning the variational inequality (1.60) we have the following existence result:

Theorem 1.5 *Let* (1.41)–(1.45) *hold and suppose* $\text{meas}(\Gamma_1) > 0$. *Then there exists at least one solution of the variational inequality* (1.60).

In order to prove Theorem 1.5 the following lemma will be useful:

Lemma 1.2 (*i*) *There exists* $C_h > 0$ (*which depends only on* Ω *and* $h > 0$) *such that*

$$\|\mathscr{L}_h \tau\|_{[L^2(\Gamma)]^{N \times N}} \leq C_h \|\tau\|_{\mathscr{H}} \qquad \forall \tau \in \mathscr{H}. \tag{1.61}$$

(*ii*) *The operator* $\mathscr{L}_h \in B(\mathscr{H}, [L^2(\Gamma)]^{N \times N})$ *is continuous from* $\mathscr{H}$ *weak to* $[L^2(\Gamma)]^{N \times N}$ *strong.*

PROOF. (i) From (1.33) one can easily deduce (1.61).

(ii) Let $I_h \in B(H, H_1)$ be given by

$$(I_h v)(x) = \int_\Omega w_h(x - y) v(y)\, dy \qquad \forall x \in \Omega, \quad v \in H \tag{1.62}$$

and let us prove that I_h is continuous from H weak to H_1 weak. Indeed let $v_n \to 0$ weakly in H and let $\varphi \in \mathscr{D}(\Omega)$. Note that $\langle I_h v_n, \varphi \rangle_H = \langle v_n, I_h \varphi \rangle_H \to 0$; hence $I_h v_n \to 0$ weakly in H. If we bear in mind that

$$\left\langle \frac{\partial}{\partial x_i}(I_h v_n), \varphi \right\rangle_H = -\left\langle I_h v_n, \frac{\partial \varphi}{\partial x_i} \right\rangle_H \to 0,$$

we deduce that $\nabla(I_h v_n) \to 0$ weakly in $\boldsymbol{H}$; hence $I_h v_n \to 0$ weakly in H_1. Since

$$(\mathscr{L}_h \tau)_{ij} = r_h^{-1} \gamma_1 (I_h \tau_{ij}) \qquad \forall \tau \in \mathscr{H}, \tag{1.63}$$

and γ_1 is compact from H_1 to $L^2(\Gamma)$, we deduce that $\mathscr{L}_h$ is also compact from $\mathscr{H}$ to $[L^2(\Gamma)]^{N \times N}$.

PROOF OF THEOREM 1.5. Let us denote by $G: K \times K \to \mathbb{R}$ the function given by

$$\begin{aligned} G(u, v) = {} & a(u, u - v) + m(u) - m(v) + \mu d(u, u) \\ & - \mu d(u, v) - \langle \tilde{f}, v - u \rangle \qquad \forall u, v \in K. \end{aligned} \tag{1.64}$$

It can be easily seen that (1.60) can be written as

$$u \in K,\ G(u, v) \leq 0 \qquad \forall v \in K. \tag{1.65}$$

In order to use a consequence of Ky-Fans' theorem (Theorem A2.2), let us remark that from the linearity of $v \to a(u, v) + \tilde{f}(u, v)$ for all $u \in K$ and the convexity of $v \to m(v)$ and $v \to \mu d(u, v)$ for all $u \in K$ we deduce that for all $u \in K$, $v \to G(u, v)$ is concave. From Lemma 1.2 we deduce that $u \to -a(u, v) + m(u) + \mu d(u, u) + \mu d(u, v) - \tilde{f}(u)$ is weakly continuous on $\boldsymbol{H}_1$. Bearing in mind that $u \to a(u, u)$ is convex and continuous we deduce that $u \to a(u, u)$ is weakly lower semicontinuous. Hence for all $v \in K$, $u \to G(u, v)$ is weakly lower semicontinuous. Let $M > 0$ such that

$$|a(u, v)| \leq 2\eta M \|u\|_{\boldsymbol{H}_1} \|v\|_{\boldsymbol{H}_1} \qquad \forall u, v \in \boldsymbol{H}_1 \tag{1.66}$$

and let $u_0 \in K$ (see (1.41)). From (1.57) we deduce that

$$a(u - v, u - v) \geq 2\eta\rho \|u - v\|^2_{\boldsymbol{H}_1} \qquad \forall u, v \in K. \tag{1.67}$$

From (1.66) and (1.67) we have

$$G(v, u_0) \geq a(v - u_0, v - u_0) + a(u_0, v - u_0) - m(u_0) - \mu d(v, u_0) - \langle \tilde{f}, v - u_0 \rangle$$

$$\geq 2\eta\rho \|v - u_0\|^2_{\boldsymbol{H}_1} - 2\eta \|u_0\|_{\boldsymbol{H}_1} \|v - u_0\|_{\boldsymbol{H}_1} - m(u_0) - \mu(gC_\gamma + 2C_h\eta M \|v\|_{\boldsymbol{H}_1});$$

hence $\lim_{\|v\|_{H1} \to +\infty} G(v, u_0) = +\infty$. We can use now Corollary A2.3 and we deduce the statement of Theorem 1.5.

Remark 1.5 *Let S: $K \to K$ be the map which associates with every $u \in K$ the solution of the following variational inequality*:

$$v \in K;\ a(v, v - w) + m(v) - m(w) + \mu d(u, v) - \mu d(u, w) \leq \langle \tilde{f}, v - w \rangle \qquad \forall w \in K. \tag{1.68}$$

We can easily observe that any solution of (1.60) *is a fixed point for S and conversely. Hence, another proof of Theorem* 1.5 *can be given using the Tihonov fixed-point theorem and proving that S has at least one fixed point.*

The following theorem states a uniqueness result of the variational inequality (1.60) for small fraction coefficient μ:

Theorem 1.6 *Let* (1.41)–(1.45) *hold and suppose* meas $\Gamma_1 > 0$. *Then there exists μ_0 (which depends only on Ω, Γ_1, and h) such that for all $\mu \in [0, \mu_0)$ the variational inequality* (1.60) *has at most one solution.*

PROOF. In order to prove the uniqueness we shall prove that G given by (1.65) is strictly monotone, i.e.

$$G(u, v) + G(v, u) > 0 \qquad \forall u, v \in K, \quad u \neq v. \tag{1.69}$$

If we denote by C_γ the norm of $\gamma_1 \in B(\boldsymbol{H}_1, [L^2(\Gamma)]^N)$, then from (1.61) and (1.66) and from the inequality

$$\int_{\Gamma_5} (|\mathscr{L}_h D(u)| - |\mathscr{L}_h D(v)|)(|u| - |v|)\, \mathrm{d}\Gamma \leq \int_{\Gamma_5} |\mathscr{L}_h D(u - v)| \cdot |u - v|\, \mathrm{d}\Gamma,$$

we deduce

$$\int_{\Gamma_5} (|\mathscr{L}_h D(u)| - |\mathscr{L}_h D(v)|)(|u| - |v|)\, d\Gamma \le M C_h C_\gamma \|u - v\|^2_{H_1} \qquad (1.70)$$

After some computation we have

$$G(u, v) + G(v, u) = a(u - v, u - v) + 2\eta\mu \int_{\Gamma_5} (|\mathscr{L}_h D(u)| - |\mathscr{L}_h D(v)|)(|u| - |v|)\, d\Gamma$$

and from (1.70) and (1.67) we obtain

$$G(u, v) + G(v, u) \ge 2\eta(\rho - \mu M C_h C_\gamma)\|u - v\|^2_{H_1}.$$

Hence if we put $\mu_0 = \rho/(M C_h C_\gamma)$, then for all $\mu \in [0, \mu_0)$ we deduce (1.69).

Remark 1.6 *Another proof of Theorem* 1.6 *can be given by imposing the operator S (given in Remark* 1.5) *to be a contraction. The same* μ_0 *is obtained.*

Remark 1.7 *As follows from the proof of Theorem* 1.6, *the critical friction coefficient* μ_0 *does not depend on the material constants* η *and* g, *nor on the external forces* b, F, f_v *or the imposed velocity* U.

The following result shows us that if the blocking phenomena occurs, then we cannot have another equilibrium solution, for all friction coefficient $\mu \ge 0$:

Lemma 1.3 *Let* (1.41)–(1.45) *hold and suppose* $U \equiv 0$ *and* meas $\Gamma_1 > 0$. *If* $u = 0$ *is the solution of* (1.60), *then it is unique for all* $\mu \ge 0$.

PROOF. Let G be given by (1.64) and let v be another solution of (1.60). From $G(0, v) \le 0$ and $G(v, 0) \le 0$ we deduce $G(0, v) + G(v, 0) \le 0$. If we make some computations we get $a(v,v) + 2\eta\mu \int_{\Gamma_5} |\mathscr{L}_h D(v)||v|\, d\Gamma \le 0$; hence $v = 0$ for all $\mu \ge 0$.

2 The blocking property of the solution

In this section we study the blocking phenomenon for two boundary-value problems (MPA) and (MPF) describing the stationary flow of the Bingham fluid through a domain $\Omega \subset \mathbb{R}^3$ with friction and without friction. Both these problems are particular forms of (1.17)–(1.25) and are modelled by variational inequalities of the second kind. We are interested here in finding the conditions for these variational inequalities to have $u = 0$ as a solution, which means the Bingham fluid is blocked in Ω.

2.1 Problem statements and blocking property

Let $\Omega \subset \mathbb{R}^3$ satisfy the requirements in Section 1.2; the mechanical problem (MPA) consists in finding the velocity function $u: \Omega \to \mathbb{R}^3$ and the stress function $T: \Omega \to \mathcal{S}_3$ such that (1.17)–(1.20) are satisfied with the following boundary conditions:

$$\left.\begin{aligned} &u = 0 && \text{on } \Gamma_6 \\ &t = F_1 && \text{on } \Gamma_7 \\ &u_\tau = 0,\ t\cdot\nu = f_{\nu_1} && \text{on } \Gamma_8 \\ &t_\tau = 0,\ u\cdot\nu = 0 && \text{on } \Gamma_9 \end{aligned}\right\} \tag{2.1}$$

where $\Gamma = \Gamma_6 \cup \Gamma_7 \cup \Gamma_8 \cup \Gamma_9$ is a partition of Γ.

The mechanical problem (MPF) consists in finding the velocity function $u: \Omega \to \mathbb{R}^3$ and the stress function $T: \Omega \to \mathcal{S}_3$ such that (1.17)–(1.25) are satisfied with $U = 0$ and μ a positive constant. We suppose that (1.42)–(1.44), (1.46) are satisfied and $F_1 \in [L^2(\Gamma_7)]^3$, $f_{\nu_1} \in L^2(\Gamma_8)$. We consider the space V defined by (1.53) and the space V_1 defined by

$$V_1 = \{v \in \boldsymbol{H}_1 \,|\, \operatorname{div} v = 0 \text{ in } \Omega,\ v|_{\Gamma_6} = 0,\ v_\tau|_{\Gamma_8} = 0,\ v\cdot\nu|_{\Gamma_9} = 0\}. \tag{2.2}$$

We also denote by V' and V_1' respectively the strong duals of V and V_1, and by $\langle\,,\,\rangle$ the duality pairing between V' and V, V_1', and V_1 respectively. Since $U = 0$ we can take $u_0 = 0$ in (1.53), hence by (1.55) and (1.49) we have

$$\langle f, v\rangle = \int_\Omega b\cdot v\,\mathrm{d}x + \int_{\Gamma_2} F\cdot v\,\mathrm{d}\Gamma + \int_{\Gamma_3} f_\nu v\cdot\nu\,\mathrm{d}\Gamma \qquad \forall v \in V. \tag{2.3}$$

Similarly, we consider the element $f_1 \in V_1'$ defined by

$$\langle f_1, v\rangle = \int_\Omega b\cdot v\,\mathrm{d}x + \int_{\Gamma_7} F_1\cdot v\,\mathrm{d}\Gamma + \int_{\Gamma_8} f_{\nu_1} v\cdot\nu\,\mathrm{d}\Gamma \qquad \forall v \in V_1 \tag{2.4}$$

and let $j, h = \boldsymbol{H}_1 \to \mathbb{R}_+$ be the seminorms defined by

$$j(v) = \int_\Omega |D(v)|\,\mathrm{d}x, \tag{2.5}$$

$$h(v) = \int_{\Gamma_5} |S||v|\,\mathrm{d}\Gamma \tag{2.6}$$

for all $v \in \boldsymbol{H}_1$.

With these notations, as follows from Section 1.3, we have that if $(u; T)$ is a smooth solution of the mechanical problem (MPA), then u satisfies the variational inequality

$$u \in V_1, \quad a(u, v-u) + gj(v) - gj(u) \geq \langle f_1, v-u\rangle \qquad \forall v \in V_1, \tag{2.7}$$

where a is defined by (1.47). Moreover, if $(u; T)$ is a smooth solution of the mechanical problem (MPF), then u satisfies the variational inequality

$$u \in V, \quad a(u, v - u) + gj(v) + \mu h(v) - gj(u) - \mu h(u) \geq \langle f, v - u \rangle \qquad \forall v \in V. \tag{2.8}$$

For this reason, we shall consider (2.7) and (2.8) as the *variational formulation* for the problems (MPA) and (MPF) respectively. In the next paragraphs we are interested in finding the conditions for which (2.7) has $u = 0_{V_1}$ as a solution and (2.8) has $u = 0_V$ as a solution; from the physical point of view this property means a 'blocking property' for the Bingham fluid in Ω. More precisely, we study in detail the following problems:

(1) Find the set of external forces such that the blocking stands for a great enough yield limit. For this it is necessary to define the blocking space $B(j)$ for variational inequalities of the form (2.7) and the space $B(j, h)$ for variational inequalities of the form (2.8).
(2) Find the critical yield limit for which the blocking stands if the external forces belong to $B(j)$ in the (MPA) case described by (2.7).
(3) Find the link between the yield limit g and the friction coefficient μ such that the blocking stands if the external forces belong to $B(j, h)$ in the (MPF) case described by (2.8).
(4) Find the link between the blocking properties for the two problems studied.

In order to discuss problems (1)–(4) we start from the study of the blocking property for abstract variational inequalities of the form (2.7), (2.8).

2.2 The blocking property for abstract variational inequalities

In this section the following notations are used:

$(V, \|\cdot\|_V)$ is the real reflexive Banach space;
$(V', \|\cdot\|_{V'})$ is the strong topological dual of V;
$\langle , \rangle$ is the canonical duality pairing from $V' \times V$ into $\mathbb{R}$;
2^V is the set of all subsets of V;
a: $V \times V \to \mathbb{R}$ is a bilinear functional on V;
j, h: $V \to \mathbb{R}_+$ are continuous seminorms on V.

We can define the operator P: $V' \times \mathbb{R}_+ \to 2^V$ which maps (f, g) in the set of solutions of the variational inequality

$$u \in V, \quad a(u, v - u) + gj(v) - gj(u) \geq \langle f, v - u \rangle \tag{2.9}$$

for all $v \in V$. By analogy with (2.7) we call (2.9) a variational inequality of (MPA) type.

Let us consider the set:

$$\begin{aligned}B(j) &= \{f \in V' \mid \text{there exists } g \geq 0 \text{ such that } P(f, g) \ni 0_V\} \\ &= \{f \in V' \mid \text{there exists } g \geq 0 \text{ such that } |\langle f, v\rangle| \leq gj(v) \quad \forall v \in V\} \quad (2.10)\end{aligned}$$

and define the map g_j: $B(j) \to \mathbb{R}_+$ by

$$\begin{aligned}g_j(f) &= \inf\{g \geq 0 \mid P(f, g) \ni 0_V\} \\ &= \inf\{g \geq 0 \mid |\langle f, v\rangle| \leq gj(v) \quad \forall v \in V\}. \quad (2.11)\end{aligned}$$

Theorem 2.1 (1) *$B(j)$ is a subspace of V', g_j is a norm on $B(j)$ and the inclusion i: $(B(j), g_j) \hookrightarrow (V', \|\cdot\|_{V'})$ is continuous:*

(2) *For all $f \in B(j)$ we have*

$$g_j(f) = \sup_{v \notin \operatorname{Ker} j} |\langle f, v\rangle|/j(v). \quad (2.12)$$

(3) *$B(j) = \{0_{V'}\}$ iff j vanishes everywhere in V.*

(4) *$P(f, g) \ni 0_V$ iff f belongs to $B(j)$ and $g \geq g_j(f)$.*

PROOF. (1) From (2.10) it is easy to see that $B(j)$ is a subspace of V'. Let $f \in B(j)$; from (2.11) we get

$$|\langle f, v\rangle| \leq g_j(f)j(v) \qquad \forall v \in V \quad (2.13)$$

and g_j is a norm on $B(j)$. Since j is a continuous seminorm on V there exists $C > 0$ such that $j(v) \leq C\|v\|_V$ for all $v \in V$, and by (2.13) we get $|\langle f, v\rangle| \leq Cg_j(f)\|v\|_V$ for all $v \in V$. It results

$$\|f\|_{V'} \leq Cg_j(f) \quad (2.14)$$

hence (1) is completely proved.

(2) It results from (2.11) and (2.13).

(3) Let $v_0 \in V$, $v_0 \neq 0_V$ such that $j(v_0) \neq 0$; we denote by $\mathrm{Sp}(v_0)$ the space generated by v_0 and let f_0: $\mathrm{Sp}(v_0) \to \mathbb{R}$ be the functional defined by $f_0(\lambda v_0) = \lambda j(v_0)$ for all $\lambda \in \mathbb{R}$. Using the Hahn–Banach theorem we get that there exists a linear functional f: $V \to \mathbb{R}$ such that $f|_{\mathrm{Sp}(v_0)} = f_0$ and

$$|\langle f, v\rangle| \leq j(v) \qquad \forall v \in V. \quad (2.15)$$

Since j is continuous we have $f \in V'$ and (2.15) implies $f \in B(j)$; it follows that $f(v_0) = f_0(v_0) = j(v_0) \neq 0$, hence $f \neq 0_{V'}$. So, we proved that if j does not identically vanish then $B(j) \neq \{0_{V'}\}$, hence $B(j) = \{0_{V'}\}$ implies that j vanishes identically. The converse implication results from (2.10).

(4) Is a consequence of (2.10), (2.11), and (2.13).

The property of the variational inequality (4.9) to have $u = 0_V$ as a solution is called the *blocking property*. As follows from the previous theorem, this

property is determined by the normed space $(B(j), g_j)$ which will be called the *blocking space* of the variational inequality (2.9).
For some applications it is useful to give another expression for (2.11):

Lemma 2.1 *Let $f \in B(j)$, $f \neq 0_{V'}$, $V_f = \{v \in V | \langle f, v \rangle = 1\}$ and*

$$\alpha_j = \inf_{v \in V_f} j(v). \tag{2.16}$$

Then $\alpha_f > 0$ and $g_j(f) = 1/\alpha_f$.

PROOF. Using (2.11) we have $|\langle f, v \rangle| \leq g_j(f) j(v)$ for all $v \in V$ and taking $v \in V_f$ we obtain $1 \leq g_j(f) j(v)$ hence $1/g_j(f) \leq j(v)$. From (2.16) we get $1/g_j(f) \leq \alpha_f$, which implies $\alpha_j > 0$ and $g_j(f) \geq 1/\alpha_f$. For the converse inequality let us observe that if $\langle f, v \rangle \neq 0$ we get $v/\langle f, v \rangle \in V_f$; hence $|\langle f, v \rangle| \leq (1/\alpha_f) j(v)$. So we have $|\langle f, v \rangle| \leq (1/\alpha_f) j(v)$ for all $v \in V$ and using (2.11) we deduce $g_j(f) \leq 1/\alpha_f$.

Let us now study the blocking property for variational inequalities of the form (2.8). With the previous notations we consider the operator Q: $V' \times \mathbb{R}_+ \times \mathbb{R}_+ \to 2^V$, which maps (f, g, μ) into the set of solutions of the variational inequality

$$u \in V, \quad a(u, v - u) + gj(v) + \mu h(v) - gj(u) - \mu h(u) \geq \langle f, v - u \rangle \qquad \forall v \in V. \tag{2.17}$$

By analogy with (2.8) we call (2.17) a variational inequality of (MPF)-type, Let $B(j, h)$ be the set defined by

$$\begin{aligned} B(j, h) &= \{f \in V' | \text{there exists } g \geq 0 \text{ and } \mu \geq 0 \text{ and such that } Q(f, g, \mu) \ni 0_V\} \\ &= \{f \in V' | \text{there exists } g \geq 0 \text{ and } \mu \geq 0 \text{ such that} \\ &\qquad gj(v) + \mu h(v) \geq |\langle f, v \rangle| \quad \forall v \in V\}. \end{aligned} \tag{2.18}$$

Using the results in Section 4.1 we may consider the blocking space of the continuous seminorm $j + h$:

$$\begin{aligned} B(j + h) = \{f \in V' | &\text{there exists } \mu \geq 0 \text{ such that} \\ &\mu(j(v) + h(v)) \geq |\langle f, v \rangle| \quad \forall v \in V\}. \end{aligned} \tag{2.19}$$

From (2.18) and (2.19) it is clear that $B(j, h) = B(j + h)$ and using Theorem 2.1(3) of Chapter 2 we obtain $B(j, h) = \{0_{V'}\}$ iff j and h vanish everywhere on V.

Let us suppose now that f belongs to $B(j, h)$; in order to obtain a result similar to Theorem 2.1(4) we characterize the set Q_f defined by

$$Q_f = \{(g, \mu) \in \mathbb{R}_+ \times \mathbb{R}_+ | Q(f, g, \mu) \ni 0_V\}. \tag{2.20}$$

For this reason let us consider the function $F_j\colon \mathbb{R}_+ \times \mathbb{R}_+ \to \mathbb{R}$,

$$F_f(g, \mu) = \inf_{\|V\|_V = 1} \{gj(v) + \mu h(v) - \langle f, v \rangle\} \tag{2.21}$$

and we have

$$Q_f = \{(g, \mu) \in \mathbb{R}_+ \times \mathbb{R}_+ \mid F_f(g, \mu) \geq 0\}. \tag{2.22}$$

Lemma 2.2 (1) *F_f is a concave upper semicontinuous function.*

(2) *For each $\mu, g \geq 0$ the functions $F_f(\cdot, \mu)\colon \mathbb{R}_+ \to \mathbb{R}_+$ and $F_f(g, \cdot)\colon \mathbb{R}_+ \to \mathbb{R}_+$ are increasing.*

(3) *There exists*

$$\lim_{\mu \to \infty} (\inf\{g \geq 0 \mid F_f(g, \mu) \geq 0\}) = l_1(f)$$

and

$$\lim_{g \to \infty} (\inf\{\mu \geq 0 \mid F_f(g, \mu) \geq 0\}) = l_2(f);$$

(4) $l_1(f) \geq |\langle f, v \rangle|/j(v)$ $\forall v \in V$ such that $h(v) = 0$ and $j(v) \neq 0$ and $l_2(f) \geq \langle f, v \rangle/h(v)$ $\forall v \in V$ such that $j(v) = 0$ and $h(v) \neq 0$.

PROOF. The first two statements are obvious. In order to prove (3) we observe that if g and μ are large enough the functions

$$\mu \to \inf\{g \geq 0 \mid F_f(g, \mu) \geq 0\}$$

and $g \to \inf\{\mu \geq 0 \mid F_f(g, \mu) \geq 0\}$ are decreasing. Indeed, if, for instance, $0 \leq \mu_1 \leq \mu_2$, then by (2.21) we have $F_f(g, \mu_1) \leq F_f(g, \mu_2)$ for all $g \geq 0$; hence

$$\{g \geq 0 \mid F_f(g, \mu_1) \geq 0\} \subset \{g \geq 0 \mid F_f(g, \mu_2) \geq 0\};$$

it follows that

$$\inf\{g \geq 0 \mid F_f(g, \mu_2) \geq 0\} \leq \inf\{g \geq 0 \mid F_f(g, \mu_1) \geq 0\}.$$

(4) Let $v \in V$ such that $h(v) = 0$, $j(v) \neq 0$ and let $g \geq 0$, $\mu \geq 0$ be such that $F_f(g, \mu) \geq 0$. It follows that $\|v\|_V \neq 0$ and by (2.21) we get

$$gj(v/\|v\|_V) + \mu h(v/\|v\|_V) \geq \langle f, v/\|v\|_V \rangle,$$

and since $h(v) = 0$ it follows that $gj(v) \geq \langle f, v \rangle$; replacing v by $-v$ we obtain $gj(v) \geq |\langle f, v \rangle|$; hence $g \geq |\langle f, v \rangle|/j(v)$. So, $\inf\{g \geq 0 \mid F_f(g, \mu) \geq 0\} \geq |\langle f, v \rangle|/j(v)$, and taking the limit when $\mu \to \infty$ we obtain the first statement; the second one can be deduced in the same way, replacing j by h and μ by g.

The properties of the set Q_f are characterized by the following theorem;

Theorem 2.2 (1) $Q_f \subset \mathbb{R}_+ \times \mathbb{R}_+$ *is a convex closed set;*

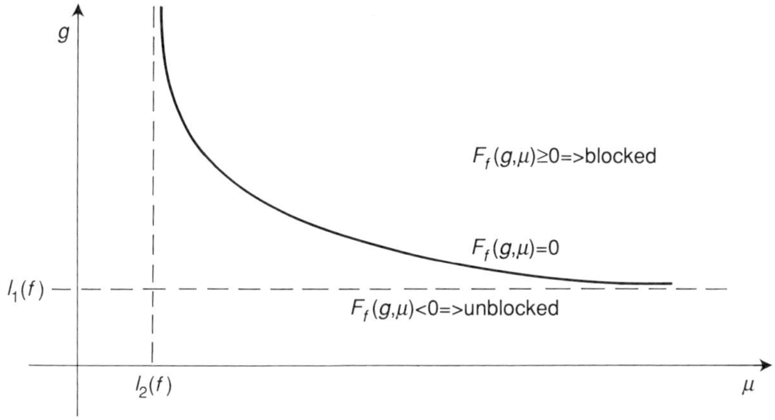

Fig. 2.1 A representation of the set Q_f defined by (2.22).

(2) *if* $(g, \mu) \in Q_f$ *then* $[g, +\infty) \times [\mu, +\infty) \subset Q_f$;
(3) $Q_f \subset [l_1(f), +\infty) \times [l_2(f), +\infty)$.

PROOF. (1) Lemma 2.2(1) and (2.22) are used.
(2) This follows from (2.22) and Lemma 2.2(2).
(3) Consider $(g, \mu) \in Q_f$; from (2.22) we have $F_f(g, \mu) \geq 0$; hence $g \geq \inf\{g \geq 0 | F_f(g, \mu) \geq 0\} \geq l_1(f)$ hence $g \in [l_1(f), +\infty)$; similarly we get $\mu \in [l_2(f), +\infty)$.

The previous theorem allows us to consider the set Q_f as in Figure 2.1. Next we impose the supplementary assumption: there exists $c > 0$ so that

$$h(v) \leq cj(v) \qquad \forall v \in V. \tag{2.23}$$

Using (2.18) and (2.10) we get $B(j, h) = B(j)$.

Remark 2.1 *Since we supposed* $f \in B(j, h)$ *we get* $f \in B(j)$ *and from* (2.10), (2.21) *it follows that for all* $\mu \geq 0$ *the set* $\{g \geq 0 | F_f(g, \mu) \geq 0\}$ *is not empty.*

We can add other properties of the set Q_f. For this, define $G_f: \mathbb{R}_+ \to \mathbb{R}_+$ by

$$G_f(\mu) = \sup_{v \notin \operatorname{Ker} j} \frac{\langle f, v\rangle - \mu h(v)}{j(v)} \qquad \forall \mu \geq 0 \tag{2.24}$$

Theorem 2.3 *For all* $\mu \geq 0$ *we have*:
(1) $G_f(\mu) = \inf\{g \geq 0 | F_f(g, \mu) \geq 0\}$
(2) $g \geq G_j(\mu)$ *iff* $F_f(g, \mu) \geq 0$
(3) G_f *is a convex decreasing function.*

PROOF. (1) Let $g \geq 0$ such that $F_f(g, \mu) \geq 0$, i.e. $gj(v) + \mu h(v) \geq \langle f, v \rangle$ for all $v \in V$. If $v \notin \operatorname{Ker} j$ we have $g \geq [\langle f, v \rangle - \mu h(v)]/j(v)$ and, using (2.24), we get $g \geq G_f(\mu)$. So, we proved that

$$F_f(g, \mu) \geq 0 \Rightarrow g \geq G_f(\mu) \tag{2.25}$$

$$G_f(\mu) \leq \inf\{g \geq 0 \mid F_f(g, \mu) \geq 0\}. \tag{2.26}$$

Moreover, from (2.24) we have $G_f(\mu)j(v) + \mu h(v) \geq \langle f, v \rangle$ for all $v \in V$ such that $j(v) \neq 0$; if $j(v) = 0$ then $h(v) = 0$ by (2.23) and since $f \in B(j, h) = B(j)$ we have $\langle f, v \rangle = 0$. So it follows that $G_f(\mu)j(v) + \mu h(v) \geq \langle f, v \rangle$ for all $v \in V$ and by (2.21) we get

$$F_f(G_f(\mu), \mu) \geq 0. \tag{2.27}$$

We have proved that

$$G_f(\mu) \geq \inf\{g \geq 0 \mid F_f(g, \mu) \geq 0\}. \tag{2.28}$$

The statement of (1) follows now from (2.26) and (2.28).

(2) From (2.27) and (2.22) we get $(G_f(\mu), \mu) \in Q_f$. Using Theorem 2.2(2) and (2.22) we get

$$g \geq G_f(\mu) \Rightarrow F_f(g, \mu) \geq 0. \tag{2.29}$$

The statement of (2) follows now from (2.25) and (2.29).

(3) This follows from (2.24).

From Theorem 2.3 it follows that the boundary of Q_f is the set

$$\{0\} \times [G_f(0), +\infty) \cup \{(g, \mu) \in \mathbb{R}_+ \times \mathbb{R}_+ \mid g = G_f(\mu)\}$$

as shown in Figure 2.2. Moreover, from (2.20), (2.22), (2.18) and the previous

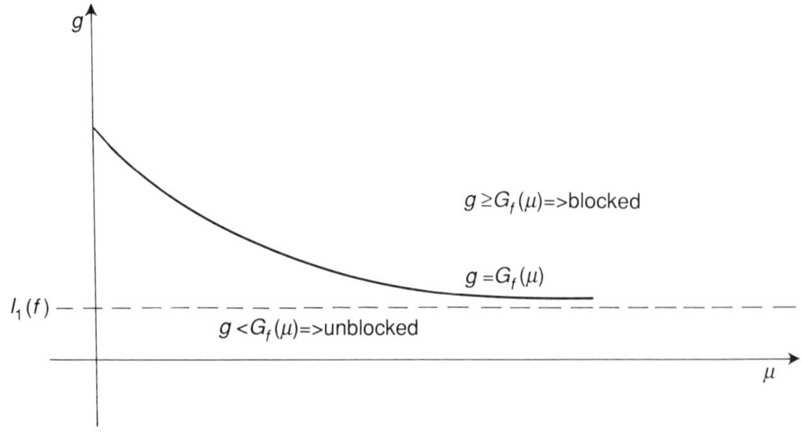

Fig. 2.2 A representation of the set Q_f in the particular case when h is continuous with respect to j.

theorem we get

$$Q(f, g, \mu) \ni 0_V \Leftrightarrow f \in B(j) \quad \text{and} \quad g \geq G_f(\mu). \tag{2.30}$$

2.3 The blocking property in the case without friction

In this section we apply the abstract results given in the previous section for the variational inequality (2.7); the notations (2.2), (2.4), and (2.5) are used. Let us consider the space $LD(\Omega)$ defined in Section 2.2 of Chapter 2 and the following subspace of $LD(\Omega)$:

$$LD_1(\Omega) = \{v \in LD(\Omega) | \operatorname{tr} D(v) = 0, v|_{\Gamma_6} = 0, v_\tau|_{\Gamma_8} = 0, v \cdot \nu|_{\Gamma_9} = 0\}.$$

Let $LD_1'(\Omega)$ be the dual of the space $LD_1(\Omega)$ and let us recall that

$$\mathscr{R} = \{r: \Omega \to \mathbb{R}^3 | r(x) = m + n \wedge x, m, n \in \mathbb{R}^3\}$$

is the set of rigid motions.

Theorem 2.4 *The blocking space of the seminorm $j: V_1 \to \mathbb{R}$ defined by (2.5) is given by*

$$B(j) = \{f_1 \in LD_1'(\Omega) | \langle f_1, r\rangle = 0 \quad \forall r \in \mathscr{R} \cap LD_1(\Omega)\}.$$

PROOF. Let $f_1 \in B(j)$; from (2.10) there exists $g \geq 0$ such that $|\langle f_1, v\rangle| \leq gj(v)$ for all $v \in V_1$; hence $|\langle f_1, v\rangle| \leq g\|v\|_{LD(\Omega)}$. Since V_1 is dense in $LD_1(\Omega)$, there exists a unique extension of f_1 to an element of $LD_1'(\Omega)$; therefore we can consider $f_1 \in LD_1'(\Omega)$. Let r be an element of $\mathscr{R} \cap LD_1(\Omega) = \mathscr{R} \cap V_1$; then $|\langle f_1, r\rangle| \leq gj(r)$ and, since $j(r) = 0$ we deduce $\langle f_1, r\rangle = 0$.

Conversely, let $f_1 \in LD_1'(\Omega)$ and $\langle f_1, r\rangle = 0$ for all $r \in \mathscr{R} \cap LD_1(\Omega)$. There exists $c_1 > 0$ such that

$$|\langle f_1, v\rangle| \leq c_1 \|v\|_{LD(\Omega)} \qquad \forall v \in LD_1(\Omega). \tag{2.31}$$

From (2.39) of Chapter 2 there exists $c_2 > 0$ such that, for all $v \in LD_1(\Omega)$, there exists $r_v \in \mathscr{R} \cap V_1$ that satisfies the inequality

$$\|v - r_v\|_{LD(\Omega)} \leq c_2 \int_\Omega |D(v)| \, dx. \tag{2.32}$$

From (2.31) and (2.32) taking $v \in V_1$ we get

$$|\langle f\, v\rangle| = |\langle f_1, v - r_v\rangle| \leq c_1 \|v - r_v\|_{LD(\Omega)} \leq c_1 c_2 j(v)$$

and using (2.10) we have $f_1 \in B(j)$.

Remark 2.2 *The condition $\langle f_1, r\rangle = 0$ for all $r \in \mathscr{R} \cap V_1$ is not unique to the blocking phenomena; it represents the necessary and sufficient condition for* (2.7) *to have solutions.*

Using the above theorem, we can impose on the external forces as few restrictive conditions as the blocking phenomena can stand:

Corollary 2.1 *If* $b \in [L^\infty(\Omega)]^3$, $F_1 \in [L^\infty(\Gamma_5)]^3$, $f_{v_1} \in L^\infty(\Gamma_6)$ *and*

$$\int_\Omega b \cdot r \, \mathrm{d}x + \int_{\Gamma_7} F_1 \cdot r \, \mathrm{d}\Gamma + \int_{\Gamma_8} f_{v_1} r \cdot v \, \mathrm{d}\Gamma = 0$$

for all $r \in \mathscr{R}$ *with* $r|_{\Gamma_6} = 0$, $r_\tau|_{\Gamma_8} = 0$, $r \cdot v|_{\Gamma_9} = 0$, *then the functional* f_1 *defined by* (2.4) *belongs to the space* $B(j)$ (*hence, for* $g \geq g_j(f)$, *inequality* (2.7) *has* 0_{V_1} *as a solution*).

PROOF. Let $b \in [L^\infty(\Omega)]^3$, $F_1 \in [L^\infty(\Gamma_7)]^3$, $f_{v_1} \in L^\infty(\Gamma_8)$; for every $v \in LD_1(\Omega)$, using (2.36) of Chapter 2, we have

$$\begin{aligned} |\langle f_1, v \rangle| &\leq \int_\Omega |b||v| \, \mathrm{d}x + \int_{\Gamma_7} |F_1||v| \, \mathrm{d}x + \int_{\Gamma_8} |f_{\lambda_1}||v| \, \mathrm{d}\Gamma \\ &\leq \|b\|_{[L^\infty(\Omega)]^3} \|v\|_{[L^1(\Omega)]^3} + \|F_1\|_{[L^\infty(\Gamma_7)]^3} \|\gamma_1(v)\|_{[L^1(\Omega)]^3} \\ &\quad + \|f_{\lambda_1}\|_{[L^\infty(\Gamma_8)]} \|\gamma_1(v)\|_{[L^1(\Gamma)]^3} \\ &\leq C(\|b\|_{[L^\infty(\Omega)]^3} + \|F_1\|_{[L^\infty\Gamma_7)]^3} + \|f_{v_1}\|_{[L^\infty(\Gamma_8)]}) \|v\|_{LD(\Omega)} \qquad (C > 0). \end{aligned}$$

Hence we can apply the above theorem and Corollary 2.1 follows.

Remark 2.3 *Corollary* 2.1 *has the following physical interpretations*:

(1) The case meas. $\Gamma_6 \geq 0$. *In this case* $\mathscr{R} \cap V_1 = \{0\}$; *hence the condition* $\langle f_1, r \rangle = 0$ *for all* $r \in \mathscr{R} \cap V$ *is satisfied; the variational problem* (MPA) *has a unique solution and if the external forces are bounded, then there exists a critical yield limit* $g_j(f_1)$ *such that the Bingham fluid flows iff* $g < g_j(f_1)$. *The fluid is blocked iff* $g \geq g_j(f_1)$. *Relation* (2.12) *gives this critical yield limit.*

(2) The case $\Gamma_6 = \varnothing$. *In this case, if the external forces are bounded and they are statically equivalent to zero for a great enough yield limit, the Bingham fluid has rigid-body behaviour.*

Remark 2.4 *Using Lemma* 2.1 *and Theorem* 2.1 *we can easily deduce a result of Glowinski* (1973) *for the problem of the laminar and stationary flow of the Bingham fluid in a cylindrical pipe with adherence boundary conditions. As is shown in the paper of Duvaut and Lions* (1972, *Ch. VI*), *this problem is described by a variational inequality of the form* (2.9) *with the following notation*: $\Omega \subset \mathbb{R}^2$ *is the cross-section of the pipe*; $V = H_0^1(\Omega)$; $j(v) = \int_\Omega |\nabla v| \, \mathrm{d}x$; $a(u, v) = \eta \int_\Omega \nabla u \cdot \nabla v \, \mathrm{d}x$; $f = c \in \mathbb{R}$ *is the drop in pressure per unit length. Using* (2.40) *in Chapter* 2 *we get* $\|v\|_{L^1(\Omega)} \leq kj(v)$ *for all* $v \in H_0^1(\Omega)$ $(k > 0)$ *hence* $\mathbb{R} \subset B(j)$; *moreover, from Theorem* 2.1 *and Lemma* 2.1 *we have* $g_j(c) = g_j(c \cdot 1) = |c| g_j(1) = |c| \cdot (1/\alpha_1)$ *where* $\alpha_1 = \inf_{v \in V_1} j(v)$, $V_1 = \{v \in H_0^1(\Omega) \mid \int_\Omega v \, \mathrm{d}x = 1\}$ (*see* (2.16)). *Using Theorem* 2.1(4) *we have* $u = 0_{H_0^1(\Omega)}$ *if* $fg \geq |c|/\alpha_1$. *This result, obtained by Glowinski* (1973) *using a different method, points out the constant* $g^*(c) = |c|/\alpha_1$, *which depends on the drop in pressure and the cross-section of the pipe, with the following physical interpretations*:

(1) *If the yield limit of the Bingham fluid satisfies the inequality $g < g^*(c)$, then the Bingham fluid flows through the pipe under the action of the drop in pressure c per unit length.*

(2) *If the yield limit of the Bingham fluid satisfies the inequality $g \geq g^*(c)$, then the Bingham fluid does not flow through the pipe under the action of the drop in pressure c per unit length.*

2.4 The blocking property in the case with friction

In this section we give the mechanical interpretations of the abstract results given in Section 2.2 for the variational inequality (2.8); the notations (1.53), (2.3), (2.5), and (2.6) are used. Let $LD_2(\Omega)$ be the space given by

$$LD_2(\Omega) = \{v \in LD(\Omega) \mid \operatorname{tr} D(v) = 0, v|_{\Gamma_1} = 0, v_\tau|_{\Gamma_3} = 0, v \cdot \nu|_{\Gamma_4 \cup \Gamma_5} = 0\}.$$

We denote the dual of $LD_2(\Omega)$ by $LD'_2(\Omega)$. We suppose $\operatorname{meas}\{x \in \Gamma_5 \mid S(x) \neq 0\} > 0$.

Theorem 2.5 *The blocking space $B(j, h)$ defined by* (2.18) *in the* (MPF) *case is given by $B(j, h) = LD'_2(\Omega)$.*

PROOF. Let f belong to $B(j, h)$. *There exist $g, \mu \geq 0$ such that* $|\langle f, v\rangle| \leq gj(v) + \mu h(v)$ for all $v \in V$. But $j(v) \leq \|v\|_{LD(\Omega)}$ and from (2.36) of Chapter 1 there exists $c > 0$ such that $h(v) \leq c\|S\|_{L^\infty(\Gamma_5)} \cdot \|v\|_{LD(\Omega)}$ for all $v \in LD(\Omega)$. Hence

$$|\langle f, v\rangle| \leq (g + \mu c\|S\|_{L^\infty(\Gamma_5)}\|v\|_{LD(\Omega)}$$

for all $v \in V$. Since V is dense in $LD_2(\Omega)$ we have a unique extension of f to an element of $LD'_2(\Omega)$; hence, we may consider f as belonging to $LD'_2(\Omega)$.

Conversely, let f be an element of $LD'_2(\Omega)$. Since h is a continuous seminorm on $LD_2(\Omega)$ and a norm on $\mathscr{R}$, using (2.40) of Chapter 2 we find that $j + h$ and $\|\cdot\|_{LD(\Omega)}$ are equivalent norms on $LD_2(\Omega)$. Therefore, for every $v \in V$ we have

$$|\langle f, v\rangle| \leq \|f\|_{LD'(\Omega)}\|v\|_{LD(\Omega)} \leq k\|f\|_{LD'(\Omega)}(j + h)(v) \qquad (k > 0)$$

hence $f \in B(j + h) = B(j, h)$.

Similarly as in Corollary 2.1, from the previous theorem we get

Corollary 2.2 *If $b \in [L^\infty(\Omega)]^3$, $F \in [L^\infty(\Gamma_2)]^3$, $f_\nu \in L^\infty(\Gamma_3)$, then $f \in V'$ defined by* (2.3) *belongs to the blocking space $B(j, h)$.*

Remark 2.5 *From Corollary* 2.2 *and Theorem* 2.2 *we deduce that if the external forces are bounded, then the set $Q_f \subset \mathbb{R}_+ \times \mathbb{R}_+$ of all pairs (g, μ) for which the Bingham fluid is blocked is convex and closed (it is schematically*

represented in Figure 2.1). *If the blocking stands for a pair* (g, μ), *then it also stands for every pair* (g', μ') *with* $g' \geq g$ *and* $\mu' \geq \mu$. *There exist two positive numbers* $l_1(f)$ *and* $l_2(f)$ *such that if* $(g, \mu) \in [0, l_1(f)) \times \mathbb{R}_+ \cup \mathbb{R}_+ \times [0, l_2(f))$, *then the fluid flows.*

If meas $\Gamma_1 > 0$, using (2.40) of Chapter 2 in the case of the continuous seminorm $p(v) = \gamma_1(v)|_{\Gamma_1}$ (see (2.36) in Chapter 2) we find that j is equivalent on $LD_2(\Omega)$ with the canonical norm $\|\cdot\|_{LD(\Omega)}$; so, since h is a continuous seminorm on $LD_2(\Omega)$ there exists $c > 0$ such that $h(v) \leq cj(v)$ for all $v \in LD^2(\Omega)$. Using Theorem 2.3 we obain that for any friction coefficient $\mu \geq 0$ there exists a critical yield limit $G_f(\mu)$ such that the fluid is blocked iff $g \geq G_f(\mu)$. This critical yield limit is a convex decreasing function and has the limit $l_1(f)$ when $\mu \to +\infty$ (the graphic representation is given in Figure 2.2). If $g \geq G_f(0)$, then the fluid is blocked for any friction coefficient $\mu \geq 0$. If $l_1(f) < g \leq G_f(0)$, there exists a critical friction coefficient $\mu_f(g)$ such that the fluid is blocked iff $\mu \geq \mu_f(g)$. If $0 \leq g < l_1(f)$, then the fluid flows for any friction coefficient $\mu \geq 0$.

Remark 2.6 *If in the* (*MPA*) *case we put* $\Gamma_6 = \Gamma_1 \cup \{x \in \Gamma_5 | S(x) > 0\}$, $\Gamma_7 = \Gamma_2$, $\Gamma_8 = \Gamma_3$, $\Gamma_9 = \Gamma_4 \cup \{x \in \Gamma_5 | S(x) = 0\}$, $F_1 = F$ *and* $f_{\nu_1} = f_\nu$, *using Lemma* 2.2(4) *we obtain*

$$l_1(f) \geq \sup_{\substack{v \in V \\ h(v)=0 \\ j(v) \neq 0}} \frac{|\langle f, v \rangle|}{j(v)} = \sup_{\substack{v \in V \cap \operatorname{Ker} h \\ j(v) \neq 0}} \frac{|\langle f, v \rangle|}{j(v)}.$$

But, since from (1.53), (2.2), *and* (2.6) *we have* $V \cap \operatorname{Ker} h = V_1$, *we get*

$$l_1(f) \geq \sup_{\substack{v \in V_1 \\ j(v) \neq 0}} \frac{|\langle f, v \rangle|}{j(v)}$$

and using (2.12) *we get* $l_1(f) \geq g_j(f)$. *This last inequality says that under the same external forces, if the problem* (*MPF*) *has a rigid solution, then the problem* (*MPA*) *also has a rigid solution. So, under the same external forces if the blocking phenomenon occurs in the* (*MPF*) *case it also occurs in the* (*MPA*) *case.*

3 A numerical approach

In this section we shall consider a numerical method for solving the variational inequality (1.56) by using penalization, regularization and discretization techniques (the last one is made by means of the finite element method). Thus the problem will be reduced to solving a non-linear finite-dimensional system, for which an iterative method will be presented. Finally, this numerical method will be applied to wire drawing through conical dies and some numerical results for the copper wire drawing are given.

In all this section (1.41)–(1.46) will be supposed to hold and we also assume that meas $\Gamma_1 > 0$.

3.1 The penalized problem

The linear space V (given by (1.53)) on which the variational problem (1.56) is formulated contains the incompressibility condition (1.20), which is cumbersome for numerical approximations. Therefore we employ a penalty method which is similar to the one used for the flow of an incompressible Stokes fluid (see Temam (1979)).

Let us consider W, a closed subspace of $\boldsymbol{H}_1$, given by

$$W = \{v \in \boldsymbol{H}_1 \,|\, v = 0 \text{ on } \Gamma_1,\ v_\tau = 0 \text{ on } \Gamma_3,\ v \cdot \nu = 0 \text{ on } \Gamma_4 \cup \Gamma_5\}, \tag{3.1}$$

in which the incompressibility condition is not involved. Let us denote by $c\colon \boldsymbol{H}_1 \times \boldsymbol{H}_1 \to \mathbb{R}$ the following bilinear form

$$c(u, v) = \int_\Omega (\operatorname{div} u) \cdot (\operatorname{div} v)\, \mathrm{d}x \qquad \forall u, v \in \boldsymbol{H}_1. \tag{3.2}$$

For each $\varepsilon > 0$ we consider the following penalized variational inequality:

$$\bar{u}_\varepsilon \in W;\ a(\bar{u}_\varepsilon, \bar{u}_\varepsilon - v) + j(\bar{u}_\varepsilon) - j(v) + \frac{1}{\varepsilon} c(\bar{u}_\varepsilon, \bar{u}_\varepsilon - v) \le \langle f, \bar{u}_\varepsilon - v \rangle \qquad \forall v \in W \tag{3.3}$$

that has a unique solution $\bar{u}_\varepsilon \in W$ (the same techniques as in the proof of Theorem 1.3 can be used).

The following lemma enables us to consider that $\bar{u}_\varepsilon$ is an approximation of $\bar{u} = u - u_0$, the solution of (1.56).

Lemma 3.1 *Let $\bar{u}_\varepsilon$ be the solution of* (3.3) *for each $\varepsilon > 0$ and let $\bar{u} = u - u_0$ be the solution of* (1.56); *then we have*

$$\bar{u}_\varepsilon \to \bar{u} \text{ strongly in } \boldsymbol{H}_1 \text{ when } \varepsilon \to 0. \tag{3.4}$$

PROOF. If we put $v = 0$ in (3.3) and we use (1.57) we have

$$2\eta\rho \|\bar{u}_\varepsilon\|^2_{\boldsymbol{H}_1} \le a(\bar{u}_\varepsilon, \bar{u}_\varepsilon) + j(\bar{u}_\varepsilon) + \frac{1}{\varepsilon} c(\bar{u}_\varepsilon, \bar{u}_\varepsilon) \le \|f\|_{W'} \|\bar{u}_\varepsilon\|_{\boldsymbol{H}_1}.$$

Hence we get

$$\|\bar{u}_\varepsilon\|_{\boldsymbol{H}_1} \le (2\eta\rho)^{-1} \|f\|_{W'}. \tag{3.5}$$

Therefore the sequence $(\bar{u}_\varepsilon)_{>0}$ is bounded and we shall denote by $\bar{u}' \in W$

the element to which a subsequence $(\bar{u}_{\varepsilon'})_{\varepsilon'>0} \subset (\bar{u}_\varepsilon)_{\varepsilon>0}$ is weakly convergent, hence

$$\bar{u}_{\varepsilon'} \to \bar{u}' \text{ weakly in } W \text{ when } \varepsilon' \to 0. \tag{3.6}$$

In order to prove that $\bar{u}' \in V$ let us remark that from

$$a(\bar{u}_\varepsilon, \bar{u}_\varepsilon) + j(\bar{u}_\varepsilon) + \frac{1}{\varepsilon} c(\bar{u}_\varepsilon, \bar{u}_\varepsilon) \leq \|f\|_{W'} \|u_\varepsilon\|_{\boldsymbol{H}_1}$$

we deduce

$$c(\bar{u}_\varepsilon, \bar{u}_\varepsilon) \leq \varepsilon \|f\|_{W'}^2 (2\eta\rho)^{-1}. \tag{3.7}$$

Having in mind that $v \to c(v, v)$ is convex and continuous we deduce that $v \to c(v, v)$ is weakly lower semicontinuous; hence $c(\bar{u}', \bar{u}') \leq \underline{\lim}_{\varepsilon' \to 0} c(\bar{u}_{\varepsilon'}, \bar{u}_{\varepsilon'})$ and from (3.7) we obtain $c(\bar{u}', \bar{u}') = 0$, hence $\operatorname{div} \bar{u}' = 0$ and $\bar{u}' \in V$.

For all $v \in V$ the inequality (3.3) becomes

$$a(\bar{u}_{\varepsilon'}, \bar{u}_{\varepsilon'} - v) + j(\bar{u}_{\varepsilon'}) - j(v) + \frac{1}{\varepsilon} c(\bar{u}_{\varepsilon'}, \bar{u}_{\varepsilon'}) \leq \langle f, \bar{u}_{\varepsilon'} - v \rangle$$

and we deduce

$$a(\bar{u}_\varepsilon, \bar{u}_\varepsilon - v) + j(\bar{u}_\varepsilon) - j(v) \leq \langle f, \bar{u}_\varepsilon - v \rangle \qquad \forall v \in V. \tag{3.8}$$

Taking into account that $w \to a(w, v) + \langle f, w \rangle$ is weakly continuous and $w \to a(w, w) + j(w)$ is convex and continuous, hence weakly lower semicontinuous, we deduce

$$\begin{aligned} a(\bar{u}', \bar{u}') + j(\bar{u}') - a(\bar{u}', v) - \langle f, \bar{u} - v \rangle - j(v) &\leq \underline{\lim}_{\varepsilon' \to 0} \big(a(\bar{u}_{\varepsilon'}, \bar{u}_{\varepsilon'}) + j(\bar{u}_{\varepsilon'}) \\ &\quad - a(\bar{u}_{\varepsilon'}, v) - \langle f, \bar{u}_{\varepsilon'} - v \rangle - j(v)\big) \\ &\leq 0. \end{aligned}$$

From the above inequality we find that $\bar{u}'$ is a solution of (1.56) but from the uniqueness result stated in Theorem 1.3 we obtain $\bar{u}' = \bar{u}$. Since $\bar{u}$ is the unique weak limit of any subsequence of $(\bar{u}_\varepsilon)_{\varepsilon>0}$ we deduce that the whole sequence $(\bar{u}_\varepsilon)_{\varepsilon>0}$ is weakly convergent to $\bar{u}$.

In order to obtain the strong convergence result let us put in (3.8) $v = \bar{u}$. From (1.57) we have

$$\begin{aligned} 2\eta\rho \|\bar{u}_\varepsilon - \bar{u}\|_{\boldsymbol{H}_1}^2 &\leq a(\bar{u}_\varepsilon - \bar{u}, \bar{u}_\varepsilon - u) \\ &\leq j(\bar{u}) - j(\bar{u}_\varepsilon) + a(\bar{u}, \bar{u} - \bar{u}_\varepsilon) + \langle f, \bar{u}_\varepsilon - \bar{u} \rangle. \end{aligned} \tag{3.9}$$

Bearing in mind that j is weakly lower semicontinuous, i.e.

$$\overline{\lim_{\varepsilon \to 0}} \, (j(\bar{u}) - j(\bar{u}_\varepsilon)) = j(\bar{u}) - \underline{\lim}_{\varepsilon \to 0} j(\bar{u}_\varepsilon) \leq 0,$$

from (3.9) we deduce $\bar{u}_\varepsilon \to \bar{u}$ strongly in $\boldsymbol{H}_1$.

3.2 The discrete and regularized problem

In order to have an internal approximation of the subspace W let us consider $(W_h)_{h>0} \subset W$, a finite-dimensional family of subspaces (obtained, for instance, by means of the finite-element method) with the following property:

$$\forall v \in W \text{ there exists } v_h \in W_h \text{ such that } v_h \to v \text{ in } \boldsymbol{H}_1 \text{ for } h \to 0. \quad (3.10)$$

Let $\bar{u}_{\varepsilon h} \in W_h$ be the unique solution of the discrete variational inequality

$$\bar{u}_{\varepsilon h} \in W_h, \quad a(\bar{u}_{\varepsilon h}, \bar{u}_{\varepsilon h} - v_h) + j(\bar{u}_{\varepsilon h}) - j(v_h) + \frac{1}{\varepsilon} c(\bar{u}_{\varepsilon h}, \bar{u}_{\varepsilon h} - v_h)$$
$$\leq \langle f, \bar{u}_{\varepsilon h} - v_h \rangle \qquad \forall v_h \in W_h. \quad (3.11)$$

The following lemma shows that $\bar{u}_{\varepsilon h}$ is a good approximation of $\bar{u}_\varepsilon$ if the family of subspace $(W_h)_{h>0}$ is a good internal approximation of W (i.e. (3.10) holds).

Lemma 3.2 *Let $(\bar{u}_{\varepsilon h})_{h>0}$ be the solution of* (3.11) *for each $h > 0$ and $\bar{u}_\varepsilon$ be the solution of* (3.3) *for a fixed $\varepsilon > 0$. If* (3.10) *holds then*

$$\bar{u}_{\varepsilon h} \to \bar{u}_\varepsilon \text{ strongly in } \boldsymbol{H}_1 \text{ when } h \to 0. \quad (3.12)$$

PROOF. If we denote by $\bar{a}$: $W \times W \to \mathbb{R}$ the form given by

$$\bar{a}(u, v) = a(u, v) + \frac{1}{\varepsilon} c(u, v) \qquad \forall u, v \in W, \quad (3.13)$$

then we can use Lemma A4.2 with $\bar{a}$ instead of a, W instead of X, and $(W_h)_{h>0}$ instead of X_h.

If we define J_ε: $W \to \mathbb{R}$ by

$$J_\varepsilon(v) = \tfrac{1}{2}a(v, v) + j(v) + \frac{1}{2\varepsilon} c(v, v) - \langle f, v \rangle \qquad \forall v \in W, \quad (3.14)$$

then the solution $\bar{u}_{\varepsilon h}$ of the variational inequality (3.11) can be characterized as the solution of the following minimum problem:

$$\bar{u}_{\varepsilon h} \in W_h, \quad J_\varepsilon(\bar{u}_{\varepsilon h}) \leq J_\varepsilon(v_h) \qquad \forall v_h \in W_h. \quad (3.15)$$

Unfortunately J_ε is not Gâteaux differentiable and hence we cannot transform (3.15) into the non-linear algebraic system $\nabla J_\varepsilon(\bar{u}_{\varepsilon h}) \cdot v_h = 0$ for all $v_h \in W_h$. Therefore we have to consider the function φ_γ: $\mathbb{R} \to \mathbb{R}$ defined by

$$\varphi_\gamma(x) = \sqrt{(x^2 + \gamma^2)} - \gamma, \quad (3.16)$$

which regularizes the function $x \to |x|$. Indeed φ_γ is differentiable and the following inequality holds:

$$\big||x| - \varphi_\gamma(|x|)\big| < \gamma \qquad \forall x \in \mathbb{R}, \quad \gamma \geq 0. \tag{3.17}$$

Let j_γ: $W \to \mathbb{R}$ be the regularized function that corresponds to j given by

$$j_\gamma(v) = g \int_\Omega \varphi_\gamma(|D(v + u_0)|)\,\mathrm{d}x + \int_{\Gamma_5} \mu S \varphi_\gamma(|v + u_0|)\,\mathrm{d}\Gamma \tag{3.18}$$

and let us define $J_{\varepsilon\gamma}$: $W \to \mathbb{R}$ by

$$J_{\varepsilon\gamma}(v) = \tfrac{1}{2}a(v, v) + \frac{1}{2\varepsilon}c(v, v) + j_\gamma(v) - \langle f, v\rangle \qquad \forall v \in W. \tag{3.19}$$

For each $\gamma > 0$ we consider the following regularized minimum problem:

$$\bar{u}_{\varepsilon h\gamma} \in V_h, \quad J_{\varepsilon\gamma}(\bar{u}_{\varepsilon h\gamma}) \leq J_{\varepsilon\gamma}(v_h) \qquad \forall v_h \in V_h, \tag{3.20}$$

which is equivalent to the following regularized variational inequality:

$$\bar{u}_{\varepsilon h\gamma} \in V_h, \quad a(\bar{u}_{\varepsilon h\gamma}, \bar{u}_{\varepsilon h\gamma} - \bar{v}_h) + j_\gamma(\bar{u}_{\varepsilon h\gamma}) - j_\gamma(v_h) + \frac{1}{\varepsilon}c(\bar{u}_{\varepsilon h\gamma}, \bar{u}_{\varepsilon h\gamma} - v_h)$$
$$\leq \langle f, \bar{u}_{\varepsilon h\gamma} - v_h\rangle \qquad \forall v_h \in V_h. \tag{3.21}$$

Since j_γ is also convex and continuous we deduce that (3.20) or (3.21) has a unique solution $\bar{u}_{\varepsilon h\gamma} \in V_h$.

The following lemma shows that $\bar{u}_{\varepsilon h\gamma}$ is a good approximation for $\bar{u}_{\varepsilon h}$ if γ is sufficiently small.

Lemma 3.3 *Let $(\bar{u}_{\varepsilon h\gamma})_{\gamma>0}$ be the solution of* (3.20) *or* (3.21) *for each $\gamma > 0$ and let $\bar{u}_{\varepsilon h}$ be the solution of* (3.11) *for fixed ε, $h > 0$. Then there exists $C > 0$ which depends only on Ω, Γ_1, η, g, μ and S such that*

$$\|\bar{u}_{\varepsilon h\gamma} - \bar{u}_{\varepsilon h}\|_{\boldsymbol{H}_1} \leq c\gamma^{1/2} \qquad \forall \gamma \geq 0. \tag{3.22}$$

PROOF. If we put in (3.11) $v_h = \bar{u}_{\varepsilon h\gamma}$ and in (3.21) $v_h = \bar{u}_{\varepsilon h}$ after some algebra we get

$$a(\bar{u}_{\varepsilon h} - \bar{u}_{\varepsilon h\gamma}, \bar{u}_{\varepsilon h} - \bar{u}_{\varepsilon h\gamma}) + \frac{1}{\varepsilon}c(\bar{u}_{\varepsilon h} - \bar{u}_{\varepsilon h\gamma}, \bar{u}_\varepsilon - \bar{u}_{\varepsilon h\gamma})$$
$$+ j(\bar{u}_{\varepsilon h}) - j_\gamma(\bar{u}_{\varepsilon h}) + j_\gamma(\bar{u}_{\varepsilon h\gamma}) - j(u_{\varepsilon h\gamma}) \geq 0. \tag{3.23}$$

From (3.17) we can easily deduce

$$|j_\gamma(v) - j(v)| \leq \gamma\left(g \text{ meas}(\Omega) + \int_{\Gamma_5} \mu S\right) \tag{3.24}$$

If we use now (3.24) and (1.57) in (3.23) we obtain

$$2\eta\rho\|\bar{u}_{\varepsilon h} - \bar{u}_{\varepsilon h\gamma}\|_{\boldsymbol{H}}^2 \leq 2\gamma\left(g \operatorname{meas}(\Omega) + \int_{\Gamma_5} \mu S\right);$$

hence (3.22) holds.

Note finally that we can compute the gradient of j_γ; after some algebra we have

$$\langle \nabla j_\gamma(u), v\rangle = g\int_\Omega (D(u+u_0)\cdot D(v)/(|D(u)|^2+\gamma^2)^{1/2})\,\mathrm{d}x$$
$$+\int_{\Gamma_5} (\mu S(u+u_0)\cdot v/(|u+u_0|^2+\gamma^2)^{1/2})\,\mathrm{d}\Gamma \qquad \forall u, v \in W. \quad (3.25)$$

Moreover, since $J_{\varepsilon\gamma}$ is a differentiable and convex function by Corollary A2.2, problem (3.20) (or (3.21)) is equivalent to the following non-linear algebraic system:

$$\bar{u}_{\varepsilon h\gamma} \in W_h, \quad \langle \nabla J_{\varepsilon\gamma}(\bar{u}_{\varepsilon h\gamma}), v_h\rangle = 0 \qquad \forall v_h \in W_h \quad (3.26)$$

where $\nabla J_{\varepsilon\gamma}$ is given by

$$\langle \nabla J_{\varepsilon\gamma}(u), v\rangle = a(u,v) + \frac{1}{\varepsilon}c(u,v) + \nabla j_\gamma(u)\cdot v - \langle f, v\rangle \qquad \forall u, v \in W. \quad (3.27)$$

3.3 A Newton iterative method

In order to solve the non-linear system (3.26) we shall use the Newton iterative method. Hence we shall consider a sequence of linear algebraic systems recursively defined by

$$\bar{u}^0_{\varepsilon h\gamma} \in W_h \quad (3.28)$$

$$\langle H(\bar{u}^n_{\varepsilon h\gamma})(\bar{u}^{n+1}_{\varepsilon h\gamma} - \bar{u}^n_{\varepsilon h\gamma})\rangle, v_h + \langle \nabla J_{\varepsilon\gamma}(\bar{u}^n_{\varepsilon h\gamma}), v_h\rangle = 0 \qquad \forall v_h \in W_h, \quad (3.29)$$

where H denotes the Hessian matrix $\nabla(\nabla J_{\varepsilon\gamma})$ given by

$$\langle H(u)v, w\rangle = a(v,w) + \frac{1}{\varepsilon}c(v,w)$$
$$+\int_\Omega \frac{g[(D(v)\cdot D(w)(|D(u+u_0)|^2+\gamma^2) - (D(u+u_0)\cdot D(v))(D(u+v_0)\cdot D(w))]}{(|D(u+u_0)|^2+\gamma^2)^{3/2}}\,\mathrm{d}x$$
$$+\int_{\Gamma_5} \frac{\mu S[(v\cdot w)(|u+u_0|^2+\gamma^2) - ((u+u_0)\cdot v)((u+u_0)\cdot w)}{(|u+u_0|^2+\gamma^2)^{3/2}}\,\mathrm{d}\Gamma. \quad (3.30)$$

In some concrete problems (see for instance the wire-drawing problem in the next section) the greatest part of the computing effort is to compute the functional $J_{\varepsilon\gamma}$, its gradient $\nabla J_{\varepsilon\gamma}$ and the hessian matrix H (there are no great differences between them from the point of view of the volume of computation). The Newton method needs only a small number of iterations for a good approximation (provided that the start function $\bar{u}^0_{\varepsilon h\gamma}$ is good enough) hence we have to compute the hessian matrix and the gradient vector only a few times. Since the volume of computation needed to solve the linear system (3.29) is much smaller than the one needed to compute the gradient $\nabla J_{\varepsilon\gamma}$, we have chosen this Newton type method from the convex optimization theory.

The Newton iterative method gives us the opportunity to remove the restrictive assumption that S is a given datum on Γ_5. Indeed we can replace S from (3.30) with $|T'| = g + 2\eta|D(u_h)|$, where u_h is the computed velocity field from the previous step.

The following lemma shows that if the start function $\bar{u}^0_{\varepsilon h\gamma}$ is chosen in a neighbourhood of $\bar{u}_{\varepsilon h\gamma}$, then $\bar{u}^n_{\varepsilon h\gamma}$ is a good approximation of $\bar{u}_{\varepsilon h\gamma}$ for n large.

Lemma 3.4 *Let $\varepsilon, h, \gamma > 0$ be fixed, $\bar{u}_{\varepsilon h\gamma}$ be given by (3.20), (3.21), or (3.26), and $(\bar{u}^n_{\varepsilon h\gamma})_{n\in N}$ the sequence obtained by solving the recursive linear system (3.28)–(3.29). Then there exists $\bar{\mathscr{U}}$, a neighbourhood of $\bar{u}_{\varepsilon h\gamma}$, such that if $\bar{u}^0_{\varepsilon h\gamma} \in \bar{\mathscr{U}}$, then*

$$\bar{u}^n_{\varepsilon h\gamma} \to \bar{u}_{\varepsilon h\gamma} \qquad \text{in } W_h \quad \text{when } n \to \infty. \tag{3.31}$$

Remark 3.1 *Since the Bingham fluid can be thought of as a non-linear perturbation of the Stokes fluid in many concrete problems it is useful to choose $\bar{u}^0_{\varepsilon h\gamma}$ as the approximation solution for the Stokes fluid. Hence $\bar{u}^0_{\varepsilon h\gamma}$ is the solution of the following linear algebraic system:*

$$\bar{u}^0_{\varepsilon h\gamma} \in W_h, \quad a(\bar{u}^0_{\varepsilon h\gamma}, v_h) + \frac{1}{\varepsilon} c(\bar{u}^0_{\varepsilon h\gamma}, v_h) = \langle f, v_h \rangle \qquad \forall v_h \in W_h, \tag{3.32}$$

which is obtained from (3.29) with $\bar{u}^{-1}_{\varepsilon h\gamma} = 0$, and $H, \nabla J_{\varepsilon\gamma}$ are given by (3.27), (3.30) with $g = 0$, $\mu = 0$.

PROOF OF LEMMA 3.4. In order to prove the local convergence of the Newton method we shall use Lemma A4.3. We have only to check that there exist $\alpha, \beta > 0$ such that

$$\alpha\|v\|^2_{H_1} \leq \langle H(u)v, v\rangle \leq \beta\|v\|^2_{H_1} \qquad \forall u, v \in W \tag{3.33}$$

and the function $u_h \to H(u_h)$ is continuous from W_h to $B(W_h)$. Indeed, since j_γ is convex then $\nabla(\nabla j_\gamma)$ is positive; hence from (1.57) we can easily deduce that the first inequality in (3.33) holds with $\alpha = 2\eta\rho$. From direct computation we have

$$\langle H(u)v, v\rangle \leq \left(2\eta + \frac{1}{\varepsilon}\right) M \|v\|_{\boldsymbol{H}_1}^2 + \frac{2g}{\gamma}\|D(v)\|_{\mathscr{H}}^2 + \frac{2}{\gamma}\int_{\Gamma_5} \mu S|v|^2 \, \mathrm{d}\Gamma$$

$$\leq \left(2\eta + \frac{1}{\varepsilon} + \frac{2g}{\gamma}\right) M \|v\|_{\boldsymbol{H}_1}^2 + \frac{2}{\gamma} C \|\mu S\|_{L^\infty(\Gamma_5)} \|v\|_{\boldsymbol{H}_1}^2$$

and hence (3.33) is proved. Now let $u_h^n \in W_h$ such that $u_h^n \to u_h \in W_h$ in $\boldsymbol{H}_1$ when $n \to \infty$ and let $K \subset \Omega$ be a compact set of the finite element triangulation. Since u_h^n are polynomial functions on K from the convergence $D(u_h^n) \to D(u_h)$ in $[L^2(K)]_s^{N \times N}$ we deduce that $D(u_h^n) \to D(u_n)$ uniformly on K. Using a similar argument we obtain that $u_h^n \to u_h$ uniformly on $\partial K \cap \Gamma_5$. From (3.30) we now easily deduce that $\langle H(u_h^n)v_h, w_h\rangle \to \langle H(u_h)v_h, w_h\rangle$ for all $v_h, w_h \in W_h$, hence $H(u_h^n) \to H(u_h)$ in $B(W_h)$.

Let $u_\varepsilon = \bar{u}_\varepsilon + u_0$, $u_{\varepsilon h} = \bar{u}_{\varepsilon h} + u_0$, $\bar{u}_{\varepsilon h\gamma} = \bar{u}_{\varepsilon h\gamma} + u_0$, $u_{\varepsilon h\gamma}^n = \bar{u}_{\varepsilon h\gamma}^n + u_0$ for all $\varepsilon, h, \gamma > 0$, $n \in \mathbb{N}$. From Lemmas 3.1–4 we deduce the following final convergence result:

Theorem 3.1 *If u is the solution of* (1.51) *or* (1.52) *then we have:*

(*i*) *$u_\varepsilon \to u$ strongly in $\boldsymbol{H}_1$ when $\varepsilon \to 0$;*
(*ii*) *$u_{\varepsilon h} \to u_\varepsilon$ strongly in $\boldsymbol{H}_1$ when $h \to 0$;*
(*iii*) *$u_{\varepsilon h\gamma} \to u_{\varepsilon h}$ strongly in $\boldsymbol{H}_1$ when $\gamma \to 0$;*
(*iv*) *there exists a neighbourhood $\mathscr{U}$ of $u_{\varepsilon h\gamma}$ such that if $u_{\varepsilon h\gamma}^0 \in \mathscr{U}$ then $u_{\varepsilon h\gamma}^n \to u_{\varepsilon h\gamma}$ strongly in $\boldsymbol{H}_1$ when $n \to \infty$.*

3.4 An application to the wire-drawing problem

The bounded domain $\Omega \subset \mathbb{R}^3$ drawn in Figure 3.1 represents the geometry to be considered in the wire-drawing problem. Due to the axisymmetry of the domain we use the cylindrical coordinates and hence the problem is reduced to a two-dimensional one in the (z, r) *plane*.

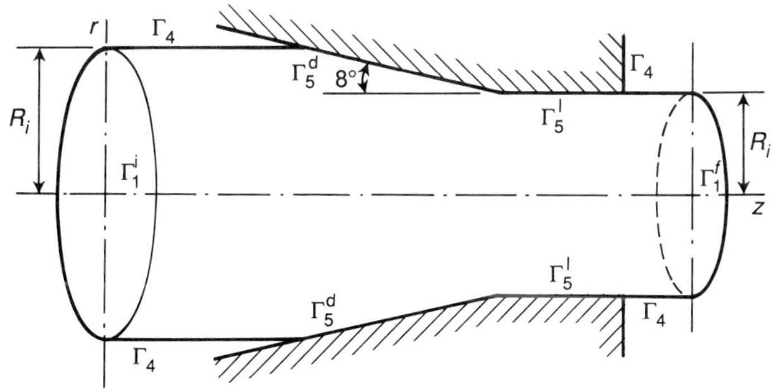

Fig. 3.1 Main geometrical elements for flow of a viscoplastic wire through a conical die.

Equations (1.17)–(1.20) are considered in the domain and the following boundary conditions are imposed:

- On Γ_1^f, at the exit of the die, the final velocity v_f is imposed (we put $U = v_f e_z$ in (1.21) on Γ_1^f).
- On Γ_1^i, at the entrance of the die, several conditions can be considered: the initial velocity obtained from the incompressibility condition $v_i R_i^2 = v_f R_f^2$ can be imposed (hence we put $U = v_i e_z$ in (1.21) on Γ_1^i), or the back stress can be given, $Tv = f = T_b e_z$ (hence we put $F = T_b e_z$ in (1.22)), or the two conditions can be combined by imposing $Tv \cdot v = T_b$ and $u_\tau = 0$ (hence we put $f_v = T_b$ in (1.23)). In the numerical computations presented further only the first condition is considered on Γ_1^i (hence $\Gamma_1 = \Gamma_1^f \cup \Gamma_1^i$ and $U = v_f e_z$ on Γ_1^f, $U = v_i e_z$ on Γ_1^i).
- On Γ_4, which is in contact with the air, we impose (1.24).
- On $\Gamma_5 = \Gamma_5^d \cup \Gamma_5^l$, the surface of the die and of the land, the friction law (1.25) is considered; the friction coefficients being different on Γ_5^d and Γ_5^l and denoted by μ_d and μ_l. Only the viscoplastic friction law was used.

In order to construct the space W_h the finite-element method was used. The two-dimensional domains from the (z, r) plane, which generates, by rotation around the Oz-axis, the domain Ω, was divided into quadrilateral elements so that the volumes of the three-dimensional elements obtained by rotation are almost equal. The mesh size near the die is smaller; in that region the strain is expected to increase.

The mesh pattern presented in Figure 3.2 consist of 85 finite elements and 108 nodes.

In order to determine $\bar{u}_{\varepsilon h\gamma}^0$ used in the iterative method (3.28)–(3.29), the system (3.29) was first solved for $\bar{u}_{\varepsilon h\gamma}^{-1} = 0$, $g = 0$, $\mu_1 = \mu_2 = 0$ (see also Remark 3.1). The restriction assumption that S is a given datum on Γ_5 was removed by replacing S in the computation of the Hessian matrix (3.30) at step $n+1$ with $g + 2\eta|D(\bar{u}_{\varepsilon h\gamma}^n)|$. For good convergence five iterations are enough.

In order to compare the computed results with the experimental data, the

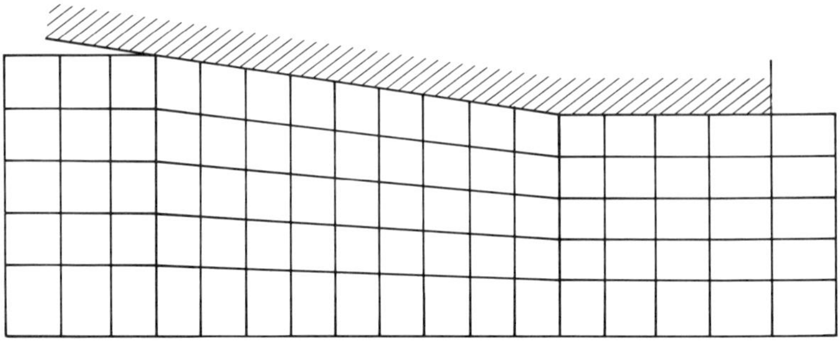

Fig. 3.2 Mesh pattern used in the finite-element discretization of the domain Ω.

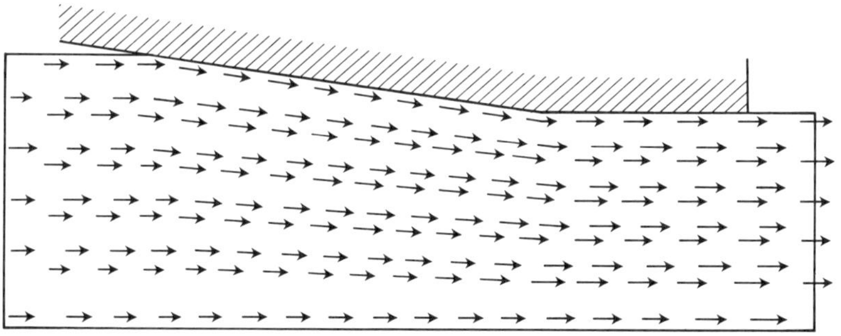

Fig. 3.3 Computed velocity field.

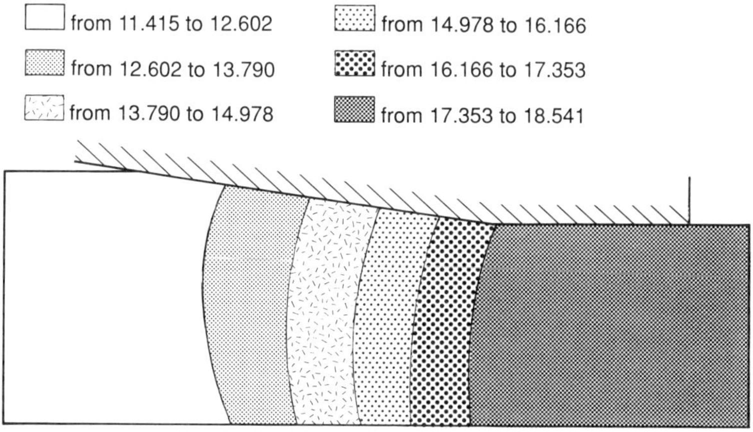

Fig. 3.4 Computed velocity component u_z (in mm/s).

various constants involved were chosen for copper wire drawing as in Cristescu (1977). The drawing speeds considered are low compared with the industrial speeds. This is because in the mathematical model the viscosity coefficient η does not depend on the strain rate. For large drawing speeds the strain rate has a great variation in the considered domain and thus η cannot be considered constant.

The geometrical data are $R_i = 0.47$ mm, $R_f = 0.373$ mm (which correspond with a reduction in area of 37%), $\theta = 8°$ and the length of the land 0.362 mm. The physical data are $v_i = 11.529$ mm/s, $v_f = 18.0$ mm/s, $\eta = 2.94$N s/mm^2, $g = 289$ N/mm^2, $\mu_1 = 0.09$, $\mu_2 = 0.08$. The penalty factor $1/\varepsilon$ was chosen to be $10^3\eta$ and the regularization factor $\gamma = 10^{-6}$.

The computed velocity field is presented in Figure 3.3.

As can be seen in Figures 3.3 and 3.4 the velocity increases with z, and due to the friction near the die the increase is delayed.

In Figure 3.5 we notice that the velocity component of on the r axis is positive at the exit of the die describing the die swelling phenomenon.

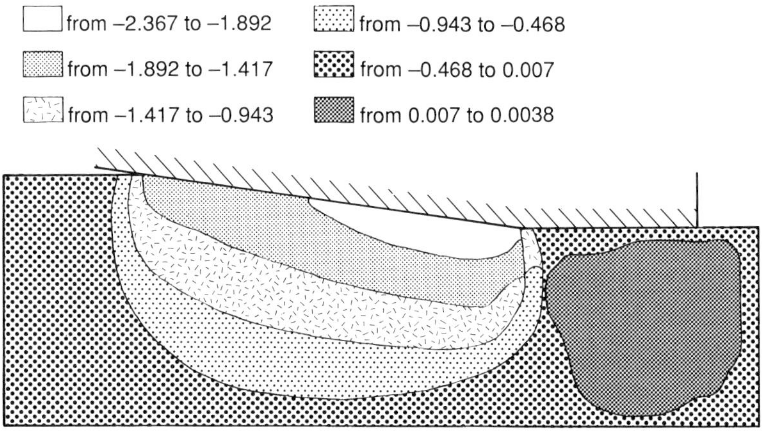

Fig. 3.5 Computed velocity component u_r (in mm/s)..

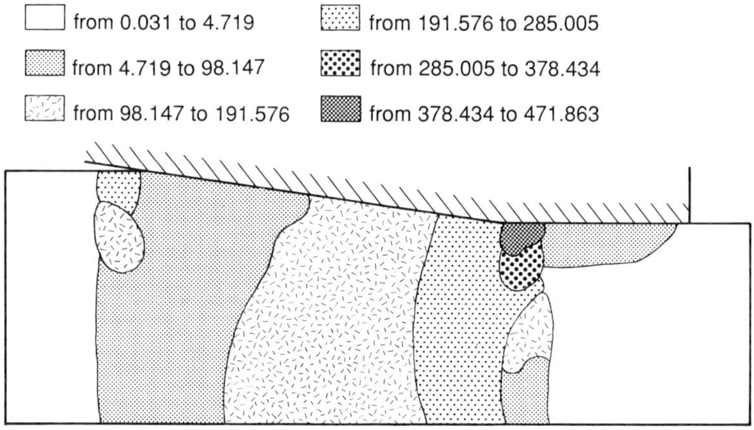

Fig. 3.6 Computed strain rate $|D(u)|$ (in s^{-1}).

In Figure 3.6 it can be observed that at the entrance and at the exit of the die the wire behaves as a rigid one and the domain with maximum strain rate appears at the exit of the conical surface of the die.

The drawing stress was computed using the following relation

$$T_d^n = \left[2\eta / \|D(u_{\varepsilon h\gamma}^n)\|_{\mathscr{H}}^2 + \frac{1}{\varepsilon} \|\operatorname{div} u_{\varepsilon h\gamma}^n\|_H^2 + g \int_\Omega |D(u_{\varepsilon h\gamma}^n)| \, dx \right.$$

$$\left. + \int_{\Gamma_5} \mu(g + 2\eta |D(u_{\varepsilon h\gamma}^n)|) |u^n| \, d\Gamma \right] \Big/ (\pi v_f R_f^2) \tag{3.34}$$

Thus a drawing stress $T_d = 304 \text{ N/mm}^2$ was obtained, corresponding to the experimental data.

Bibliographical notes

From the mechanical point of view the Bingham model may be considered as a strict generalization of the Newtonian viscous incompressible fluid. The reader interested in the description of this model may find details in the works of Oldroyd (1947*a,b,c*), Eirich (1956), Reiner (1960, Ch. VIII), Cristescu and Suliciu (1982, Ch. VIII).

From the mathematical point of view the Bingham model is retained because it is the simplest generalization of the Newtonian fluid model which leads to variational inequalities such that some of the well-known results for the Navier–Stokes equations can be extended.

The first investigation in the study of the Bingham fluid by means of functional methods were made by Duvaut and Lions (1970). Existence and uniqueness results concerning various evolutionary and stationary problems in the study of the Bingham fluid with adherence can be found in the work of Duvaut and Lions (1972). Numerical approaches for problems of this kind were considered by various authors, such as Fortin (1972) (non-stationary flow with homogeneous boundary conditions), Glowinski (1973) (the blocking property for the laminar flow through a pipe). Glowinski *et al.* (1976) (laminar flow with homogeneous boundary conditions).

The friction law (1.25) in the case $S = |T'|$ was introduced by Cristescu (1975) and applied hereafter in many metal-forming problems. The results given in Section 1.4 (Theorem 1.1 included) belong to Ionescu (1985). Problems of the form (1.17)–(1.25) can also be investigated by a different method. So, in the papers of Ionescu and Sofonea (1984), Ionescu *et al.* (1985*b*), it is shown that such types of mechanical problems lead to multivalued equations of the form $Pu \ni f$, for which a variational method can be applied in order to prove the existence and uniqueness of the solution.

The blocking phenomenon was pointed out in the study of laminar and stationary flow of the Bingham fluid through a cylindrical pipe (see Remark 2.4). So considering the adherence condition, in Duvaut and Lions (1972, p. 317) it is shown that if the yield limit g is great enough, the Bingham fluid does not flow, hence it is blocked in the pipe. In the works of Glowinski (1973) and Glowinski *et al.* (1976), Ch. 4, the critical yield limit for the blocking property is computed as a function of the drop in pressure per unit length. The same result was obtained in a different way by Sofonea (1982*a*) and, by analogy, the blocking property for abstract variational inequality of the form (2.9) was introduced and studied in Sofonea (1982*b*). The influence of the friction was studied by Ionescu and Sofonea (1986). Section 2 was written following the ideas of this paper.

The results presented in Sections 3.1–3.4 were written following Ionescu *et al.* (1985*a*), and finally the numerical results given in Section 3.5 were performed by Ionescu and Vernescu (1988).

APPENDIX

In this appendix we summarize the main concepts and results necessary in the reading of Chapters 1–5. Some preliminaries on linear and non-linear analysis and evolution equations in real Banach spaces are presented. Some numerical methods in solving the problems of the form studied in Chapters 3–5 are also prescribed. Most of the material is standard and can be found in many books or surveys. For this reason some important results have been given without proof while referring to the bibliography for proofs as well as for more information in topics presented here.

1 Elements of linear analysis

In this section we summarize some basic concepts of functional analysis. Our intention is to present as simply as possible some of the classical notions and results which are frequently used in this book. For this, some of the concepts were introduced in a simplified way.

1.1 Normed linear spaces and linear operators

A real linear space X is called a *normed linear space* if with every $x \in X$ there is associated a real number $\|x\|_X$, the norm of x, such that

$$\|x\|_X \geq 0 \qquad \text{and} \qquad \|x\| = 0 \quad \text{iff } x = 0,$$

$$\|x + y\|_X \leq \|x\|_X + \|y\|_X \qquad \text{(triangle inequality)},$$

$$\|\alpha x\|_X = |\alpha| \, \|x\|_X,$$

for all $x, y \in X$ and $\alpha \in \mathbb{R}$.

The convergence $\|x_n - x\|_X \to 0$ in a normed linear space X will be denoted by $x_n \to x$ or $\lim x_n = x$ and we say that the sequence $(x_n) \subset X$ converges (strongly) to the element x in X. The adjective 'strong' is introduced here to distinguish it from other types of convergences which will be introduced in the next section.

A sequence $(x_n) \subset X$ is called a *Cauchy sequence* iff it satisfies the Cauchy's convergence criterion

$$\lim \|x_n - x_m\|_X = 0 \qquad \text{when} \quad n, m \to \infty.$$

One can easily prove that every convergent sequence is a Cauchy sequence. Since the converse statement is not always true, the following definition is justified: a normed space X is called a *Banach space* if it is complete, i.e. every Cauchy sequence (x_n) of X converges (strongly) to an element $x \in X$.

Let us now consider two linear normed spaces $(X, \|\cdot\|_X)$ and $(Y, \|\cdot\|_Y)$. On the product linear space $X \times Y$ we shall define the norm

$$\|(x; y)\|_{X \times Y} = (\|x\|_X^2 + \|y\|_Y^2)^{1/2}.$$

If we endow $X \times Y$ with this norm and X, Y, are Banach spaces then $X \times Y$ is also a Banach space.

A linear operator $T\colon D(T) \subset X \to Y$ is called a *closed linear operator* if its graph

$$G(T) = \{(x; Tx) \in X \times Y \,|\, x \in D(T)\}$$

is a closed subset of $X \times Y$. One can easily see that T is a closed operator iff the following condition holds:

$$\forall (x_n) \subset D(T), \quad \lim x_n = x, \quad \lim Tx_n = y \qquad \text{implies} \quad x \in D(T) \quad \text{and} \quad Tx = y.$$

The inverse T^{-1} of a closed linear operator $T\colon D(T) \subset X \to Y$, if it exists, is a closed linear operator.

Let X be a Banach space and let $T\colon D(T) \in X \to X$ be a linear closed operator. If we denote by $[D(T)]$ the linear space $D(T)$ endowed with the graph norm of T given by

$$\|x\|_T = \|x\|_X^2 + \|Tx\|_X^2)^{1/2} \qquad \forall x \in D(T),$$

then $[D(T)]$ is a Banach space.

Let us now consider a linear operator $T\colon D(T) \subset X \to Y$ where X and Y are linear normed spaces. The operator T is *continuous* iff there exists $\beta > 0$ such that

$$\|Tx\|_Y \leq \beta \|x\|_X \qquad \forall x \in D(T). \tag{1.1}$$

From (1.1) one can deduce that a linear operator $T\colon D(T) \subset X \to Y$ admits a *continuous inverse* T^{-1} iff there exists $\gamma > 0$ such that

$$\|Tx\|_Y \geq \gamma \|x\|_X \qquad \forall x \in D(T).$$

A continuous linear operator $T\colon X \to Y$ is called a *bounded linear operator*. The linear space of all bounded linear operators on X into Y will be denoted by $B(X, Y)$ and for the sake of simplicity we shall write $B(X)$ instead of $B(X, X)$. The linear space $B(X, Y)$ will be endowed with the norm

$$\|T\|_{B(X,Y)} = \sup_{\|x\|_X = 1} \|Tx\|_Y. \tag{1.2}$$

In this way $B(X, Y)$ becomes a normed linear space and it can be proved that if Y is a Banach space then $B(X, Y)$ is also a Banach space.

The completeness of a Banach space enables us to obtain the following consequence of the *open mapping theorem*:

Theorem 1.1 *Let X, Y be two Banach spaces and $T \in B(X, Y)$ be one to one. Then $T^{-1} \in B(Y, X)$.*

1.2 Duality and weak topologies

In the special case when Y is the real number field topologized in the usual way, $B(X, Y)$ is called the *strong dual* of X and it is denoted by X'. The elements of X' are called *continuous functionals*. As it follows from (1.2) the norm on X' is given by

$$\|f\|_{X'} = \sup_{\|x\|_X = 1} |f(x)| \tag{1.3}$$

and X' is a Banach space. For any $x \in X$ and $x' \in X'$ we shall denote by $\langle x', x \rangle_{X',X}$ or $x'(x)$ the value of the functional x' in the point x. We have

Theorem 1.2 (Hahn–Banach). *Let X be a normed linear space, M a linear subspace of X and $f_1 \in M'$. Then there exists $f \in X'$ such that f is an extension of f_1 and $\|f\|_{X'} = \|f_1\|_{M'}$.*

As a consequence of the above theorem we recall:

Corollary 1.1 *Let X be a normed linear space and let $x_0 \neq 0$ be an element of X. Then there exists an element $f_0 \in X'$ such that $\|f_0\|_{X'} = 1$, $\langle f_0, x_0 \rangle_{X',X} = \|x_0\|_X$.*

Corollary 1.2 *Let X be a normed linear space, M a closed subspace of X and $x_0 \in X \backslash M$. Then there exists an element $f_0 \in X'$ such that $\langle f_0, x_0 \rangle_{X',X} \neq 0$ and $\langle f_0, x \rangle_{X',X} = 0$ for all $x \in M$.*

Now let X be a normed linear space and let us denote by $X'' = (X')'$ the dual of the Banach space X', which will be called the *bidual* of X. Each element $x_0 \in X$ defines a continuous linear functional f_0 on X' (i.e. it belongs to X'') by the equality $\langle f_0, x' \rangle_{X'',X'} = \langle x', x_0 \rangle_{X',X}$. The mapping $x_0 \to f_0 = J(x_0)$ of X into X'' is linear and isometric (i.e. $\|J(x_0)\|_{X''} = \|x_0\|_X$ for all $x_0 \in X$). So, the normed linear space X may be considered as a linear subspace of the Banach space X'' by the embedding $x_0 \to J(x_0)$ and we have the following definition: a normed linear space X is said to be *reflexive if* X may be identified with its bidual X'' by the canonical embedding J (i.e. if $J(X) = X''$).

We say that a sequence $(x_n) \subset X$ is *weakly convergent* to the element $x \in X$

and we write '$x_n \to x$ weakly in X' or '$x_n \rightharpoonup x$' if

$$\langle x', x_n \rangle_{X',X} \to \langle x', x \rangle_{X',X} \qquad \forall x' \in X'.$$

The following statements hold:

(1) if $x_n \to x$ strongly in X then $x_n \to x$ weakly in X;
(2) if $x_n \to x$ weakly in X then $(\|x_n\|_X)$ is a bounded sequence and

$$\|x\|_X \leq \underline{\lim} \, \|x_n\|_X$$

(3) if $x_n \to x$ weakly in X and $x'_n \to x'$ (strongly) in X' then

$$\langle x'_n, x_n \rangle_{X',X} \to \langle x', x \rangle_{X',X}.$$

Theorem 1.3 (Eberlein–Smulian). *Let X be a Banach space. Then X is reflexive iff for all bounded sequences $(x_n) \subset X$ there exists $(x'_n) \subset (x_n)$ a subsequence of (x_n) and $x \in X$ such that $x'_n \to x$ weakly in X.*

We say that a sequence $(x'_n) \subset X'$ is *weakly* convergent* to the element $x' \in X'$ and we write '$x'_n \to x'$ weakly* in X'' or '$x'_n \overset{*}{\rightharpoonup} x'$' if

$$\langle x'_n, x \rangle_{X',X} \to \langle x', x \rangle_{X',X} \qquad \forall x \in X.$$

The following statements hold:

(1) if $x'_n \to x'$ strongly in X' then $x'_n \to x'$ weakly in X':
(2) if $x'_n \to x'$ weakly in X' then $x'_n \to x'$ weakly* in X';
(3) if $x'_n \to x'$ weakly in X' then (x'_n) is a bounded sequence of X' and

$$\|x'\|_{X'} \leq \underline{\lim} \, \|x'_n\|_{X'}$$

(4) if $x'_n \to x'$ weakly* in X' and $x_n \to x$ strongly in X then

$$\langle x'_n, x_n \rangle_{X',X} \to \langle x', x \rangle_{X',X}.$$

We finish this section by the following definition:

A Banach space X is called *separable* if there exists a countable dense subset of X. We have

Theorem 1.4 *Let X be a separable Banach space and let $(f_n) \subset X'$ be a bounded sequence. Then there exists $(f'_n) \subset (f_n)$, a subsequence of (f_n), and $f \in X'$ such that $f'_n \to f$ weakly* in X'.*

1.3 Hilbert spaces

Let X be a real linear space. We say that the bilinear form $\langle \cdot, \cdot \rangle_X : X \times X \to \mathbb{R}$ is an inner product on X if it is symmetric (i.e. $\langle x, y \rangle_X = \langle y, x \rangle_X$ for all $x, y \in X$) and positive definite (i.e. $\langle x, x \rangle_X \geq 0$ and $\langle x, x \rangle_X > 0$ for $x \neq 0$).

Let X be endowed with the norm generated by the inner product $\langle \cdot, \cdot \rangle_X$, defined by

$$\|x\|_X = \langle x, x \rangle_X^{1/2} \qquad \forall x \in X.$$

A normed linear space X which has the norm generated by an inner product, as above, is called a *pre-Hilbert space*. If in addition X is a Banach space, then we say that X is a *Hilbert space*.

If X is a pre-Hilbert space one can easily prove that the Cauchy–Schwartz inequality holds:

$$|\langle x, y \rangle_X| \leq \|x\|_X \|y\|_X \qquad \forall x, y \in X.$$

Theorem 1.5 (Projection on a convex closed set). *Let X be a real Hilbert space and $K \subset X$ be a closed convex subset of X. Then for all $x \in X$ there exists a unique element $y = P_K x \in K$ such that*

$$\|x - y\|_X = \min_{z \in K} \|x - z\|_X.$$

In the following we consider a Hilbert space X. The map $P_K: X \to K$ defined by Theorem 1.5 is called the *projection map* on the closed convex subset K. One can prove that $y = P_K x$ if and only if the following variational inequality holds:

$$y \in K, \quad \langle y - x, z - y \rangle_X \geq 0 \qquad \forall z \in K.$$

Moreover, the projection map P_K is a contraction operator, i.e.

$$\|P_K x - P_K y\|_X \leq \|x - y\|_X \qquad \forall x, y \in X.$$

If in particular $K = M$ is a closed subspace of X, then $P_M \in B(X)$ and $\|P_M\|_{B(X)} = 1$.

For $x, y \in X$ we say that x is *orthogonal* to y and we write $x \perp y$ if $\langle x, y \rangle_X = 0$. Two closed linear subspaces $M, N \subset X$ are called *orthogonal* if for all $x \in M$, $y \in N$ we have $\langle x, y \rangle_X = 0$. We write $X = M \oplus N$ if M and N are orthogonal and for all $x \in X$ there exists a unique pair $(x', x'') \in M \times N$ such that $x = x' + x''$.

Let M be a closed linear subspace of X. We denote by $M^\perp$ the following subspace of X:

$$M^\perp = \{x'' \in X \,|\, \langle x', x'' \rangle_X = 0 \qquad \forall x' \in M\}.$$

Since for all $x \in X$ there exists $x' = P_M x \in M$ and $x'' = x - P_M x \in M^\perp$ such that $x = x' + x''$ we have $X = M \oplus M^\perp$.

Theorem 1.6 (the Riesz representation theorem). *Let X be a Hilbert space and $x' \in X'$. Then there exists a uniquely determined element $x = x_{x'} \in X$ such that*

$$\langle x', y \rangle_{X', X} = \langle x, y \rangle_X \qquad \forall y \in X$$

and

$$\|x\|_X = \|x'\|_{X'}.$$

From the previous theorem it follows that every Hilbert space is a reflexive Banach space, the totality X' of bounded linear functionals on X also constitutes a Hilbert space, and there exists a norm-preserving one-to-one linear correspondence $x' \to I(x') = x_{x'}$ between X' and X. By this correspondence X' may be identified with X as an abstract set. This identification is usually used but not always. Let us see an example in which this identification will not be used.

Let $(Y, \| \ \|_Y, \langle \cdot, \cdot \rangle_Y)$ be a Hilbert space such that Y is continuously and densely embedded in X (i.e. $Y \subset X$, Y is a dense subset of X and there exists $a > 0$ such that $\|y\|_X \le a\|y\|_Y$ for all $y \in Y$). For all $x \in X$ we remark that $x' = I^{-1}(x) \in X'$ given by $y \to \langle x, y \rangle_X$ is a bounded linear functional on Y. If we denote by Tx the above element of Y', then we have

$$\langle Tx, y \rangle_{Y',Y} = \langle x, y \rangle_X \qquad \forall x \in X, y \in Y.$$

One can prove that T is injective, continuous and $T(X)$ is a dense subset of Y'. Using the operator T we get

$$Y \subset X \approx X' \subset Y'. \tag{1.4}$$

Since we have already identified X with X' we can no longer identify Y with Y' such that (1.4) holds. The Hilbert space X from (1.4) will be called the *pivot space*.

Now let X be a Hilbert space and $A: D(A) \subset X \to X$ be a linear operator with $D(A)$ a dense subspace of X. Let us denote by

$$D(A^*) = \{x \in X \mid \exists y \in X \text{ such that } \langle Az, x \rangle_X = \langle z, y \rangle_X \quad \forall z \in D(A)\}. \tag{1.5}$$

Since $D(A)$ is a dense subspace of X we find that y from above is determined by x. If we denote by $A^*x = y$ then $A^*: D(A^*) \subset X \to X$ is called the *adjoint operator* of A and is a closed operator. We also recall that A is called a *symmetric operator* if $A \subset A^*$ (i.e. $D(A) \subset D(A^*)$ and $A^*|_{D(A)} = A$) and a *self-adjoint operator* if $A = A^*$.

2 Elements of non-linear analysis

In this section we summarize the basic results concerning convex functions, elliptic variational inequalities and theory of non-linear monotone operators in Hilbert spaces. Some of the concepts and results presented here hold also in more general cases but for simplicity we sometimes revert to the Hilbertian case.

2.1 Convex functions

Let X be a real linear space. The function $\varphi: X \to (-\infty, +\infty]$ is called *convex* if for all $u, v \in X$ and $t \in (0, 1)$, the following inequality holds:

$$\varphi(tu + (1-t)v) \geq t\varphi(u) + (1-t)\varphi(v). \tag{2.1}$$

The function φ is called *strictly convex* if (2.1) is a strict inequality whenever u and v are distinct points of X; moreover, if X is a linear normed space, the function $\varphi: X \to (-\infty, +\infty]$ is said to be *uniformly convex* if there exists $m > 0$ such that

$$t\varphi(u) + (1-t)\varphi(v) - \varphi(tu + (1-t)v) \leq \frac{m}{2} t(1-t)\|u - v\|_X^2 \tag{2.2}$$

for all $u, v \in X$ and $t \in (0, 1)$. The function $\varphi: X \to (-\infty, +\infty]$ is said to be *concave* if the function $-\varphi$ is convex. All the previous definitions could be formulated in the case of functions defined on convex sets $K \subset X$ but, taking $\varphi(u) = +\infty$ if $u \notin K$ we may assume that φ is defined everywhere in X.

A function $\varphi: X \to (-\infty, +\infty]$ is called *proper* if it is not the constant function $+\infty$, i.e. there exists $u_0 \in X$ such that $\varphi(u_0) < +\infty$. Given a function $\varphi: X \to (-\infty, +\infty]$ we define the *effective domain* and the *epigraph* of φ by

$$D(\varphi) = \{u \in X \,|\, \varphi(u) < +\infty\},$$

$$\text{epi } \varphi = \{(u; \alpha) \in X \in \mathbb{R} \,|\, \varphi(u) \leq \alpha\}.$$

It is easy to see that φ is convex iff $D(\varphi)$ is a convex set of X and $\varphi: D(\varphi) \to \mathbb{R}$ is a convex function. Moreover, φ is convex iff its epigraph is a convex subset of $X \times \mathbb{R}$.

Now let X be a topological space and $\varphi: X \to (-\infty, +\infty]$. The function φ is called *lower-semicontinuous* at $u_0 \in X$ if $\liminf_{u \to u_0} \varphi(u) = \varphi(u_0)$. We recall that

$$\liminf_{u \to u_0} \varphi(u) = \sup_{V \in \mathscr{V}(u)} \inf_{v \in V} \varphi(v),$$

where $\mathscr{V}(u)$ is a base of neighbourhoods of u in X. A function which is lower semicontinuous at each point *of* $X(K)$ is called *lower semicontinuous* (on K).

The lower semicontinuity property can be characterized as follows:

Lemma 2.1 *Let X be a topological space and $\varphi: X \to (-\infty, +\infty]$. Then the following conditions are equivalent*:

(*i*) *φ is lower semicontinuous on X;*
(*ii*) the level sets $\{u \in X \,|\, \varphi(u) \leq \lambda\}$ *are closed, for all $\lambda \in \mathbb{R}$;*
(*iii*) *the epigraph of φ is closed in $X \times \mathbb{R}$.*

Since in normed vector spaces every convex set is simultaneously closed in the weak and strong topology, from Lemma 2.1 we deduce the following corollary:

Corollary 2.1 *Let X be a real normed space. A proper convex function $\varphi: X \to (-\infty, +\infty]$ is lower semicontinuous on X if and only if it is lower semicontinuous with respect to the weak topology on X.*

Let K be a subset of a set X; we define the *indicator function* of the set K by

$$\psi_K(v) = \begin{cases} 0 & \text{if } v \in K \\ +\infty & \text{if } v \notin K. \end{cases} \tag{2.3}$$

Using Lemma 2.1 it is easy to prove the following result:

Lemma 2.2 *Let X be a linear topological space and $K \subset X$. The set K is a non-empty closed convex set of X if and only if its indicator function ψ_K is a proper convex lower semicontinuous function on X.*

Using a consequence of the Hahn–Banach theorem, the following property of convex lower semicontinuous functions can be proved:

Lemma 2.3 *Let X be a real normed space and $\varphi: X \to (-\infty, +\infty]$ a proper convex and lower semicontinuous function on X. Then φ is bounded from below by an affine function, i.e. there exists $f \in X'$ and $g \in \mathbb{R}$ such that $\varphi(v) \geq \langle f, v\rangle_{X',X} + g$ for all $v \in X$.*

An important property of lower semicontinuous functions is given by a well-known theorem of the Weierstrass type:

Theorem 2.1 *Let X be a reflexive Banach space and $\varphi: X \to (-\infty, +\infty]$ a proper lower semicontinuous convex function such that $\varphi(x) \to +\infty$ when $\|x\|_X \to +\infty$ (i.e. φ is coercive). Then φ is bounded from below and attains its infimum on X. The set of minimizers of φ is closed and it is a convex set in X if φ is a convex function. Moreover, if φ is a strictly convex function, then φ attain its infimum on X in only one point.*

Let us now consider a real Hilbert space X, with the inner product $\langle\, ,\rangle_X$ and the norm $\|\cdot\|_X$; we also denote by 2^X the set of all subsets of X.

A function $\varphi: X \to (-\infty, +\infty]$ is said to be *Gâteaux differentiable at* $u \in X$ if there exists an element $\nabla\varphi(u) \in X$ such that

$$\lim_{t \to 0} \frac{\varphi(u + tv) - \varphi(u)}{t} = \langle \nabla\varphi(u), v\rangle_X \qquad v \in X. \tag{2.4}$$

The element $\nabla\varphi(u) \in X$ is called the *gradient of* φ at u; φ is said to be

Gâteaux differentiable on X if it is Gâteaux differentiable in every point of X; in this case the operator $u \to \nabla\varphi(u)$: $X \to X$ will be called the gradient operator of φ.

The convexity of Gâteaux differentiable functions can be characterized as follows:

Lemma 2.4 *Let $\varphi: X \to \mathbb{R}$ be a Gâteaux differentiable function. Then φ is a convex function if and only if*

$$\varphi(v) - \varphi(u) \geq \langle \nabla\varphi(u), v - u \rangle_X \qquad \forall u, v \in X. \tag{2.5}$$

PROOF. If φ is a convex function, then for all $u, v \in X$ and $t \in (0, 1)$ from (2.1) we deduce $t(\varphi(v) - \varphi(u)) \geq \varphi(u + t(v - u)) - \varphi(u)$. Dividing by t and passing to the limit when $t \downarrow 0$ from (2.4) we get (2.5). Conversely, let (2.5) hold and $u, v \in X$, $t \in (0, 1)$. Let $w = (1 - t)u + tv$; from (2.5) we get $\varphi(v) - \varphi(w) \geq \langle \nabla\varphi(w), v - w \rangle_X$, $\varphi(u) - \varphi(w) \geq \langle \nabla\varphi(w), u - w \rangle_X$. Multiplying these inequalities by t and $(1 - t)$ respectively and adding the results we get (2.1), hence φ is a convex function.

From the previous lemma it is easy to deduce the following result:

Corollary 2.2 *Let $\varphi: X \to \mathbb{R}$ be a convex and Gâteaux differentiable function. Then u is a minimizer of φ on X if and only if $\nabla\varphi(u) = \theta_X$.*

Inequality (2.5) suggests a generalization of the gradient operator for convex functions defined on real Hilbert spaces. More precisely, given a function $\varphi: X \to (-\infty, +\infty]$ we define the multivalued mapping $\partial\varphi: X \to 2^X$ by

$$\partial\varphi(u) = \{f \in X \mid \varphi(v) - \varphi(u) \geq \langle f, v - u \rangle_X \text{ for all } v \in X\}. \tag{2.6}$$

$\partial\varphi$ is called the *subdifferential mapping* of φ; the elements $f \in \partial\varphi(u)$ are called *subgradients* of φ at u and the inequality in (2.6) is called the *subgradient inequality*. It is immediately clear from (2.6) that $\partial\varphi(u)$ is always a closed convex subset of X, which may also happen to be empty, e.g. if $\varphi(u) = +\infty$ and φ is proper. The set of those $u \in X$ for which $\partial\varphi(u) \neq \emptyset$ is called the *domain* of $\partial\varphi$ and is denoted by $D(\partial\varphi)$. Clearly $D(\partial\varphi)$ is a subset of $D(\varphi)$. The function φ is said to be *subdifferentiable* at u if $u \in D(\partial\varphi)$ and *subdifferentiable* if $D(\partial\varphi) = X$. Again using (2.6) we deduce that every subdifferentiable function on X is a proper convex and lower semicontinuous function on X.

The 'geometric' significance of the subdifferentiability of φ is best illustrated in connection with its epigraph. Using (2.6) we get that φ is subdifferentiable at u and f is a subgradient of φ at u if and only if the graph of the function $v \to \varphi(u) + \langle f, v - u \rangle_X$ is a non-vertical supporting hyperplane at epi φ at the point $(u; \varphi(u)) \in X \times \mathbb{R}$.

Note that from (2.6) we get

$$\varphi(v) \geq \varphi(u) \qquad \forall v \in X \quad \text{iff } \theta_X \in \partial\varphi(u). \tag{2.7}$$

In the case of convex functions the link between the gradient operator and subdifferential is given by the following lemma:

Lemma 2.5 *Let $\varphi: X \to (-\infty, +\infty]$ be a convex Gâteaux differentiable function. Then φ is subdifferentiable, $\partial\varphi$ is a univalued operator on X and $\partial\varphi(u) = \{\nabla\varphi(u)\}$ for every $u \in X$.*

PROOF. Let $u \in X$; from Lemma 2.6 and (2.6) we get $\nabla\varphi(u) \in \partial\varphi(u)$; hence $u \in D(\partial\varphi)$, i.e. φ is subdifferentiable in u. Let $f \in \partial\varphi(u)$; using (2.6) we get $\varphi(u + tv) - \varphi(u) \geq \langle f, tv \rangle_X$ for all $v \in X$ and $t \in \mathbb{R}$. Dividing by $t > 0$ and using (2.4) we get $\langle \nabla\varphi(u), v \rangle_X \geq \langle f, v \rangle_X$ for all $v \in X$, i.e. $f = \nabla\varphi(u)$.

2.2 Elliptic variational inequalities

Let X be a real normed space, X' its dual space and $\langle \, , \rangle$ the duality pairing between X' and X. Let also $a: X \times X \to \mathbb{R}$ be a bilinear form on X, $\varphi: X \to (-\infty, +\infty]$ be a proper function and $f \in X'$. A lot of boundary-value problems in partial differential equations as well as various problems in mechanics of a continuous media lead to mathematical problems of the following form: find u such that

$$u \in X, \quad a(u, v - u) + \varphi(v) - \varphi(u) \geq \langle f, v - u \rangle \qquad \forall v \in X. \tag{2.8}$$

Problem (2.8) is called *elliptic variational inequality of the second kind* on X. Many problems of different domains lead to similar mathematical problems: find u such that

$$u \in K, \quad a(u, v - u) \geq \langle f, v - u \rangle \qquad \forall v \in K, \tag{2.9}$$

where K is a non-empty subset of X. Problem (2.3) is called *elliptic variational inequality of the first kind.*

Let us observe that if $\varphi \equiv 0$ (or $K = X$), then (2.8) (respectively (2.9)) reduces to the following problem: find u such that

$$u \in X, \quad a(u, v) = \langle f, v \rangle \qquad \forall v \in X. \tag{2.10}$$

We refer to (2.10) as a *variational equation.*

Finally let us remark that all the problems (2.8)–(2.10) are of the form

$$u \in K, \quad \varphi(u) + G(u, v) \leq \varphi(v) \qquad \forall v \in K, \tag{2.11}$$

in which $K \subset X$ and $G: K \times K \to \mathbb{R}$; (2.11) will be called the *Ky-Fan inequality.*

In order to give some classical existence results for problems (2.8)–(2.11) we consider the following assumptions:

$$\text{there exists } M > 0 \text{ such that } \quad |a(u, v)| \leq M \|u\|_X \|v\|_X \qquad \forall u, v \in X \tag{2.12}$$

(i.e. a is a continuous form on X);

$$\text{there exists } m > 0 \text{ such that } \quad a(v,v) \geq m\|v\|_X^2 \quad \forall v \in X \tag{2.13}$$

(i.e. a is a coercive form on X);

$$\varphi\colon X \to (-\infty, +\infty] \text{ is a proper convex lower semicontinuous function;} \tag{2.14}$$

$$K \text{ is a non-empty closed convex subset of } X. \tag{2.15}$$

We start with a result concerning the Ky-Fan inequality, from which we shall derive other existence results:

Theorem 2.2 *Let X be a topological linear space and K a compact convex set of X. Let* (2.14) *hold and $G\colon K \times K \to \bar{\mathbb{R}}$ with the following properties:*

$$G(v, v) \leq 0 \qquad \forall v \in K \tag{2.16}$$

$$v \to G(u, v) \text{ is a concave function} \qquad \forall u \in K \tag{2.17}$$

$$u \to G(u, v) \text{ is lower semicontinuous on } K \qquad \forall v \in K. \tag{2.18}$$

We also suppose that there exists K_0, a compact subset of X, and $w_0 \in K_0 \cap K \cap D(\varphi)$ such that $\varphi(v) + G(v, w_0) > \varphi(w_0)$ for all $v \in K \setminus K_0$. Then, there exists at least a solution of the variational inequality (2.11).

In some concrete problems it is more convenient to use the following form of Theorem 2.2:

Corollary 2.3 *Let X be a real reflexive Banach space and $K \subset X$. Let us suppose that* (2.14)–(2.17) *hold. Let us also assume that*

$$u \to G(u, v) \text{ is weakly lower semicontinuous on } K, \text{ for all } v \in K \tag{2.19}$$

K is bounded or there exists $w_0 \in K \cap D(\varphi)$ such that

$$\lim_{\|u\|_X \to +\infty} G(u, w_0) + \varphi(u) = +\infty. \tag{2.20}$$

Then there exists at least a solution of the variational inequality (2.11).

PROOF. We shall use Theorem 2.2 with X endowed with the weak topology. If K is bounded, then from (2.15) we find that it is a compact set in the weak topology. Hence, we put $K_0 \equiv K$ in Theorem 2.2, and the conclusion will follow. If K is not bounded we choose n_0 large enough such that $G(u, w_0) + \varphi(u) > \varphi(w_0)$ for all $u \in K$ with $\|u\|_X > n_0$. If we bear in mind that $K_0 = \overline{B(0, n_0)}$ is a compact subset of X in the weak topology, then we may use Theorem 2.2 to get that there exists at least a solution of (2.11).

From Corollary 2.3 one can easily deduce existence results for elliptic variational inequalities:

Corollary 2.4 *Let X be a real reflexive Banach space, a, a bilinear form on X, $\varphi: X \to (-\infty, +\infty]$ and $f \in X'$. Let us assume that* (2.12)–(2.14) *hold. Then there exists a unique solution of the variational inequality* (2.8).

PROOF. Take $K = X$ and $G: X \to \mathbb{R}$ defined by

$$G(u,v) = a(u, u - v) - \langle f, v - u \rangle. \tag{2.21}$$

Let us verify the assumptions of Corollary 2.3; indeed, (2.17) and (2.16) are obviously satisfied; (2.19) results from (2.13) and Corollary 2.1. Finally (2.20) results from (2.12)–(2.14) and Lemma 2.3 taking w_0 as an arbitrary element of $D(\varphi)$. Applying Corollary 2.3 we obtain the existence of an element $u \in D(\varphi)$ such that $\varphi(u) + G(u, v) \leq \varphi(v)$ for all $v \in D(\varphi)$. Using (2.21) we get that u satisfies (2.8). The uniqueness of the solution in (2.8) results from (2.13).

Corollary 2.5 *Let X be a real reflexive Banach space, a, a bilinear form on X, K a subset of X and $f \in X'$. Let us assume that* (2.12), (2.13), *and* (2.15) *are satisfied. Then, there exists a unique solution of the variational inequality* (2.9).

PROOF. Since (2.15) is satisfied, from Lemma 2.2 we get that the indicator function ψ_K defined by (2.3) satisfies (2.14). Corollary 2.5 follows now from Corollary 2.4 taking $\varphi = \psi_K$ in (2.8).

Corollary 2.6 *Let X be a real reflexive Banach space, a is a bilinear form on X and $f \in X'$. Let us assume that* (2.12) *and* (2.13) *are satisfied. Then there exists a unique solution of the variational equation* (2.10).

PROOF. Corollary 2.4 can be applied in the case $\varphi \equiv 0$ or Corollary 2.5 in the case $K = X$.

In the case when a is a *symmetric* form, i.e.

$$a(u, v) = a(v, u) \qquad \forall u, v \in X, \tag{2.22}$$

then problems (2.8)–(2.10) are equivalent to minimum problems of the form

$$u \in K, \quad J(u) \leq J(v) \qquad \forall v \in K, \tag{2.23}$$

where $J: K \to (-\infty, +\infty]$ is a proper function. More precisely, we have:

Lemma 2.6 *Let X be a real normed space, a is a bilinear symmetric form on X*

such that $a(v, v) \geq 0$ *for all* $v \in X$ *(i.e. a is positive) and* $f \in X'$. *The following statements hold:*

(i) *if* φ *is a convex proper function then* u *is a solution of the variational inequality* (2.8) *if and only if* u *is a solution of* (2.23), *where* $K = X$ *and*

$$J(v) = \tfrac{1}{2}a(v, v) + \varphi(v) - \langle f, v\rangle \qquad \forall v \in X; \tag{2.24}$$

(ii) *if* K *is a convex non-empty set of* X, *then* u *is a solution of the variational inequality* (2.9) *if and only if* u *is a solution of* (2.23), *where*

$$J(v) = \tfrac{1}{2}a(v, v) - \langle f, v\rangle \qquad \forall v \in K; \tag{2.25}$$

(iii) u *is a solution of the variational equation* (2.10) *if and only if* u *is a solution of* (2.23) *where* $K = X$ *and* J *is given by* (2.25).

PROOF. (i) Let u be a solution of (2.8). For every $v \in X$ from (2.24) and (2.22) we deduce

$$J(v) - J(u) = \tfrac{1}{2}a(u - v, u - v) + a(u, v - u) + \varphi(v) - \varphi(u) - \langle f, v - u\rangle.$$

Using (2.13) and (2.8) we get (2.23). Conversely, let u be a solution of (2.23), (2.24); for every $v \in X$ and $t \in (0, 1)$ we have

$$J(u + t(v - u)) - J(u) = \frac{t^2}{2} a(u - v, u - v) + ta(u, v - u) + \varphi(u + t(v - u)) - \varphi(u) - t\langle f, v - u\rangle \geq 0.$$

Using (2.1) we get

$$\frac{t^2}{2} a(u - v, u - v) + ta(u, v - u) + t(\varphi(v) - \varphi(u)) \geq t\langle f, v - u\rangle.$$

Dividing by $t > 0$ and passing to the limit when $t \downarrow 0$ we get (2.8)

Proofs of (ii) and (iii) result from (i) taking $\varphi = \psi_K$ and $\varphi \equiv 0$ respectively.

2.3 Maximal monotone operators in Hilbert spaces

Throughout this section X will be a real Hilbert space with the inner product denoted by $\langle\, ,\, \rangle_X$ and the norm $\|\cdot\|_X$; an element of the product space $X \times X$ will be written as $(x; y)$, and we denote by 2^X the set of all subsets of X. We consider a multivalued operator $A: X \to 2^X$ and we set

$$D(A) = \{x \in X \mid Ax \neq \varnothing\}, \qquad R(A) = \bigcup_{x \in X} Ax;$$

the set $D(A)$ is called the *domain* of A and the set $R(A)$ is called the *range* of A. If for every $x \in D(A)$ the set $Ax \subset X$ contains a single element then we say that A is a *single-valued operator.*

Let now A and B be two multivalued operators on X and $\lambda, \mu \in \mathbb{R}$; the operator $\lambda A + \mu B$ is defined on X by

$$x \to \lambda Ax + \mu Bx = \{\lambda y + \mu z \,|\, y \in Ax, z \in Bx\}$$

for all $x \in D(\lambda A + \mu B) = D(A) \cap D(B)$.

Every multivalued operator on X can be viewed as a subset of $X \times X$, i.e. it is not distinguished from its graph

$$A = \{(x; y) \in X \times X \,|\, x \in D(A), y \in Ax\}.$$

So, we can always consider the *inverse* of A denoted by A^{-1} and defined by $y \in A^{-1}x$ iff $x \in Ay$.

A subset $A \subset X \times X$ is called *monotone* if

$$\langle y_1 - y_2, x_1 - x_2 \rangle_X \geq 0 \qquad \forall (x_1; y_1), (x_2; y_2) \in A. \tag{2.26}$$

A monotone subset of $X \times X$ is said to be *maximal monotone* if it is not properly contained in any other monotone subset of $X \times X$. From the above definition we get that $A \subset X \times X$ is a maximal monotone subset iff A is monotone and for all pairs $(x; f) \in X \times X$ such that $\langle f - g, x - y \rangle_X \geq 0$ for all $(y; g) \in A$ we get $(x; f) \in A$.

There is a close connection between the property of maximal monotonicity and the surjectivity property of the operator $I_X + \lambda A$ with $\lambda > 0$. The fundamental result in this direction is the celebrated *theorem of Minty*:

Theorem 2.3 *Let A be a monotone subset of $X \times X$. Then A is maximal monotone if and only if for each $\lambda > 0$ (equivalently for some $\lambda > 0$) $R(I_X + \lambda A) = X$.*

From Theorem 2.3 it is easy to obtain the following result:

Lemma 2.7 *Let A be a maximal monotone set in $X \times X$ and $\lambda > 0$. Then, for every $f \in X$ there exists a unique element $x \in D(A)$ such that $x + \lambda Ax \ni f$.*

This lemma shows that if A is a maximal monotone set in $X \times X$, then $(I_X + \lambda A)^{-1}$ is well defined on H and it is a single-valued operator, for every $\lambda > 0$. Let $J_\lambda = (I_X + \lambda A)^{-1}$; then for every $f, x \in H$ we have

$$x = J_\lambda f \Leftrightarrow x + \lambda Ax \ni f. \tag{2.27}$$

J_λ is called the *resolvent operator* of A and it is a monotone non-expansive operator (i.e. $\|J_\lambda x - J_\lambda v\|_X \leq \|x - y\|_X$ for all $x, y \in X$). We also define the *Yosida approximant* of A by $A_\lambda = (1/\lambda)(I_X - J_\lambda)$ for every $\lambda > 0$; it is also a monotone and Lipschitz operator on X.

Using (2.26) it is clear that the sum of two maximal monotone operators is a monotone operator but some examples can be given in order to prove that it is not always a maximal monotone operator on X. Some sufficient conditions for the maximality of sum of maximal monotone operators can be given. The simplest case is the case of a single-valued perturbation:

Lemma 2.8 *Let $A: X \to 2^X$ be a maximal monotone operator and $B: X \to X$ be a Lipschitz monotone operator. Then the sum $A + B: X \to 2^X$ is also a maximal monotone operator.*

Let us now investigate some monotony properties of single-valued operators. A single valued operator $A: D(A) \subset X \to X$ is said to be:

(1) *monotone*, if

$$\langle Ax_1 - Ax_2, x_1 - x_2 \rangle_X \geq 0 \qquad \forall x_1, x_2 \in D(A); \tag{2.28}$$

(2) *strictly monotone*, if

$$\langle Ax_1 - Ax_2, x_1 - x_2 \rangle_X > 0 \qquad \forall x_1, x_2 \in D(A), \quad x_1 \neq x_2; \tag{2.29}$$

(3) *strongly monotone*, if there exists $m > 0$ such that

$$\langle Ax_1 - Ax_2, x_1 - x_2 \rangle_X \geq m\|x_1 - x_2\|_X^2 \qquad \forall x_1, x_2 \in D(A); \tag{2.30}$$

(4) *hemicontinuous*, if $D(A) = X$ and

$$\lim_{t \downarrow 0} \langle A(1 - t)x + ty), z \rangle_X = \langle Ax, z \rangle_X \qquad \forall x, y, z \in X; \tag{2.31}$$

(5) *coercive*, if $\lim_n \langle Ax_n, x_n \rangle_X / \|x_n\|_X = +\infty$ for all $(x_n)_n \subset D(A)$ such that $\lim_n \|x_n\|_X = +\infty$;

(6) *potential*, if $D(A) = X$ and there exists a Gâteaux differentiable function on X such that $A = \nabla\varphi$, i.e.

$$\langle Au, v \rangle_X = \lim_{t \to 0} \frac{\varphi(u + tv) - \varphi(u)}{t} \qquad \forall u, v \in X; \tag{2.32}$$

in this case φ is called a *potential* of A;

(7) *Gâteaux differentiable* if $D(A) = X$ and there exists a linear and continuous operator $\nabla A: X \to B(X)$ such that

$$\lim_{t \to 0} \frac{A(u + tv) - Au}{t} = (\nabla A(u))v \qquad \forall u, v \in X; \tag{2.33}$$

in this case ∇A is called the *Gâteaux differential* of A.

Moreover, if $D(A)$ is a dense subspace of X and $A: D(A) \subset X \to X$ is a linear operator, we use the notation

$$D(A^*) = \{v \in X \mid \text{there exists } w \in X \text{ such that } \langle Au, v \rangle_X = \langle u, w \rangle_X \text{ for all } u \in D(A)\} \tag{2.34}$$

and $A^*: D(A^*) \subset X \to X$ the *adjoint operator* of A, i.e. $A^*v = w$ for all $v \in D(A)$, see also (1.5).

The following results hold:

Theorem 2.4 *Let $A: X \to X$ be a monotone and hemicontinuous operator. Then A is maximal monotone. If in addition A is coercive, then A is surjective, i.e. $R(A) = X$.*

Lemma 2.9 *Let $D(A)$ be a dense subspace of X and $A: D(A) \subset X \to X$ be a linear monotone operator. Then A is maximal monotone if and only if A is closed and A^* is a monotone operator.*

Lemma 2.10 *Let $\varphi: X \to \mathbb{R}$ be a Gâteaux differentiable functional and $\nabla\varphi: X \to X$ the gradient operator of φ. Then:*

(i) *φ is a convex function if and only if $\nabla\varphi$ is a monotone operator;*
(ii) *φ is a strictly convex function if and only if $\nabla\varphi$ is a strictly monotone operator;*
(iii) *φ is a uniformly convex function if and only if $\nabla\varphi$ is a strongly monotone operator.*

From the above-mentioned results we can give some remarkable examples of maximal monotone operators:

(1) every linear and positive operator $A: X \to X$ is a maximal monotone operator. Indeed, since A is linear (2.31) is satisfied and Theorem 2.4 can be applied;
(2) if K is a closed convex set of X and P_K denotes the projection map on K, then P_K is a maximal monotone operator. Indeed, since P_K is a Lipschitz and monotone operator Theorem 2.4 can be used;
(3) more generally, the proximity map $\text{prox}_\varphi: X \to X$ attached to a convex proper lower semicontinuous function, $\varphi: X \to (-\infty, +\infty]$ is a monotone and non-expansive operator (see Moreau, 1965); hence, using again Theorem 2.4, we get that prox_φ is a maximal monotone operator;
(4) if $T: X \to X$ is a Lipschitz continuous operator with the Lipschitz constant denoted by L, then the operator $A: X \to X$ defined by $A = I_X - (1/L)T$ is a maximal monotone operator. In order to prove this, Theorem 2.4 can be used again;
(5) if $\varphi: X \to (-\infty, +\infty]$ is a convex proper lower semicontinuous function, then the subdifferential mapping $\partial\varphi: X \to 2^X$ is a maximal monotone operator. Indeed, if $u, v \in X$ and $f \in \partial\varphi(u)$, $g \in \partial\varphi(v)$ from (2.6), we get $\varphi(v) - \varphi(u) \geq \langle f, v - u\rangle_X$, $\varphi(u) - \varphi(v) \geq \langle g, u - v\rangle_X$; hence $\langle f - g, u - v\rangle_X \geq 0$ and by (2.26) $\partial\varphi$ is a monotone operator. Moreover, for every $f \in X$ there exists $u \in X$ such that $\langle u, v - u\rangle_X + \varphi(v) - \varphi(u) \geq \langle f, v - u\rangle_X$ for all $v \in X$ (see Corollary 2.4) hence, by (2.6) we get $u + \partial\varphi(u) \ni f$. Using Theorem 2.3 it follows that $\partial\varphi$ is a maximal monotone operator.

We finish this section by presenting the Kerner–Vainberg potentiality theorem:

Theorem 2.5 *Let $P: X \to X$ be a Gâteaux differentiable operator such that $u \to \langle (\nabla P(u))v, w\rangle_X$ is a continuous function on X, for all $v, w \in X$. Then P is a potential operator if and only if*

$$\langle (\nabla P(u))v, w\rangle_X = \langle v, (\nabla P(u))w\rangle_X \qquad \forall u, v, w \in X. \tag{2.35}$$

In this case a potential of P is given by

$$J(v) = J(0) + \int_0^1 \langle P(tv), v\rangle_X \,\mathrm{d}t \qquad \forall v \in X. \tag{2.36}$$

Remark 2.1 *Under the assumptions of Theorem* 2.5 *one can prove that P is a strongly monotone operator* (*i.e.* (2.30) *holds*) *iff $\nabla P(\cdot)$ is a positively defined operator on X, i.e. there exists $m > 0$ such that*

$$\langle (\nabla P(u))v, v\rangle_X \geq m\|v\|_X^2 \qquad \forall u, v \in X. \tag{2.37}$$

3 Evolution equations in Banach spaces

In this section we summarize the results used in Chapters 3 and 4 in the study of some evolution equations. So, existence and uniqueness results for ordinary differential equations as well as some applications of C_0 linear semigroup theory are presented. Finally some useful results concerning non-linear evolution equations in Hilbert spaces involving monotone operators are also given.

3.1 Ordinary differential equations in Banach spaces

Let X be a real Banach space and $T > 0$. An operator A: $[0, T] \times X \to X$ is said to be:

(*i*) *locally Lipschitz continuous* with respect to the second argument, uniformly with respect to the first argument if for all $x \in X$ there exists $r > 0$ and $L > 0$ such that

$$\|A(t, y) - A(t, z)\|_X \leq L\|y - z\|_X \qquad \forall y, z \in B(x, r) \quad \text{and} \quad t \in [0, T];$$

(ii) *Lipschitz continuous* with respect to the second argument, uniformly with respect to the first argument if there exists $L > 0$ such that

$$\|A(t, x) - A(t, y)\|_X \leq L\|x - y\|_X \qquad \forall x, y \in X \quad \text{and} \quad t \in [0, T].$$

The following well-known locally existence result can be proved:

Theorem 3.1 *Let A: $[0, T] \times X \to X$ be a continuous operator, locally Lipschitz continuous with respect to the second argument, uniformly with respect to the* *first argument. Then for every* $x \in X$ *there exist* $T_0 \in (0, T]$ *and a unique function* $u \in C^1([0, T_0], X)$ *which is a solution of the Cauchy problem*

$$\dot{u}(t) = A(t, u(t)) \qquad \forall t \in (0, T_0] \tag{3.1}$$

$$u(0) = x. \tag{3.2}$$

The following global existence result also holds:

Theorem 3.2 *Let* $A: \mathbb{R}_+ \times X \to X$ *be a continuous operator which is Lipschitz continuous with respect to the second argument, uniformly with respect to the first argument. Then for every* $x \in X$ *there exists a unique function* $u \in C^1(\mathbb{R}_+, X)$ *which is a solution of the Cauchy problem*

$$\dot{u}(t) = A(t, u(t)) \qquad \forall t > 0. \tag{3.3}$$

$$u(0) = x. \tag{3.4}$$

The following theorem states that the conclusion of the above theorem holds for weaker conditions on the operator A:

Theorem 3.3 *Let* X *be a real Hilbert space and* $A: \mathbb{R}_+ \times X \to X$ *a continuous operator such that there exists* $D > 0$ *with*

$$\langle A(t, x_1) - A(t, x_2), x_1 - x_2 \rangle_X \leq D\|x_1 - x_2\|_X^2 \qquad \forall t \geq 0 \quad \text{and} \quad x_1, x_2 \in X. \tag{3.5}$$

Then for all $x \in X$ *there exists a unique solution* $u \in C^1(\mathbb{R}_+, X)$ *of the Cauchy problem* (3.3), (3.4).

3.2 Linear evolution equations

In the Section 3.1 we have given some existence results for the Cauchy problem (3.3), (3.4) in the case when A is a continuous non-linear operator. In this subsection we are interested in existence results for the Cauchy problem in the case when A is not continuous but it is linear. An important role in the study of these equations is played by the semigroups of linear operator theory. Hence we start by recalling some basic properties of C_0 semigroups.

Let X be a real Banach space. A one-parameter family $(T(t))_{t \geq 0}$ of bounded linear operators in X is called a C_0 *semigroup* if the following conditions hold:

$$T(0) = I_X \tag{3.6}$$

$$T(t) \cdot T(s) = T(t + s) \qquad \forall t, s \geq 0 \tag{3.7}$$

$$\lim_{t \downarrow 0} T(t)x = x \qquad \forall x \in X. \tag{3.8}$$

The following classical lemma holds:

Lemma 3.1 *If $(T(t))_{t>0}$ is a C_0 semigroup, then for all $x \in X$ the function $t \to T(t)x$ is continuous from $[0, +\infty)$ to X and there exist $M \geq 1$, $\omega \in \mathbb{R}$ such that*

$$\| T(t)\|_{B(X)} \leq M \exp(\omega t) \qquad \forall t \in [0, +\infty). \tag{3.9}$$

Let $(T(t))_{t \geq 0}$ be a C_0 semigroup and let us denote by $D(A)$ the following subspace of X:

$$D(A) = \{x \in X \mid \text{there exists} \lim_{t \downarrow 0}(T(t)x - x)/t \in X\}. \tag{3.10}$$

We can consider now A: $D(A) \subset X \to X$, the linear operator defined by

$$Ax = \lim_{t \downarrow 0}(T(t)x - x)/t \qquad \forall x \in D(A) \tag{3.11}$$

which is called the *infinitesimal generator* of the semigroup $(T(t))_{t \geq 0}$. The basic properties of the operator A can be summarized as follows:

Lemma 3.2 *Let $(T(t))_{t \geq 0}$ be a C_0 semigroup and let A: $D(A) \subset X \to X$ be its infinitesimal generator. Then:*

(*i*) *for all $x \in X$, $t \in [0, +\infty)$ we have $\int_0^t T(s)x \, \mathrm{d}s \in D(A)$ and*

$$A\left(\int_0^t T(s)x \, \mathrm{d}s\right) = T(t)x - x; \tag{3.12}$$

(*ii*) *$D(A)$ is a dense subspace of X;*

(*iii*) *for all $x \in D(A)$, $t \in [0, +\infty)$ we have $T(t)x \in D(A)$ and the function $t \to T(t)x$ belongs to $C^1(\mathbb{R}_+, X)$. Moreover*

$$\frac{\mathrm{d}}{\mathrm{d}t} T(t)x = T(t)Ax = AT(t)x \qquad \forall x \in D(A) \quad \textit{and} \quad t \in [0, +\infty) \tag{3.13}$$

(*iv*) *the generator A: $D(A) \subset X \to X$ is closed.*

Since the infinitesimal generator A is a closed operator, the space $[D(A)]$ endowed with the graph norm of A, i.e.

$$\|x\|_A^2 = \|x\|_X^2 + \|Ax\|_X^2 \qquad \forall x \in D(A) \tag{3.14}$$

is a Banach space. Similarly we define $D(A^2) = \{x \in D(A) \mid Ax \in D(A)\}$ and we endow $[D(A^2)]$ with the graph norm

$$\|x\|_{A^2}^2 = \|x\|_X^2 + \|Ax\|_X^2 + \|A^2x\|_X^2 \qquad \forall x \in D(A^2). \tag{3.15}$$

From Lemma 3.2 we can easily deduce:

Lemma 3.3 *Let $(T(t))_{t\geq 0}$ be a C_0 semigroup and let $A\colon D(A) \subset X \to X$ be its infinitesimal generator. Then the restriction $(\tilde{T}(t))_{t\geq 0}$ of $(T(t))_{t\geq 0}$ on $D(A)$ is a C_0 semigroup on the Banach space $[D(A)]$ and its infinitesimal generator is $A\colon D(A^2) \subset [D(A)] \to [D(A)]$. Hence for all $x \in D(A)$ the function $t \to \tilde{T}(t)x = T(t)x$ belongs to $C(\mathbb{R}_+, [D(A)])$. Moreover, for all $t \in [0, +\infty)$ we have*

$$x \to \int_0^t T(s)x \,\mathrm{d}s \text{ belongs to } B(X, [D(A)]) \tag{3.16}$$

$$x \to \int_0^t \tilde{T}(s)x \,\mathrm{d}s \text{ belongs to } B([D(A)], [D(A^2)]). \tag{3.17}$$

PROOF. For all $x \in [D(A^2)]$ and $t \geq 0$ using (3.13) and (3.14) we get

$$\begin{aligned}\|\tilde{T}(t)x\|_A^2 &= \|T(t)x\|_X^2 + \|T(t)Ax\|_X^2\\ &\leq \|T(t)\|_{B(X)}^2(\|x\|_X^2 + \|Ax\|_X^2)\\ &= \|T(t)\|_{B(X)}^2\|x\|_A^2;\end{aligned}$$

hence $\tilde{T}(t) \in B([D(A)])$. Using again (3.13) and (3.14) we deduce

$$\|\tilde{T}(t)x - x\|_A^2 = \|T(t)x - x\|_X^2 + \|T(t)Ax - Ax\|_X^2$$

and from (3.8) we get $\lim_{t\downarrow 0} \tilde{T}(t)x = x$ in $[D(A)]$. Hence $(\tilde{T}(t))_{t\geq 0}$ is a C_0 semigroup on $[D(A)]$. Let us prove that

$$\lim_{t\downarrow 0} \left\| \frac{\tilde{T}(t)x - x}{t} - Ax \right\|_A = 0 \qquad \forall x \in D(A^2).$$

Indeed, if $x \in D(A^2)$, from (3.13), (3.14) and (3.11) we have

$$\left\| \frac{\tilde{T}(t)x - x}{t} - Ax \right\|_A^2 = \left\| \frac{T(t)x - x}{t} - Ax \right\|_X^2 + \left\| \frac{T(t)Ax - Ax}{t} - A(Ax) \right\|_X^2 \to 0$$

for $t\downarrow 0$. Conversely, if there exists $\lim_{t\downarrow 0}(T(t)x - x)/t$ in $[D(A)]$, then there exists $\lim_{t\downarrow 0}(T(t)x - x)/t$ in X and $\lim_{t\downarrow 0}(T(t)Ax - Ax)/t$ in X, hence $x \in D(A)$ and $Ax \in D(A)$, i.e. $x \in D(A^2)$. It follows that the infinitesimal generator of $(\tilde{T}(t))_{t\geq 0}$ is $A\colon D(A^2) \subset [D(A)] \to [D(A)]$.

Now let $x \in [D(A)]$ and $t \geq 0$. Using (3.14) and (3.12) we have

$$\begin{aligned}\left\| \int_0^t T(s)x \,\mathrm{d}s \right\|_A &\leq \left\| \int_0^t T(s)x \,\mathrm{d}s \right\|_X + \|T(t)x - x\|_X\\ &\leq \left[\int_0^t \|T(s)\|_{B(X)} \,\mathrm{d}s + \|T(t) - I_X\|_{B(X)} \right] \|x\|_X;\end{aligned}$$

hence (3.16) is proved. In a similar way (3.17) can be obtained.

Subsequently we denote by $\rho(A)$ the *resolvent set* of the operator A, i.e. the set of real (or complex) numbers λ such that $\lambda I_X - A$ is invertible and $(\lambda I_X - A)^{-1} \in B(X)$. For every $\lambda \in \rho(A)$ we denote by $R(\lambda, A)$ the resolvent of A, i.e. $R(\lambda, A) = (\lambda I_X - A)^{-1}$. The following theorem (called the Hille–Yosida theorem) gives necessary and sufficient conditions for a linear operator to be the infinitesimal generator of a C_0 semigroup:

Theorem 3.4 *Let A: $D(A) \subset X \to X$ be a linear operator. A is the infinitesimal generator of a C_0 semigroup $(T(t))_{t \geq 0}$ which satisfies* (3.9) *iff $D(A)$ is dense in X, A is closed, $(\omega, +\infty) \subset \rho(A)$ and*

$$\|R(\lambda, A)^n\|_{B(X)} \leq M/(\lambda - \omega)^n \qquad \forall \lambda > \omega \quad \text{and} \quad n \in \mathbb{N}. \tag{3.18}$$

In the Hilbertian case, using the definition of the adjoint operator given in Section 1.3, the following consequences of Theorem 3.4 can be obtained:

Lemma 3.4 *Let X be a real Hilbert space and A: $D(A) \subset X \to X$ be a densely definite closed linear operator. If there exists $\omega \in \mathbb{R}$ such that*

$$\langle Au, u \rangle_X \leq \omega \|u\|_X^2 \qquad \forall u \in D(A) \tag{3.19}$$

$$\langle A^*u, u \rangle_X \leq \omega \|u\|_X^2 \qquad \forall u \in D(A^*) \tag{3.20}$$

then A is the infinitesimal generator of a C_0 semigroup $(T(t))_{t \geq 0}$ which satisfies (3.9) *with $M = 1$.*

PROOF. Let $\lambda > \omega$ and let us denote by $B = \lambda I_X - A$, B: $D(A) \subset X \to X$. From (3.19) we get that for all $u \in D(A)$ we have $(\lambda - \omega)\|u\|_X \leq \|Bu\|_X$. Hence B is injective and if we denote by $R(B)$ the range of B we get that B^{-1}: $R(B) \to D(A)$ and

$$\|B^{-1}u\|_X \leq \frac{1}{\lambda - \omega} \|u\|_X \qquad \forall u \in R(B). \tag{3.21}$$

Let us prove that $R(B)$ is a closed subspace of X. Indeed, let $(u_n)_n \subset D(A)$ and $f \in X$ such that $v_n = Bu_n \overset{n}{\to} f$ in X. From (3.21) we deduce

$$\|u_m - u_n\|_X \leq \frac{1}{\lambda - \omega} \|v_m - v_n\|_X \qquad \forall m, n \in \mathbb{N};$$

hence $(u_n)_n$ is a Cauchy sequence in X and there exists $u \in X$ such that $u_n \to u$ in X. If we have in mind that $u_n \to u$, $Au_n \to \lambda u - f$ in X, since A is closed we deduce $u \in D(A)$ and $\lambda u - f = Au$, i.e. $Bu = f$ hence $R(B)$ is closed in X. Suppose now that $R(B) \subsetneq X$; we deduce that there exists $v \in X$, $v \neq \theta_X$ such that $\langle v, Bu \rangle_X = 0$ for all $u \in D(A)$, i.e. $\langle Au, v \rangle_X = \langle u, \lambda v \rangle_X$ for all $u \in D(A)$. From (2.34) we deduce $v \in D(A^*)$ and $A^*v = \lambda v$. Using (3.20) we have $\lambda\|v\|_X^2 \leq \omega\|v\|_X^2$, i.e. $\lambda \leq \omega$, a contradiction. Hence $R(B) = X$ and by (3.21) we get $\|R(\lambda, A)\|_{B(X)} \leq 1/(\lambda - \omega)$ for all $\lambda > \omega$, and (3.18) follows with $M = 1$.

From Lemma 3.4 it easily follows that if $A = -A^*$ then A is the infinitesimal generator of a C_0 semigroup of contraction on the Hilbert space X (i.e. $M = 1$ and $\omega = 0$ in (3.9)).

Now let $(X, \langle , \rangle_X, \| \ \|_X)$ be a real Hilbert space and $A: D(A) \subset X \to X$ be a linear and densely defined operator. If $A^*: D(A^*) \subset X \to X$ is the adjoint of A, then we denote by $[D(A^*)]$ the space $D(A^*)$ endowed with the graph norm of A^* and by $[D(A^*)]'$ its strong dual, $\langle , \rangle_*$ being the duality between $[D(A^*)]'$ and $[D(A^*)]$. If we identify X with its dual we get the continuous inclusions $[D(A^*)] \subset X \subset [D(A^*)]'$ (see section 1.3).

The following theorem gives a construction technique for extended generators of C^0 semigroups of linear contractions.

Theorem 3.5 *Let $A: D(A) \subset X \to X$ be the generator of a C^0 semigroup of linear contractions $(T(t))_{t \geq 0}$ acting on X. If we define $\tilde{A}: X \subset [D(A^*)]' \to [D(A^*)]'$ given by*

$$\langle \tilde{A}u, v \rangle_* = \langle u, A^*v \rangle_X \qquad \forall u \in X, \quad v \in D(A^*), \tag{3.22}$$

then $\tilde{A}$ is the generator of a C^0 semigroup $(\tilde{T}(t))_{t \geq 0}$ of linear contractions acting on $[D(A^*)]'$ with $[D(\tilde{A})] \simeq X$, $\tilde{A}|_{D(A)} = A$ and $\tilde{T}(t)|_X = T(t)$.

PROOF. We shall denote by $\|\cdot\|_{A^*}$ and $\|\cdot\|_*$ the norms on $[D(A^*)]$ and $[D(A^*)]'$ respectively. Since A^* is the generator of a C^0 semigroup of linear contractions (see for instance Pazy, 1983, p. 41) we get that the restriction $A^*: D(A^{*2}) \subset [D(A^*)] \to [D(A^*)]$ has the same property and if we denote by $R(\lambda, A^*)$ the resolvent of A^* we have

$$\|R(\lambda, A^*)w\|_{A^*} \leq \frac{1}{\lambda} \|w\|_{A^*} \qquad \forall w \in D(A^*) \quad \text{and} \quad \lambda > 0.$$

Using this inequality we can easily deduce that

$$\|\lambda u - \tilde{A}u\|_* \geq \lambda \|u\|_* \qquad \forall u \in X. \tag{3.23}$$

Let us prove now that $\lambda I - \tilde{A}$ is a closed operator where I is the identity map on $[D(A^*)]'$. To see this let $u_n \in X$ and $f \in [D(A^*)]'$ such that $u_n \to u \in [D(A^*)]'$, $\lambda u_n - \tilde{A}u_n \to f$ in $[D(A^*)]'$. From (3.22) we get

$$\langle u_n, \lambda w - A^*w \rangle_X \to \langle f, w \rangle_* \qquad \forall w \in D(A^*)$$

and since $\lambda I - A^*$ is surjective we have obtained that u_n is weakly convergent in X, hence $u \in X$ and $f = \lambda u - \tilde{A}u$.

We prove now that $\lambda I - \tilde{A}$ is surjective. Since it is a closed operator which has a continuous inverse, the range of $\lambda I - \tilde{A}$ is a closed subspace of X. If this subspace does not coincide with $[D(A^*)]'$, then there exists $v \in D(A^*)$, $v \neq 0$ such that $\langle \lambda u - \tilde{A}u, v \rangle_* = 0$ for all $u \in X$. From (3.22) we deduce $\lambda v - A^*v = 0$, hence $v = 0$, a contradiction.

Now, by using (3.23) and the fact that $\lambda I - \tilde{A}$ is one to one it follows that $\tilde{A}$ is the generator of a C^0 semigroup of contractions $(\tilde{T}(t))_{t \geq 0}$.

From (3.22) one can easily deduce that $\|\tilde{A}u\|_* \leq \|u\|$, hence $\|u\|_{\tilde{A}} \leq 2\|u\|_X$ for all $u \in X$. Thus, since the identity operator from X to $[D(\tilde{A})]$ is a continuous bijection, it has a continuous inverse, and we obtain that $X \simeq [D(\tilde{A})]$. From (3.22) it results $\tilde{A}|_{D(A)} = A$ and we get that for all $x \in D(A)$ the functions $t \to T(t)x$ and $t \to \tilde{T}(t)x$ satisfy the same Cauchy problem, which has a unique solution. By consequence $T(t)x = \tilde{T}(t)x$ for all $x \in D(A)$ and using the density of $D(A)$ in X we get $\tilde{T}(t)|_X = T(t)$.

Let us return now to the case of a real Banach space X. The following theorem shows what happens when an infinitesimal generator A is perturbed by a bounded linear operator on X:

Theorem 3.6 *Let $A{:}D(A) \subset X \to X$ be an infinitesimal generator of a C_0 semigroup $(T(t))_{t \geq 0}$ which satisfies (3.9) and let $B \in B(X)$. Then $A + B$: $D(A) \subset X \to X$ is the infinitesimal generator of a C_0 semigroup $(S(t))_{t \geq 0}$ satisfying*

$$\|S(t)\|_{B(X)} \leq M \exp(\omega + M\|B\|_{B(X)}t) \qquad \forall t \geq 0.$$

Let $A{:}D(A) \subset X \to X$ be a linear operator and let us consider the following Cauchy problem:

$$\frac{du}{dt} = Au(t) \qquad \forall t > 0. \tag{3.24}$$

$$u(0) = x. \tag{3.25}$$

We call u a *classical solution* for (3.24), (3.25) if

$$u \in C^0(\mathbb{R}_+, X) \cap C^1((0, +\infty), X) \qquad u(t) \in D(A) \quad \forall t > 0$$

and (3.24)–(3.25) hold.

The following theorem gives the link between the C_0 semigroup theory presented above and the Cauchy problem (3.24), (3.25):

Theorem 3.7 *Let A: $D(A) \subset X \to X$ be a dense definite linear operator with $\rho(A) \neq \emptyset$. Then the Cauchy problem (3.24), (3.25) has a unique classical solution $u \in C^1(\mathbb{R}_+, X)$ for all $x \in D(A)$ iff A is the infinitesimal generator of a C_0 semigroup $(T(t))_{t \geq 0}$. Moreover, the solution u of (3.24), (3.25) is given by $u(t) = T(t)x$ and $u \in C^0(\mathbb{R}_+, [D(A)])$.*

The above theorem gives us the opportunity to call $u(t) = T(t)x$ *the mild solution* of (3.24), (3.25), for all $x \in X$.

3.3 Lipschitz perturbation of linear evolution equations

Let us now consider the following affine Cauchy problem on the real Banach space X:

$$\frac{du}{dt} = Au(t) + f(t), \qquad t \in (0, T] \tag{3.26}$$

$$u(0) = x \tag{3.27}$$

where A is the infinitesimal generator of a C_0 semigroup $(T(t))_{t \geq 0}$ and $T > 0$.

We call u a *classical solution* for (3.26)–(3.27) if

$$u \in C^0([0, T], X) \cap C^1([0, T], X), u(t) \in D(A) \quad \forall t \in (0, T]$$

and (3.26), (3.27) hold.

If $f \in L^1(0, T, X)$ the inhomogeneous problem (3.26), (3.27) can be 'solved' by the formula

$$u(t) = T(t)x + \int_0^t T(t-s)f(s)\,ds \qquad \forall t \in [0, T]. \tag{3.28}$$

The function $u \in C^0([0, T]), X)$ given by (3.28) will be called the *mild solution* of (3.26), (3.27). This definition is justified by the following lemma:

Lemma 3.5 *Let $f \in C^0([0, T], X)$ and $v \in C^0([0, T], X)$ be given by*

$$v(t) = \int_0^t T(t-s)f(s)\,ds \qquad \forall t \in [0, T]. \tag{3.29}$$

(*i*) *if v is differentiable in $t = t_0 \in [0, T]$, then $v(t_0) \in D(A)$ and*

$$\frac{dv}{dt}(t_0) = Av(t_0) + f(t_0);$$

(*ii*) *if $x \in D(A)$ and u is a mild solution such that $u \in C^1([0, T], X)$, then u is a classical solution of* (3.26), (3.27).

The following lemma gives sufficient conditions for a mild solution to be a classical one:

Lemma 3.6 *If one of the two following conditions hold*:

$$f \in C^1([0, T], X) \tag{3.30}$$

$$f \in C^0([0, T], X) \cap L^1(0, T, [D(A)]) \tag{3.31}$$

and $x \in D(A)$, then the non-homogeneous Cauchy problem (3.26), (3.27) *has a unique classical solution $u \in C^1([0, T], X) \cap C^0([0, T], [D(A)])$.*

A function $u \in W^{1,1}(0, T, X)$ is called a *strong solution* of (3.26), (3.27) if $u(t) \in D(A)$ a.e. on $[0, T]$, (3.26) is satisfied a.e. $t \in [0, T]$ and (3.27) holds. Moreover, we have:

Lemma 3.7 *Let $f \in W^{1,1}(0, T, X)$ and $x \in D(A)$. Then the mild solution u of (3.26), (3.27) is a strong solution and*

$$u \in W^{1,1}(0, T, X) \cap L^1(0, T, [D(A)]). \tag{3.32}$$

Moreover, if X is reflexive then $u(t) \in D(A)$ for all $t \in [0, T]$ and

$$u \in W^{1,\infty}(0, T, X) \cap L^\infty(0, T, [D(A)]). \tag{3.33}$$

PROOF. Let us prove the last statement of the lemma. From Theorem 3.4 of Chapter 2 we deduce that

$$\|f(t_1) - f(t_2)\|_X \leq \int_{t_1}^{t_2} \|\dot{f}(s)\|_X \, ds \tag{3.34}$$

for all $t_1, t_2 \in [0, T]$. Let $h > 0$ be small enough and $t \in [0, T - h]$. From (3.28) we have

$$\begin{aligned} u(t+h) - u(t) &= T(t)(T(h) - I_X)x + \int_0^{t+h} T(s)f(t+h-s) \, ds \\ &\quad - \int_0^t T(s)f(t-s) \, ds \\ &= T(t) \int_0^h T(s)Ax \, ds + \int_0^t T(s)(f(t+h-s) - f(t-s)) \, ds \\ &\quad + \int_t^{t+h} T(s)f(t+h-s) \, ds. \end{aligned}$$

Using (3.24) and (3.9) we have

$$\begin{aligned} \|u(t+h) - u(t)\|_X &\leq hM^2 \exp(2\omega T)\|Ax\|_X \\ &\quad + M \exp(\omega T) \int_0^t \int_{t-s}^{t+h-s} \|\dot{f}(\tau)\|_X \, d\tau \, ds \\ &\quad + hM \exp(\omega T) \max_{\tau \in [0,T]} \|f(\tau)\|_X. \end{aligned}$$

But

$$\begin{aligned} \int_0^t \int_{t-s}^{t+h-s} \|\dot{f}(\tau)\|_X \, d\tau \, ds &= \int_0^h \int_0^t \|\dot{f}(\tau + t - s)\|_X \, ds \, d\tau \\ &\leq h \int_0^T \|\dot{f}(s)\|_X \, ds. \end{aligned}$$

From the above inequalities we deduce that there exists $C > 0$ such that

$$\|u(t+h) - u(t)\|_X \leq Ch \qquad \forall t \in [0, T - h], \tag{3.35}$$

which shows that $t \to u(t)$ is a Lipschitz function, hence it is an absolutely continuous one. Using now Theorem 3.6(2) of Chapter 2 we deduce that $u \in W^{1,\infty}(0, T, X)$ and from Lemma 3.5(i) we have that u is a strong solution for (3.26), (3.27).

Since $\lambda I_X - A\colon [D(A)] \to X$ with $\lambda > \omega$ is an isomorphism between $[D(A)]$ and X we deduce that $[D(A)]$ is also a reflexive Banach space. From Lemma 3.5 we obtain that $u(t) \in D(A)$ a.e. $t \in [0, T]$ and $\|Au(t)\|_X \leq C + \|f(t)\|_X$ a.e. $t \in [0, T]$. If we bear in mind that A is weakly closed and $t \to \|Au(t)\|_X$ is bounded we deduce that $u(t) \in D(A)$ for all $t \in [0, T]$ and there exists $V > 0$ such that $\|u(t)\|_A \leq V$ for all $t \in [0, T]$. Since $[D(A)]$ is densely embedded in X and $u \in C^0([0, T], X)$ we deduce that u is weakly continuous from $[0, T]$ to $[D(A)]$, hence u is strongly measurable and $u \in L^\infty(0, T, [D(A)])$.

Let now $f\colon [0, T] \times X \to X$ be a continuous non-linear function and let us consider the following semilinear equation

$$\frac{\mathrm{d}u}{\mathrm{d}t} = Au(t) + f(t, u(t)), \qquad t \in (0, T] \tag{3.36}$$

$$u(0) = x \tag{3.37}$$

where A is the infinitesimal generator of a C_0-semigroup $(T(t))_{t \geq 0}$. By analogy with the non-homogeneous linear case we call $u \in C^0([0, T], X)$ a *mild solution* of (3.36), (3.37) if u is a solution of the following integral equation

$$u(t) = T(t)x + \int_0^t T(t-s)f(s, u(s))\,\mathrm{d}s \qquad \forall t \in [0, T]. \tag{3.38}$$

Theorem 3.8 (i) *If the continuous function f is locally Lipschitz continuous with respect to the second argument uniformly with respect to $t \in [0, T]$, then* (3.36)–(3.37) *has a unique local mild solution (i.e. there exists $T_0 \in (0, T]$, such that* (3.38) *is satisfied for all $t \in [0, T_0]$).*

(ii) *If f is continuous and Lipschitz continuous (globally) with respect to the second argument, uniformly with respect to $t \in [0, T]$, then* (3.36)–(3.37) *has a unique mild solution.*

A function $u \in W^{1,1}(0, T, X)$ is called a *strong solution* of (3.36), (3.37) if $u(t) \in D(A)$ a.e. $t \in [0, T]$, (3.36) is satisfied a.e. $t \in [0, T]$, and (3.37) holds. Concerning the existence and uniqueness of strong solution for the Cauchy problem (3.36), (3.37) we have the following result:

Theorem 3.9 *Let X be a real reflexive Banach space and $x \in D(A)$. Assume that there exist $g \in L^1(0, T, \mathbb{R})$, $g \geq 0$ a.e. on $(0, T)$ and $L > 0$ such that*

$$\|f(t_1, x_1) - f(t_2, x_2)\|_X \leq \left|\int_{t_1}^{t_2} g(\tau)\,\mathrm{d}\tau\right| + L\|x_1 - x_2\|_X \tag{3.39}$$

for all $t_1, t_2 \in [0, T]$, $x_1, x_2 \in X$. *Then there exists a unique strong solution* u *of* (3.36), (3.37) *such that* $u(t) \in D(A)$ *for all* $t \in [0, T]$ *and*

$$u \in W^{1,\infty}(0, T, X) \cap L^{\infty}(0, T, [D(A)]). \tag{3.40}$$

PROOF. Let u be the mild solution of (3.36), (3.37) (see Theorem 3.8(ii)) and let $h > 0$ be small enough and $t \in [0, T - h]$. From (3.38) we have

$$\begin{aligned} u(t+h) - u(t) = & \int_t^{t+h} T(s)Ax \, ds + \int_0^t T(s)(f(t+h-s, u(t+h-s)) \\ & - f(t-s, u(t-s)) \, ds \\ & + \int_t^{t+s} T(s)f(t+h-s, u(t+h-s)) \, ds. \end{aligned}$$

From (3.39) we have

$$\|f(s, u(s))\|_X \leq \|f(0, 0_X)\|_X + \int_0^T g(\tau) \, d\tau + L \max_{\tau \in [0,T]} \|u(\tau)\|_X = N.$$

Using this inequality and (3.39) we have

$$\begin{aligned} \|u(t+h) - u(t)\|_X \leq & \, hM \exp(\omega T)\|Ax\|_X + hMN \exp(\omega T) \\ & + M \exp(\omega T) \int_0^t \int_{t-s}^{t+h-s} g(\tau) \, d\tau \, ds \\ & + M \exp(\omega T) L \int_0^t \|u(\tau + h) - u(\tau)\|_X \, ds. \end{aligned}$$

If we bear in mind that

$$\int_0^t \int_{t-s}^{t+h-s} g(\tau) \, d\tau \, ds \leq h \int_0^T \dot{g}(s) \, ds$$

and if we let $\theta(t) = \|u(t+h) - u(t)\|_X$, then there exists $B_1, B_2 > 0$ such that

$$\theta(t) \leq hB_1 + B_2 \int_0^t \theta(s) \, ds,$$

and using Corollary 4.1 we deduce that there exists $C > 0$ such that (3.35) holds. If we let $\bar{f}(t) = f(t, u(t))$ for all $t \in [0, T]$, then from (3.35) and (3.39) we have

$$\|\bar{f}(t_1) - \bar{f}(t_2)\|_X \leq \int_{t_1}^{t_2} (g(\tau) + C) \, d\tau \qquad \forall t_1, t_2 \in [0, T],$$

hence $\bar{f}\colon [0, T] \to X$ is an absolutely continuous function. Using now Theorem 3.6(i) of Chapter 2 we deduce $\bar{f} \in W^{1,1}(0, T, X)$ and from Lemma 3.7 we obtain the statement of the theorem.

In the following we shall prove another type of perturbation theorem.

Let us consider $A:D(A) \subset X \to X$, the infinitesimal generator of a C^0 semigroup $(S(t))_{t \geq 0}$ and let $(Y, \| \|_Y)$ be such that

$$Y \text{ is continuously and densely embedded in } [D(A)]; \tag{3.41}$$

$$X, Y \text{ are reflexive Banach spaces;} \tag{3.42}$$

$$S(t)(Y) \subset Y \qquad \forall t \geq 0. \tag{3.43}$$

Let $\tilde{S}(t) = S(t)|_Y$ and let us suppose that

$$\tilde{S}(t) \text{ is a } C^0 \text{ semigroup acting on } Y. \tag{3.44}$$

Let us also consider the non-linear perturbation F: $[0, T] \times Y \to Y$ such that for all $y \in Y$ we have

$$t \to F(t, y) \text{ is continuous from } [0, T] \text{ to } X. \tag{3.45}$$

We shall also suppose that there exists p: $Y \to \mathbb{R}$ a continuous seminorm acting on Y and $P > 0$ such that

$$p(S(t)y) \leq Pp(y) \qquad \forall y \in Y, \quad t \in [0, T]. \tag{3.46}$$

Let $C > 0$ and L: $\mathbb{R}_+ \to \mathbb{R}_+$ a non-decreasing function such that

$$\|F(t, y)\|_Y \leq C + L(r)\|y\|_Y \tag{3.47}$$

$$p(f(t, y)) \leq C + L(r) \tag{3.48}$$

$$\|F(t, y) - F(t, \bar{y})\|_X \leq L(r)\|y - \bar{y}\|_X \tag{3.49}$$

for all $t \in [0, T]$, $y, \bar{y} \in Y$ such that $p(y), p(\bar{y}) \leq r$.

Theorem 3.10 *Let us suppose that* (3.41)–(3.49) *hold and let* $f, h \in C^0([0, T], X)$ *such that*

$$h(t) = \int_0^t S(t-s)f(s)\,\mathrm{d}s \tag{3.50}$$

If $y_0 \in Y$ *and one of the following regularity properties holds*:

$$f \in C^1([0, T], X) \qquad \text{and} \qquad h \in C^0([0, T], Y) \tag{3.51}$$

$$f \in L^1(0, T, Y), \tag{3.52}$$

then there exists $T_{\max} = T_m(y_0) \in (0, T]$ *and a unique local solution* $y \in C^1([0, T_{\max}), X) \cap C^0([0, T_{\max}), Y)$ *of the following Cauchy problem*:

$$\dot{y}(t) = Ay(t) + F(t, y(t)) + f(t) \tag{3.53}$$

$$y(0) = y_0. \tag{3.54}$$

Moreover, if $T_{\max} < T$, *then*

$$\lim_{t \uparrow T_{\max}} p(y(t)) = +\infty. \tag{3.55}$$

In order to prove Theorem 3.10, the following lemma will be useful:

Lemma 3.8 *There exists a non-increasing function* $\delta\colon [0, +\infty) \to (0, +\infty)$ *such that for all* $x \in X$, $t_0 \in (0, T]$, *there exists a unique solution*

$$y \in C^1([t_0, t_0 + \delta_0], X) \cap C^0([t_0, t_0 + \delta_0], Y)$$

of the Cauchy problem

$$\dot{y}(t) = Ay(t) + F(t, y(t)) + f(t), \qquad t \in (t_0, t_0 + \delta_0] \tag{3.56}$$

$$y(t_0) = x \tag{3.57}$$

where $\delta_0 = \min(T - t_0; \delta(p(x)))$.

PROOF. Let $V = \|h\|_{0,T,Y}$ and let $N, C_1 > 0$ be such that

$$\|S(t)\|_{B(X)}, \|\tilde{S}(t)\|_{B(Y)} \leq N \qquad \forall t \in [0, T] \tag{3.58}$$

$$p(y) \leq C_1 \|y\|_Y \qquad \forall y \in Y. \tag{3.59}$$

Let us consider $a\colon \mathbb{R}_+ \to (0, +\infty)$ given by $a(s) = 3\max(C_1 \bar{\bar{V}}; Ps)$ where $\bar{\bar{V}} = V(1 + N)$ and let $\delta\colon \mathbb{R}_+ \to (0, +\infty)$ be given by

$$\delta(s) = \alpha/(C + L(a(s))) \tag{3.60}$$

where $\alpha = \min(C_1 \bar{V}/P; 1/(2N))$. Let $r = a(p(x))$, $Q = 2(N\|x\|_Y + \nabla)$, $Z = C^0([t_0, t_0 + \delta_0], X)$ and D be a subset of Z given by

$$\begin{aligned} D = \{y \in Z \mid y(t_0) = x,\ y(t) \in Y,\ \|y(t)\|_Y \leq Q,& \\ p(y(t)) \leq r \qquad \forall t \in [t_0, t_0 + \delta_0]\}.& \end{aligned} \tag{3.61}$$

Let us prove now that for all $y \in D$ the function $t \to F(t, y(t))$ is continuous from $[0, T]$ to X. Indeed if $t_n \to t$ then from

$$\begin{aligned} \|F(t, y(t)) - F(t_n, y(t_n))\|_X &\leq \|F(t, y(t)) - F(t_n, y(t))\|_X \\ &\quad + \|F(t_n, y(t)) - F(t_n, y(t_n))\|_X \\ &\leq L(r)\|y(t_n) - y(t)\|_X + \|F(t, y(t)) - F(t_n, y(t))\|_X \end{aligned}$$

and (3.45) we get $F(t_n, y(t_n)) \to F(t, y(t))$ in X. Since $t \to F(t, y(t))$ is bounded in Y and continuous from $[0, T]$ to X and Y is a reflexive Banach space, we get that $t \to F(t, y(t))$ is weakly continuous from $[0, T]$ to Y hence it is a strongly Y-measurable function.

From the equality

$$\int_{t_0}^{t} S(t-s)f(s)\,\mathrm{d}s = h(t) - S(t-t_0)h(t_0) \in Y$$

we get

$$\left\| \int_{t_0}^{t} S(t-s)f(s)\,\mathrm{d}s \right\|_Y \leq V(1+N) = \bar{V} \qquad \forall t \in [t_0, t_0+\delta_0]. \tag{3.62}$$

Now let $\mathscr{F}: D \subset Z \to Z$ be given by

$$(\mathscr{F}(y))(t) = \tilde{S}(t-t_0)x + \int_{t_0}^{t} \tilde{S}(t-s)F(s, y(s))\,\mathrm{d}s + \int_{t_0}^{t} S(t-s)f(s)\,\mathrm{d}s. \tag{3.63}$$

Let us prove now that $\mathscr{F}(D) \subset D$. In order to do this let $y \in D$ and let $z = \mathscr{F}(y)$. For all $t \in [t_0, t_0+\delta_0]$ we have

$$p(z(t)) \leq p(\tilde{S}(t-t_0)x) + p\left(\int_{t_0}^{t} \tilde{S}(t-s)F(s, y(s))\,\mathrm{d}s\right) + p\left(\int_{t_0}^{t} S(t-s)f(s)\,\mathrm{d}s\right)$$

and using (3.46), (3.48), (3.59), and (3.62) we get

$$p(z(t)) \leq Pp(x) + P(C + L(r))\delta_0 + C_1\bar{V}.$$

If we bear in mind that $r/3 = a(p(x))/3 = \max(Pp(x); C_1\bar{V})$, from (3.60) we get

$$p(z(t)) \leq 2r/3 + P(C+L(r))\delta_0 \leq 2r/3 + C_1\bar{V} \leq r.$$

Using (3.58), (3.47) and (3.62) we deduce

$$\|z(t)\|_Y \leq N\|x\|_Y + \delta_0 N(C+L(r))Q + \bar{V}$$

and from $\delta_0 \leq 1/(2N(C+L(r)))$ and $Q/2 = N\|x\|_Y + \bar{V}$ we obtain $\|z(t)\|_Y \leq Q$; hence we have just proved that $z \in D$.

If we consider $y, \bar{y} \in D$, then from (3.49) we get $\|\mathscr{F}(y) - \mathscr{F}(\bar{y})\|_Z \leq NL(r)\delta_0\|y-\bar{y}\|_Z$ and from (3.60) we deduce

$$\|\mathscr{F}(y) - \mathscr{F}(\bar{y})\|_Z \leq \tfrac{1}{2}\|y-\bar{y}\|_Z \qquad Vy, \bar{y} \in D.$$

In order to use the Banach fixed-point theorem it is sufficient to prove that D is strongly closed in Z. Let $(y_n) \subset D$ such that $y_n \to y \in Z$ strongly in Z. For all $t \in [0, T]$ we have $y_n(t) \to y(t)$ strongly in X, but since $(y_n(t))$ is bounded in Y we deduce that $y_n(t) \to y(t)$ weakly in Y (we can pass to a subsequence, eventually); hence $\|y(t)\|_Y \leq \underline{\lim}\|y_n(t)\|_Y \leq Q$. Since p is convex and continuous we deduce that it is also weakly semicontinuous, hence

$p(y(t)) \leq \underline{\lim}\, p(y_n(t)) \leq r$. We have just proved that $y \in D$; hence D is strongly closed in Z.

We can now conclude that there exists a unique fixed point y for $\mathscr{F}$, i.e. $y = \mathscr{F}(y)$. If we have in mind that $t \to F(t, y(t))$ belongs to

$$C^0([t_0, t_0 + \delta_0], X) \cap L^\infty([t_0, t_0 + \delta_0], [D(A)])$$

and we use one of the regularities (3.51), (3.52), then from Lemma 3.6 we deduce that $y \in C^1([t_0, t_0 + \delta_0], X) \cap C^0([t_0, t_0 + \delta_0], Y)$ is the unique solution of (3.56)–(3.57).

PROOF OF THEOREM 3.10. Using Lemma 3.8, we get that there exists $T_{\max} = T_m(y_0)$ such that $[0, T_{\max})$ is the maximal interval of existence for the solution

$$y \in C^1([0, T_{\max}), X) \cap C^0([0, T_{\max}), Y)$$

for the Cauchy problem (3.53), (3.54). Let us suppose that $T_{\max} < T$ and $\lim_{t \uparrow T_{\max}} p(y(t)) \neq +\infty$. Then there exists $(t_n)_n \subset (0, T_{\max})$, an increasing sequence, and $r_0 > 0$ such that $t_n \to T_{\max}$ and $p(y(t_n)) \leq r_0$ for all $n \in \mathbb{N}$. If $\delta_1 = \min(T - T_{\max}, \delta(r_0))$, then we deduce from Lemma 3.8 that the solution y defined on $[0, t_n]$ can be prolonged on $[0, t_n + \delta_1]$ for all $n \in \mathbb{N}$. Hence it can be defined on $[0, T_{\max} + \delta_1)$, a contradiction.

3.4 Non-linear evolution equations in Hilbert spaces

In this section we present some results concerning first-order evolution equations of the form

$$\frac{du}{dt} + A(t)u(t) \ni f(t) \qquad \text{on } (0, T) \tag{3.64}$$

$$u(0) = x \tag{3.65}$$

where $(A(t))_{t \in [0,T]}$ is a family of non-linear maximal monotone operators in a real Hilbert space X. A function u: $[0, T] \to X$ is called a *solution* of (3.64), (3.65) if u is absolutely continuous, $u(t) \in D(A(t))$ for all $t \in [0, T]$, u satisfies (3.64) a.e. on $[0, T]$ and $u(0) = x$.

The following existence and uniqueness result can be proved:

Theorem 3.11 *Let X be a real Hilbert space and $(A(t))_{t \in [0,T]}$ a family of maximal monotone operators in X such that*

$$D(A(t)) = D \quad \textit{is independent of } t \qquad \forall t \in [0, T], \tag{3.66}$$

there exists $M > 0$ such that

$$\| J_\lambda A(t)x - J_\lambda A(s)x \|_X \leq M\lambda |t - s| (1 + \|x\|_X + \|A_\lambda(t)x\|_X) \tag{3.67}$$

$\forall x \in X$, $\lambda > 0$ and $t, s \in [0, T]$.

Then for every $f \in W^{1,1}(0, T, X)$ *and* $x \in D$ *there exists a unique solution* $u \in W^{1,\infty}(0, T, X)$ *of the problem* (3.64), (3.65).

Remark 3.1 *In* (3.67), *we mean by* $J_\lambda A(t)$ *and* $J_\lambda A(s)$ *the resolvent operators attached to the maximal monotone operators* $A(t)$ *and* $A(s)$ *respectively (see Section* 2.3); *in the same formula* $A_\lambda(t)$ *represents the Yosida approximant of* $A(t)$.

Using Theorem 3.11 the following existence and uniqueness result for single-valued equations involving maximal monotone operators can be proved.

Lemma 3.9 *Let* X *be a real Hilbert space,* $D(B) \subset X$ *a dense subspace of* X *and* $\mathscr{G}: [0, T] \times X \to X$, $B: D(B) \subset X \to X$ *two operators satisfying the following conditions*:

$$\langle \mathscr{G}(t, x_1) - \mathscr{G}(t, x_2), x_1 - x_2 \rangle_X \geq 0 \qquad \forall t \in [0, T] \quad \textit{and} \quad x_1, x_2 \in X; \tag{3.68}$$

there exists $\mathscr{L} > 0$ *such that*

$$\|\mathscr{G}(t_1, x_1) - \mathscr{G}(t_2, x_2)\|_X \leq \mathscr{L}(|t_1 - t_2| + \|x_1 - x_2\|_X) \tag{3.69}$$

$\forall t_1, t_2 \in [0, T]$ *and* $x_1, x_2 \in X$.

B is a linear operator such that

$$D(B^*) = D(B) \qquad \textit{and} \qquad B^* = -B, \tag{3.70}$$

where $B^*: D(B^*) \subset X \to X$ *is the adjoint of* B.

Then, for every $f \in W^{1,1}(0, T, X)$ *and* $x \in D(B)$ *there exists a unique function* $u \in W^{1,\infty}(0, T, X)$ *such that* $u(t) \in D(B)$ *for all* $t \in [0, T]$ *and*

$$\dot{u}(t) + \mathscr{G}(t, u(t)) = Bu(t) + f(t) \qquad \textit{a.e. on } (0, T) \tag{3.71}$$

$$u(0) = x. \tag{3.72}$$

PROOF. From (3.70) we get

$$\langle Bx, x \rangle_X = 0 \qquad \forall x \in D(B) \tag{3.73}$$

Hence, by (2.28) it follows that the linear operator $B^* = -B: D(B) \subset X \to X$ is a monotone operator; moreover it is a closed operator and since $(B^*)^* = (-B)^* = B$ is also a monotone operator, by Lemma 2.9 we get that B^* is a maximal monotone operator. From (3.68), (3.69) and Lemma 2.8 we get that for all $t \in [0, T]$ the operator $A(t): D(B) \subset X \to X$ defined by

$$A(t)x = \mathscr{G}(t, x) + B^*x \qquad \forall x \in D(B) \tag{3.74}$$

is also a maximal monotone operator.

Let now $\lambda > 0$, $t, s \in [0, T]$ and $x \in X$; let $y(t) = J_\lambda A(t)x$ and $y(s) = J_\lambda A(s)x$, where $J_\lambda A(t)$, $J_\lambda A(s)$ are the resolvent operators of $A(t)$ and $A(s)$ respectively.

Using (2.27) and (3.74) we have

$$y(t) + \lambda(\mathscr{G}(t, y(t)) + B^*y(t)) = x$$
$$y(s) + \lambda(\mathscr{G}(s, y(s)) + B^*y(s)) = x.$$

It follows that

$$\|y(t) - y(s)\|_X^2 + \lambda < \mathscr{G}(t, y(t)) + B^*y(t) - \mathscr{G}(s, y(s)) - B^*y(s), y(t) - y(s)\rangle_X = 0.$$

Using now (3.73), (3.70), (3.68), and (3.69), we get

$$\begin{aligned}\|y(t) - y(s)\|_X^2 &= -\lambda\langle\mathscr{G}(t, y(t)) - \mathscr{G}(s, y(s)), y(t) - y(s)\rangle_X \\ &\leq -\lambda\langle\mathscr{G}(t, y(t)) - \mathscr{G}(s, y(t)), y(t) - y(s)\rangle_X \\ &\leq \lambda\|\mathscr{G}(t, y(t)) - \mathscr{G}(s, y(t))\|_X\|y(t) - y(s)\|_X \\ &\leq \mathscr{L}\lambda|t - s| \cdot \|y(t) - y(s)\|_X.\end{aligned}$$

It follows that $\|y(t) - y(s)\|_X \leq \mathscr{L}\lambda|t - s|$, i.e.

$$\|J_\lambda A(t)x - J_\lambda A(s)x\|_X \leq \mathscr{L}\lambda|t - s|.$$

Hence, since (3.66) and (3.67) are satisfied, Lemma 3.9 follows from Theorem 3.11.

4 Some numerical methods and complements

In this section we give the numerical methods for elliptic problems and for ordinary differential equations in Hilbert spaces which were used in Chapters 3 and 5. This section contains only a few basic results concerning the numerical methods; for a more detailed study of these methods see the bibliographical notes at the end of this chapter. We also present here some useful technical results.

4.1 Numerical methods for elliptic problems

Let X be a real Hilbert space, $a: X \times X \to \mathbb{R}$ be a bilinear bounded and coercive form, i.e. there exists $M, \alpha > 0$ such that

$$|a(u, v)| \leq M\|u\|_X\|v\|_X \tag{4.1}$$

$$a(u, u) \geq \alpha\|u\|_X^2 \tag{4.2}$$

for all $u, v \in X$. Let us consider $f \in X'$ and the following linear variational equation:

$$u \in X, \quad a(u, v) = f(v) \qquad \forall v \in X. \tag{4.3}$$

Let $X_h \subset X$ be a closed subspace of X, and $f_h \in X'_h$, and let us consider the following variational equation which represents the internal approximation of (4.3):

$$u_h \in X_h, \quad a(u_h, v_h) = f_h(v_h) \qquad \forall v_h \in X_h. \tag{4.4}$$

The following lemma (called the first Strang lemma) evaluates the error introduced by considering (4.4) instead of (4.3):

Lemma 4.1 *Let* (4.1), (4.2) *hold,* $f \in X'$, $f_h \in X'_h$ *and let* u *and* u_h *be the solutions of* (4.3) *and* (4.4) *respectively. Then we have*

$$\|u - u_h\|_X \leq \alpha^{-1}\left[(1 + M) \max_{v_h \in X_h} \|u - v_h\|_X + \sup_{\substack{v_h \in X_h \\ \|v_h\|_X = 1}} |f(v_h) - f_h(v_h)|\right] \tag{4.5}$$

Let a be symmetric and let $j\colon X \to \mathbb{R}$ be a continuous and convex functional. We consider the following variational inequality:

$$u \in X, \quad a(u, v - u) + j(v) - j(u) \geq f(v - u) \qquad \forall v \in X. \tag{4.6}$$

We say that a family $(X_h)_{h>0}$ of closed subspaces of X is an *internal approximation* of X if

for all $v \in X$ and all $h > 0$ there exists $v_h \in X_h$ such that $v_h \to v$ strongly in X when $h \to 0$. (4.7)

Let us consider now for each $h > 0$ the internal approximation of the variational inequality (4.6):

$$u_h \in X_h, a(u_h, v_h - u_h) + j(v_h) - j(u_h) \geq f(v_h - u_h) \qquad \forall v_h \in X_h. \tag{4.8}$$

The following lemma shows that $(u_h)_{h>0}$ is a good approximation of u if $(X_h)_{h>0}$ is an internal approximation for X:

Lemma 4.2 *Let* (4.1), (4.2) *hold and let us suppose that* a *is symmetric and* $j\colon X \to \mathbb{R}$ *is convex and continuous. If* (4.7) *holds then*

$$u_h \to u \textit{ strongly in } X \textit{ when } h \to 0. \tag{4.9}$$

As proved in Lemma 2.6, if a is a symmetric and coercive form on X and j is a convex function X, then (4.6) is equivalent to the minimum problem

$$u \in X, \quad J(u) \leq J(v) \qquad \forall v \in X, \tag{4.10}$$

where J is the functional on X defined by $J(v) = \frac{1}{2}a(v, v) + j(v) - f(v)$. Let us suppose that J is twice Gâteaux differentiable and let us denote by $P = \nabla J$ the gradient of J and by $H = \nabla P$ the Hessian of J. In order to approximate the problem (4.10) we shall consider Newton's method, which consists of constructing a sequence $(u^n)_{n\in\mathbb{N}}$ recursively defined by

$$u^0 \in X, \quad H(u^n)(u^{n+1} - u^n) = -P(u^n) \qquad \forall n \in \mathbb{N}. \tag{4.11}$$

The following lemma shows that if u^0 is chosen close enough to u then u^n is a good approximate solution of (4.10).

Lemma 4.3 *Let J: $X \to \mathbb{R}$ be twice Gâteaux differentiable and let $P = \nabla J$ and $H = \nabla P$. Let us suppose that $v \to H(v)$ is continuous from X to $B(X)$ and there exist α, $M > 0$ such that*

$$\alpha\|w\|_X^2 \leq \langle H(v)w, w\rangle_X \leq M\|w\|_X^2 \qquad \forall v, w \in X. \tag{4.12}$$

Then there exists U, a neighbourhood of u in X (the solution of (4.10)), such that if $u^0 \in U$ then

$$u^n \to u \text{ strongly in } X \text{ when } n \to +\infty. \tag{4.13}$$

4.2 Euler's method for ordinary differential equations in Hilbert spaces

Let X be a real Hilbert space and let A: $[0, +\infty) \times X \to X$ be a continuous Lipschitz operator, i.e. there exist $L_1, L_2 > 0$ such that

$$\|A(t_1, x_1) - A(t_2, x_2)\|_X \leq L_2|t_1 - t_2| + L_1\|x_1 - x_2\|_X \tag{4.14}$$

for all $t_1, t_2 \in \mathbb{R}_+$, $x_1, x_2 \in X$.

Let $\tau > 0$ be the time step and let us consider the sequence $(y^n)_{n\in\mathbb{N}} \subset X$ recursively defined by

$$y^0 = x_0 \tag{4.15}$$

$$y^{n+1} = y^n + \tau A(n\tau, y^n) \tag{4.16}$$

which is the Euler approximation of the Cauchy problem

$$\dot{x}(t) = A(t, x(t)), \qquad t > 0 \tag{4.17}$$

$$x(0) = x_0. \tag{4.18}$$

The following lemma evaluates the error of Euler's method:

Lemma 4.4 *Let* (4.14) *hold. If $x \in C^1(\mathbb{R}_+, X)$ is the solution of the Cauchy problem* (4.17), (4.18), *and $(y^n)_{n\geq 0}$ given by* (4.15), (4.16) *is the Euler approximation of x, then we have*

$$\|x(n\tau) - y^n\|_X \leq \tau(L_2/L_1 + C_T)(\exp(n\tau L_1) - 1) \tag{4.19}$$

for all $n \in \mathbb{N}$ such that $n\tau \leq T$ and C_T is given by

$$C_T = \|\dot{x}\|_{0,T,X}. \tag{4.20}$$

The error estimate (4.19) is not useful for T large or for n large. The following results give error estimates for Euler's method over an infinite time interval:

Lemma 4.5 *Let* (4.14) *hold and let us suppose in addition that there exists* $c > 0$ *such that*

$$\langle A(t, x_1) - A(t, x_2), x_1 - x_2 \rangle_X \leq -c\|x_1 - x_2\|_X^2 \qquad \forall t \in \mathbb{R}_+, \quad x_1, x_2 \in X. \tag{4.21}$$

Let $x \in C^1(\mathbb{R}_+, X)$ *be the solution of the Cauchy problem* (4.17), (4.18), *and let* $\tau > 0$ *be the time step. We consider* B: $\tau\mathbb{N} \times X \to X$ *and* $(z_n)_{n\in\mathbb{N}}$ *be such that*

$$\|A(n\tau, y) - B(n\tau, y)\|_X \leq z_n \qquad \forall n \in \mathbb{N} \quad \text{and} \quad y \in X. \tag{4.22}$$

Let $(y^n)_{n\in\mathbb{N}} \subset X$ *be recursively defined by*

$$y^0 \in X, \quad y^{n+1} = y^n + \tau B(n\tau, y^n) \qquad \forall n \in \mathbb{N}. \tag{4.23}$$

If $\tau < \tau_0 = c/L_1^2$, *then* $q = \tau L_1^2(1 - \exp(\tau c))/c + \exp(-\tau c) < 1$ *and for all* $n \in \mathbb{N}$ *we have*

$$\|x(n\tau) - y^n\|_X \leq q^n\|x_0 - y^0\|_X + \tau L_2/(c - \tau L_1^2)$$
$$+ (1 - \exp(-c\tau))/c \sum_{i=0}^{n-1} [(1 + c/L_1)z_i + \tau L_1\|\dot{x}(i\tau)\|_X]q^{n-1-i}. \tag{4.24}$$

$$\|\dot{x}(n\tau) - (y^{n+1} - y^n)/\tau\|_X \leq L_1\|x(n\tau) - y^n\|_X + z_n. \tag{4.25}$$

PROOF. Let $n \in \mathbb{N}$ be fixed. For all $t \in [n\tau, (n+1)\tau]$ let

$$z(t) = y^n + (y^{n+1} - y^n)(t - n\tau)/\tau,$$

and note that

$$z(n\tau) = y^n, \quad \dot{z}(t) = B(n\tau, y^n), \qquad t \in [n\tau, n\tau + \tau]. \tag{4.26}$$

Let $\theta(s) = \|x(t) - z(t)\|_X^2$, $t = n\tau + s$; then from (4.17), (4.18), (4.14), (4.21), (4.22), and (4.26) we get

$$\begin{aligned}
\tfrac{1}{2}\dot{\theta}(s) &\leq \langle A(t, x(t)) - B(n\tau, y^n), x(t) - z(t)\rangle_X \\
&= \langle A(t, x(t)) - A(t, z(t)), x(t) - z(t)\rangle_X \\
&\quad + \langle A(t, z(t)) - A(n\tau, y^n), x(t) - z(t)\rangle_X \\
&\quad + \langle A(n\tau, y^n) - B(n\tau, y^n), x(t) - z(t)\rangle_X \\
&\leq -c\theta(s) + \sqrt{(\theta(s))}[sL_2 + sL_1\|y^{n+1} - y^n\|_X/\tau + z_n];
\end{aligned}$$

hence we have

$$\dot{\theta}(s) \leq -2c\theta(s) + \sqrt{\theta(s)}(\tau L_2 + L_1\|y^{n+1} - y^n\|_X + z_n), \qquad s \in [0, \tau].$$

We can now use Lemma 4.14 of Section 4.4 to deduce that

$$\begin{aligned}\|x(n\tau + \tau) - y^{n+1}\|_X \\ \leq \exp(-c\tau)\|x(n\tau) - y^n\|_X + (1 - \exp(-c\tau))(L_1\|y^{n+1} - y^n\|_X + \tau L_2 + z_n]/c\end{aligned} \tag{4.27}$$

From (4.14), (4.21)–(4.23) we can easily deduce:

$$\|y^{n+1} - y^n\|_X \leq \tau[L_1\|x(n\tau) - y^n\|_X + z_n + \|\dot{x}(n\tau)\|_X]. \tag{4.28}$$

If we use (4.28) in (4.27) we have

$$\begin{aligned}\|x(n\tau + \tau) - y^{n+1}\|_X \leq q\|x(n\tau) - y^n\|_X + (1 - \exp(-c\tau)) \\ \times [\tau L_2 + z_n(1 + \tau L_1) + \tau L_1\|\dot{x}(n\tau)\|_X]/c\end{aligned} \tag{4.29}$$

and recursively we obtain (4.29).

We also have

$$\begin{aligned}\|\dot{x}(n\tau) - (y^{n+1} - y^n)/\tau\|_X &\leq \|A(n\tau, x(n\tau)) - A(n\tau, y^n)\|_X \\ &\quad + \|A(n\tau, y^n) - B(n\tau, y^n)\|_X \\ &\leq L_1\|x(n\tau) - y^n\|_X + z_n;\end{aligned}$$

hence (4.25) holds.

A larger τ_0 for which similar inequalities to (4.24), (4.25) hold for $\tau < \tau_0$ can be obtained if $c\tau_0$ is the smallest positive solution of the equation $\exp(-x) + L_1^2(x - \exp(-x)) - 1 = 0$.

Lemma 4.6 *Let X, A and x be as in Lemma 4.5 and let $Y \subset X$ be a closed subspace. We denote by $P: X \to Y$ the projector map on Y and let $B: \mathbb{R}_+ \times X \to X$ be given by $B(t, z) = PA(t, z)$ for $t \in \mathbb{R}_+$, $z \in X$. If $y \in C^1(\mathbb{R}_+, Y)$ is the solution of the Cauchy problem*

$$\dot{y}(t) = B(t, y(t)), \qquad y(0) = y_0 \in Y, \quad t > 0, \tag{4.30}$$

then we have

$$\|x(t) - y(t)\|_X \leq \|x_0 - y_0\| \exp(-ct) + DL_1/c + \sqrt{(D\dot{D}/c)} \tag{4.31}$$

$$\|\dot{x}(t) - \dot{y}(t)\|_X \leq L_1\|x(t) - y(t)\|_X + \dot{D} \qquad \forall t \in \mathbb{R}_+, \tag{4.32}$$

where

$$D = \sup_{t \in \mathbb{R}_+} \inf_{z \in Y} \|x(t) - z\|_X, \qquad \dot{D} = \sup_{t \in \mathbb{R}_+} \inf_{z \in Y} \|\dot{x}(t) - z\|_X. \tag{4.33}$$

PROOF. Let $\theta(t) = \|x(t) - y(t)\|_X^2$; then from (4.30), (4.17), and (4.18) we get $\frac{1}{2}\dot{\theta}(t) = \langle A(t, x(t)) - B(t, y(t)), x(t) - y(t)\rangle_X$. For all $z \in Y$ we have

$$\begin{aligned}\langle A(t, y(t)) - B(t, y(t)), x(t) - y(t)\rangle_X &= \langle A(t, y(t)) - B(t, y(t)), x(t) - z\rangle_X \\ &\leq \|x(t) - z\|_X \inf_{v \in Y} \|A(t, y(t)) - v\|_X \\ &\leq \|x(t) - z\|_X (D + L_1\sqrt{\theta(t)});\end{aligned}$$

hence we have just obtained

$$\langle A(t, y(t)) - B(t, y(t)), x(t) - y(t)\rangle_X \leq D(\dot{D} + L_1\sqrt{\theta(t)}). \tag{4.34}$$

Bearing in mind that

$$\begin{aligned}\tfrac{1}{2}\dot{\theta}(t) = &\langle A(t, x(t)) - A(t, y(t)), x(t) - y(t)\rangle_X \\ &+ \langle A(t, y(t)) - B(t, y(t)), x(t) - y(t)\rangle_X,\end{aligned}$$

from (4.34) and (4.22) we obtain $\frac{1}{2}\dot{\theta}(t) \leq -c\theta(t) + L_1 D\sqrt{\theta(t)} + D\dot{D}$, and using Lemma 4.15 of Section 4.4 we deduce (4.31). If we remark that

$$\|\dot{x}(t) - \dot{y}(t)\|_X \leq \|A(t, x(t)) - B(t, x(t))\|_X + \|B(t, x(t)) - B(t, y(t))\|_X,$$

then from (4.14) we can easily obtain (4.32).

Lemma 4.7 *Let X, Y, A, B, and x be as in Lemma* 4.6 *and let* (y^n) *be given by* (4.23) *with* $y^0 \in Y$. *If* $0 < \tau < \tau_0 = c/L_1^2$, *then*

$$q = \tau L_1^2(1 - \exp(-c\tau))/c + \exp(-c\tau) < 1$$

and for all $n \in \mathbb{N}$ *we have*

$$\begin{aligned}\|x(n\tau) - y^n\|_X \leq q^n\|x_0 - y^0\|_X &+ \tau(L_2 + L_1\dot{Z})/(c - L_1^2\tau) \\ &+ 2L_1(D + c\dot{D}L_1^2)/(c - L_1^2\tau)\end{aligned} \tag{4.35}$$

$$\|\dot{x}(n\tau) - (y^{n+1} - y^n)/\tau\|_X \leq L_1\|x(n\tau) - y^n\|_X + \dot{D}, \tag{4.36}$$

where $D, \dot{D}$ *are given by* (4.33) *and* $\dot{Z} = \sup_{t \in \mathbb{R}_+} \|\dot{x}(t)\|_X$.

PROOF. Let $y \in C^1(\mathbb{R}_+, Y)$ be the solution of (4.30) with $y_0 = y^0 \in Y$. Using Lemma 4.6 we get (4.31), (4.32), hence for $t = i\tau$ we have

$$\|\dot{y}(i\tau)\|_X \leq L_1\|x_0 - y^0\|_X \exp(-c\tau i) + DL_1^2/c + \dot{D} + L_1\sqrt{(D\dot{D}/c)} + \dot{Z} \tag{4.37}$$

for all $i \in \mathbb{N}$. If we now use Lemma 4.5, in which we replace X by Y and A

by B, we deduce

$$\|y(n\tau) - y^n\|_X \leq \tau L_2/(c - L_1^2\tau) + \tau L_1(1 - \exp(-c\tau))/c \sum_{i=1}^{n-1} \|\dot{y}(i\tau)\|_X q^{n-1-i} \tag{4.38}$$

$$\|\dot{y}(n\tau) - (y^{n+1} - y^n)/\tau\|_X \leq L_1\|y(n\tau) - y^n\|_X. \tag{4.39}$$

If we use (4.37) in (4.38), after some algebra we get

$$\|y(n\tau) - y^n\|_X \leq (q^n - \exp(-c\tau n))\|x_0 - y^0\|_X + \tau/(c - L_1^2\tau)[L_2 + L_1(DL_1^2/c + \dot{D} + L_1\sqrt{(D\dot{D}/c)} + \dot{Z})]. \tag{4.40}$$

Using now (4.31) for $t = n\tau$ and (4.40) we get (4.35). Bearing in mind that

$$\|\dot{x}(n\tau) - (y^{n+1} - y^n)/\tau\|_X \leq \|A(n\tau, x(n\tau)) - B(n\tau, x(n\tau))\|_X + \|B(n\tau, x(n\tau)) - B(n\tau, y^n)\|_X,$$

we can easily deduce (4.36).

4.3 A numerical method for a non-linear evolution equation

Let $(V, \langle\,,\,\rangle, \|\cdot\|)$ be a real Hilbert space, $X, Y \subset V$ two orthogonal subspaces, $V = X \oplus Y$, $E \in B(V)$, and $B\colon \mathbb{R}_+ \times X \times V \to V$ a continuous operator. Let us consider the following problem:

Find $x \in C^1(\mathbb{R}_+, X)$, $y \in C^1(\mathbb{R}_+, Y)$ such that

$$\dot{y}(t) = E\dot{x}(t) + B(t, x(t), y(t)), \qquad t > 0 \tag{4.41}$$

$$x(0) = x_0 \in X, \qquad y(0) = y_0 \in Y. \tag{4.42}$$

Let us suppose that E is a positive definite and symmetric operator, i.e. there exists $d > 0$ such that

$$\langle Eu, u\rangle \geq d\|u\|^2 \tag{4.43}$$

$$\langle Eu, v\rangle = \langle u, Ev\rangle \qquad \forall u, v \in V. \tag{4.44}$$

Let $Z = X \times Y$ be endowed with the scalar product

$$((z_1, z_2)) = \langle Ex_1, x_2\rangle + \langle E^{-1}y_1, y_2\rangle \tag{4.45}$$

for all $z_i = (x_i; y_i) \in Z$, $i = 1, 2$, which generate the norm $|||\cdot|||$. Let us remark that Z is isomorphic with V and let us consider $F\colon \mathbb{R}_+ \times Z \to Z$ given by

$$((F(t, z_1), z_2)) = -\langle B(t, x_1, y_1), x_2\rangle + \langle E^{-1}B(t, x_1, y_1), y_2\rangle \tag{4.46}$$

for all $z_i = (x_i, y_i) \in Z$, $i = 1, 2$.

Using the above notations we can deduce the following lemma, which gives the equivalence between (4.41), (4.42) and a Cauchy problem on Z:

Lemma 4.8 *The couple $(x; y) \in C^1(\mathbb{R}_+, X \times Y)$ is a solution of* (4.41), (4.42) *iff $z = (x; y) \in C^1(\mathbb{R}_+, Z)$ is a solution of the following Cauchy problem*

$$\dot{z}(t) = F(t, z(t)), \qquad t > 0 \tag{4.47}$$

$$z(0) = z_0 = (x_0; y_0) \in Z. \tag{4.48}$$

PROOF. From (4.41) we get $0 = \langle \dot{y}(t), \bar{x} \rangle = \langle E\dot{x}(t), \bar{x} \rangle + \langle B(t, x(t), y(t)), \bar{x} \rangle$ for all $\bar{x} \in X$ and $\langle E^{-1}\dot{y}(t), \bar{y} \rangle = \langle E^{-1}B(t, x(t), y(t)), \bar{y} \rangle$ for all $\bar{y} \in Y$. If we add the last two equalities and we bear in mind (4.45) and (4.46), we obtain

$$((\dot{z}(t), \bar{z}) = ((F(t, z(t)), \bar{z})) \qquad \forall \bar{z} \in Z, \quad t > 0, \tag{4.49}$$

and (4.47) follows. Conversely, let us remark that from (4.47) we get (4.49). If we denote by $w(t) = E\dot{x}(t) - \dot{y}(t) + B(t, x(t), y(t))$ and we put $\bar{z} = (\bar{x}; 0)$ in (4.49), then we deduce $\langle w(t), \bar{x} \rangle = 0$ for all $\bar{x} \in X$, hence $w(t) \in Y$ for all $t > 0$. If we now put $\bar{z} = (0, w(t))$ in (4.49) we get $\langle E^{-1}w(t), w(t) \rangle = 0$; hence $w(t) = 0$ for all $t > 0$ and (4.41) follows.

Let us suppose in the following that there exists $L_1, L_2 > 0$ such that

$$\|B(t_1, x_1, y_1) - B(t_2, x_2, y_2)\| \le L_1|t_2 - t_2| + L_2(\|x_1 - x_2\| + \|y_1 - y_2\|) \tag{4.50}$$

for all $t_i \in \mathbb{R}_+$, $x_i \in X$, $y_i \in Y$, $i = 1, 2$. From (4.50) one can deduce that there exists $C > 0$ which depends only on $\|E\|_{B(V)}$ and d such that

$$|||F(t_1, z_1) - F(t_2, z_2)||| \le C(L_2|||z_1 - z_2||| + L_1|t_1 - t_2|) \tag{4.51}$$

for all $z_i \in Z$, $t_i \in \mathbb{R}_+$, $i = 1, 2$.

From Theorem 3.2, Lemma 4.8, and (4.51), one can easily deduce the following existence and uniqueness result for the problem (4.41), (4.42):

Lemma 4.9 *For all $(x_0; y_0) \in X \times Y$, there exists a unique pair $(x; y) \in C^1(\mathbb{R}_+, X \times Y)$ which is a solution of* (4.41), (4.42). *Moreover if we denote by $(x_i; y_i)$, $i = 1, 2$ the solution of* (4.41), (4.42) *for the initial data (x_{0i}, y_{0i}), $i = 1, 2$, then we have*

$$\|x_1(t) - x_2(t)\| + \|y_1(t) - y_2(t)\| \le C\exp(CL_2t)(\|x_{01} - x_{02}\| + \|y_{01} - y_{02}\|) \tag{4.52}$$

for all $t \in \mathbb{R}_+$, where C depends only on d and $\|E\|_{B(V)}$.

Let $X_h \subset X$ be a closed subspace of X, $T > 0$, $M \in N$, and $\tau = T/M$ be the time step. Let us denote by $(x_h^n)_{n=\overline{0,M}}$, $(y_h^n)_{n=\overline{0,M}}$ the approximate solution of the Cauchy problem (4.41), (4.42) recursively defined as follows:

$$x_h^0 \in X_h, \qquad y_h^0 \in V \tag{4.53}$$

$$\langle Ex_h^{n+1}, x_h \rangle = \langle Ex_h^n, x_h \rangle - \tau \langle B(n\tau, x_h^n, y_h^n), x_h \rangle \qquad \forall x_h \in X_h \tag{4.54}$$

$$y_h^{n+1} = y_h^n + E(x_h^{n+1} - x_h^n) - \tau B(n\tau, x_h^n, y_h^n), \qquad n = \overline{0, M-1}. \tag{4.55}$$

Remark 4.1 *If X_h is a finite-dimensional subspace of X then* (4.53)–(4.55) *represents a sequence of M linear algebraic systems.*

Theorem 4.1 *If we suppose that* (4.43), (4.44), *and* (4.50) *hold, then for all $n = \overline{0, M}$ we have the following error estimate*:

$$\begin{aligned} \|x(n\tau) - x_h^n\| + \|y(n\tau) - y_h^n\| \\ \le C \exp(CL_2T)[\|x_0 - x_h^0\| + \|y_0 - y_h^0\| + W(T)] \\ + \tau C(\exp(CL_2T) - 1)[L_1/L_2 + \dot{X}(T) + \dot{Y}(T)] \end{aligned} \tag{4.56}$$

where

$$W(T) = \sup_{t \in [0, T]} \left(\inf_{X_h \in X_h} \|x(t) - x_h\| \right) \tag{4.57}$$

$$\dot{X}(T) = \|\dot{x}\|_{0,T,V}, \qquad \dot{Y}(T) = \|\dot{y}\|_{0,T,V} \tag{4.58}$$

and $C > 0$ is a constant which depends only on d and $\|E\|_{B(V)}$.

In order to prove Theorem 4.1, we shall consider the sequences

$$(x^n)_{n=\overline{0,M}} \subset X, \qquad (y^n)_{n=\overline{0,M}} \subset Y,$$

which are solutions of the following recursive system:

$$x^0 = x_0 \in X, \qquad y^0 = y_0 \in Y \tag{4.59}$$

$$y^{n+1} - Ex^{n+1} = y^n - Ex^n + \tau B(n\tau, x^n, y^n), \qquad n = \overline{0, M-1}. \tag{4.60}$$

Lemma 4.10 *For all $n = \overline{0, M}$ we have*

$$\|x(n\tau) - x^n\| + \|y(n\tau) - y^n\| \le C\tau[L_1/L_2 + \dot{X}(T) + \dot{Y}(T)(\exp(CL_2T) - 1)] \tag{4.61}$$

PROOF. Let $z^n = (x^n; y^n) \in Z$; then from (4.59), (4.60), and (4.46), we get

$$z^{n+1} = z^n + \tau F(n\tau, z^n), \qquad n = \overline{0, M-1}, \quad z^0 = z_0 = (x_0; y_0); \tag{4.62}$$

hence $(z^n)_{n=\overline{0,M}}$ represents the approximate solution of (4.47), (4.48) obtained by using the explicit Euler method. From Lemma 4.4 we deduce that for all $n = \overline{0, M}$, we have

$$\|z(n\tau) - z^n\| \leq \tau(L_1 + L_2\dot{Z}(T))(\exp(CTL_2) - 1)/L_2 \tag{4.63}$$

where z is the solution of (4.47) and $\dot{Z} = \|\dot{z}\|_{0,T,Z}$.

We can now easily deduce (4.61) from (4.63).

Lemma 4.11 *For all $n = \overline{0, M}$ we have*

$$\|x^n - x_h^n\| + \|y^n - y_h^n\| \leq C[P(M) + \|x_0 - x_h^0\| + \|y_0 - y_h^0\|]\exp(CL_2T) \tag{4.64}$$

where $P(M) = \sup_{n=\overline{0,M}}(\inf_{x_h \in X_h}\|x^n - x_h\|)$.

PROOF. Let $g^n \in X'$, $g_h^n \in X_h'$ be given by

$$g^n(\bar{x}) = -\tau\langle B(n\tau, x^n, y^n), \bar{x}\rangle \qquad \forall \bar{x} \in X, \tag{4.65}$$

$$g_h^n(x_h) = -\tau\langle B(n\tau, x_h^n, y_h^n), x_h\rangle \qquad \forall x_h \in X_h. \tag{4.66}$$

Let $f^n(\bar{x}) = \langle Ex_0, \bar{x}\rangle + \sum_{i=0}^n g^i(\bar{x})$ for all $\bar{x} \in X$, $f_h^n(x_h) = \langle Ex_h^0, x_h\rangle + \sum_{i=0}^n g_h^i(x_h)$ for all $x_h \in X_h$ then from (4.60) we get

$$\langle Ex^{n+1}, \bar{x}\rangle = f^n(\bar{x}) \qquad \forall \bar{x} \in X \tag{4.67}$$

and from (4.54) we deduce

$$\langle Ex_h^{n+1}, x_h\rangle = f_h^n(x_h) \qquad \forall x_h \in X_h. \tag{4.68}$$

If we now use Lemma 4.1, from (4.67), (4.68) we obtain

$$\|x_h^{n+1} - x^{n+1}\| \leq C\left[\inf_{x_h \in X_h}\|x^{n+1} - x_h\| + \sup_{x_h \in X_h, \|x_h\| = 1}|f_h^n(x_h) - f^n(x_h)|\right]$$

and from (4.50) one can deduce

$$\|x^{n+1} - x_h^{n+1}\| \leq C\left[P(M) + \tau L_2\left(\sum_{i=0}^n \|x^i - x_h^i\| + \|y^i - y_h^i\|\right)\right.$$
$$\left. + \|x_0 - x_h^0\| + \|y_h^0 - y_0\|\right]. \tag{4.69}$$

Let us denote by $a_n = \|x^n - x_h^n\|$, $b_n = \|y^n - y_h^n\|$ and $d_n = \sum_{i=0}^n (a_i + b_i)$ and let us remark that (4.69) becomes

$$a_{n+1} \leq C(P(M) + \tau L_2 d_n + d_0). \tag{4.70}$$

From (4.55) and (4.60) one can easily deduce that

$$b_{n+1} \leq C(a_{n+1} + \tau L_2 d_n + d_0). \tag{4.71}$$

From (4.70) we get $d_{n+1} \leq d_n(1 + C\tau L_2) + C(d_0 + P(M))$ and recursively we obtain

$$d_{n+1} \leq \frac{1}{\tau L_2}(d_0 + P(M))(\exp(CL_2T) - 1) + d_0 \exp(CL_2T)$$

for all $n = \overline{0, M-1}$. If we use now the inequalities (4.70), (4.71) we get (4.64).

PROOF OF THEOREM 4.1. If we remark that

$$P(N) \leq W(T) + \sup_{n=\overline{0,M}} \|x(n\tau) - x^n\|,$$

then from (4.61) and (4.64) we obtain (4.56).

4.4 Some technical results

Gronwall-type inequalities represent a very efficient technique to obtain *a priori* estimates and to prove global existence theorems for evolution equations. We give here some results related to the classical Gronwall–Bellman lemma, used in Chapters 3 and 4:

Lemma 4.12 *Let $m, n \in L^1(0, T, \mathbb{R})$ be such that $m, n \geq 0$ a.e. on $(0, T)$ and let a be a real positive constant. Let also ϕ: $[0, T] \to \mathbb{R}$ be a continuous function such that*

$$\phi(s) \leq a + \int_0^s m(t)\,\mathrm{d}t + \int_0^s n(t)\phi(t)\,\mathrm{d}t \qquad \forall s \in [0, T]. \tag{4.72}$$

Then

$$\phi(s) \leq \left(a + \int_0^s m(t)\,\mathrm{d}t\right)\exp\left(\int_0^s n(t)\,\mathrm{d}t\right) \qquad \forall s \in [0, T]. \tag{4.73}$$

PROOF. Let ψ: $[0, T] \to \mathbb{R}$ be the function defined by

$$\psi(s) = a + \int_0^s m(t)\,\mathrm{d}t + \int_0^s n(t)\phi(t)\,\mathrm{d}t \qquad \forall s \in [0, T]. \tag{4.74}$$

ψ is an absolutely continuous function on $[0, T]$, $\phi(s) \leq \psi(s)$ for all $s \in [0, T]$ and hence $\dot{\psi}(s) = m(s) + n(s)\phi(s) \leq m(s) + n(s)\psi(s)$ a.e. on $(0, T)$. Multiplying this inequality by

$$\exp\left(-\int_0^s n(t)\,\mathrm{d}t\right)$$

we obtain

$$\frac{\mathrm{d}}{\mathrm{d}s}\left(\psi(s)\exp\left(-\int_0^s n(t)\,\mathrm{d}t\right)\right) \le m(s) \qquad \text{a.e. on } (0, T). \tag{4.75}$$

Since

$$s \to \psi(s)\exp\left(-\int_0^s n(t)\,\mathrm{d}t\right)$$

is absolutely continuous on $[0, T]$, using (3.6) of Chapter 2 and (4.75) we get

$$\psi(s)\exp\left(-\int_0^s n(t)\,\mathrm{d}t\right) - \psi(0) \le \int_0^s m(t)\,\mathrm{d}t \qquad \forall s \in [0, T].$$

From (4.74) we get $\psi(0) = a$; hence we obtain

$$\psi(s) \le \left(a + \int_0^s m(t)\,\mathrm{d}t\right)\exp\left(\int_0^s n(t)\,\mathrm{d}t\right) \qquad \forall s \in [0, T]$$

and using (4.72) and (4.74) we get (4.73).

In the particular case $m = 0$, Lemma 4.12 reduces to the classical Gronwall–Belmann lemma:

Corollary 4.1 *Let $n \in L^1(0, T, \mathbb{R})$, $n \ge 0$ a.e. on $(0, T)$, $a \ge 0$ and ϕ: $[0, T] \to \mathbb{R}$ be a continuous function such that $\phi(s) \le a + \int_0^s n(t)\phi(t)\,\mathrm{d}t$ for all $s \in [0, T]$. Then $\phi(s) \le a\exp(\int_0^s n(t)\,\mathrm{d}t)$ for all $s \in [0, T]$.*

Lemma 4.13 *Let $m, n \in L^1(0, T, \mathbb{R})$ such that $m, n \ge 0$ a.e. on $(0, T)$ and let a be a real positive constant. Let also ϕ: $[0, T] \to \mathbb{R}$ be a continuous function such that*

$$\tfrac{1}{2}\phi^2(s) \le \tfrac{1}{2}a^2 + \int_0^s m(t)\phi(t)\,\mathrm{d}t + \int_0^s n(t)\phi^2(t)\,\mathrm{d}t \qquad \forall s \in [0, T]. \tag{4.76}$$

Then

$$|\phi(s)| \le \left(a + \int_0^s m(t)\,\mathrm{d}t\right)\exp\left(\int_0^s n(t)\,\mathrm{d}t\right) \qquad \forall s \in [0, T]. \tag{4.77}$$

PROOF. Let $\varepsilon > 0$ and ψ_ε: $[0, T] \to \mathbb{R}$ be the function given by

$$\psi_\varepsilon(s) = \tfrac{1}{2}(a+\varepsilon)^2 + \int_0^s m(t)\phi(t)\,\mathrm{d}t + \int_0^s n(t)\phi^2(t)\,\mathrm{d}t \qquad \forall s \in [0, T]. \tag{4.78}$$

Using (4.76) we get

$$\psi_\varepsilon(s) > 0 \tag{4.79}$$

and

$$\phi^2(s) \le 2\psi_\varepsilon(s), \qquad |\phi(s)| \le \sqrt{(2\psi_\varepsilon(s))} \qquad \forall s \in [0, T]. \tag{4.80}$$

ψ_ε is absolutely continuous on $[0, T]$ and using (4.80) we have $\dot{\psi}_\varepsilon(s) = m(s)\phi(s) + n(s)\phi^2(s) \le m(s)\sqrt{(2\psi_\varepsilon(s))} + 2n(s)\psi_\varepsilon(s)$ a.e. on $(0, T)$; hence, by (4.80) we get

$$\frac{\dot{\psi}_\varepsilon(s)}{2\sqrt{(\psi_\varepsilon(s))}} \le \frac{m(s)}{\sqrt{2}} + n(s)\sqrt{(\psi_\varepsilon(s))} \qquad \text{a.e. on } (0, T). \tag{4.81}$$

Since $s \to \sqrt{(\psi_\varepsilon(s))}$ is also an absolutely continuous function on $[0, T]$ and

$$\frac{\mathrm{d}}{\mathrm{d}s}\sqrt{(\psi_\varepsilon(s))} = \frac{\dot{\psi}_\varepsilon(s)}{2\sqrt{(\psi_\varepsilon(s))}}$$

a.e. on $(0, T)$, from (3.6) of Chapter 2 and (4.81) we deduce

$$\sqrt{(\psi_\varepsilon(s))} \le \sqrt{(\psi_\varepsilon(0))} + \int_0^s \frac{m(t)}{\sqrt{2}}\,\mathrm{d}t + \int_0^s n(t)\sqrt{(\psi_\varepsilon(t))}\,\mathrm{d}t \tag{4.82}$$

for all $s \in [0, T]$. Using Lemma 4.12 from (4.82) we get

$$\sqrt{(2\psi_\varepsilon(s))} \le \left(a + \varepsilon + \int_0^s m(t)\,\mathrm{d}t\right)\exp\left(\int_0^s n(t)\,\mathrm{d}t\right) \qquad \forall s \in (0, T).$$

Using again (4.80) and passing to the limit when $\varepsilon \to 0$ we obtain (4.77).

In the particular case $n = 0$, Lemma 4.13 takes the following form.

Corollary 4.2 *Let $m \in L^1(0, T, \mathbb{R})$, $m \ge 0$ a.e. on $(0, T)$, $a \ge 0$ and $\phi\colon [0, T] \to \mathbb{R}$ be a continuous function such that $\frac{1}{2}\phi^2(s) \le \frac{1}{2}a^2 + \int_0^s m(t)\phi(t)\,\mathrm{d}t$ for all $s \in [0, T]$. Then $|\phi(s)| \le a + \int_0^s m(t)\,\mathrm{d}t$ for all $s \in [0, T]$.*

The following three lemmas were used in Chapter 3.

Lemma 4.14 *Let $p\colon [0, T] \to [0, +\infty)$ be a continuous function and $\alpha \in \mathbb{R}$. If $\theta\colon [0, T] \to [0, +\infty)$ is an absolutely continuous function such that*

$$\dot{\theta}(t) \le 2\alpha\theta(t) + 2\sqrt{(\theta(t)p(t))} \qquad \textit{a.e. } t \in [0, T] \tag{4.83}$$

then

$$\sqrt{(\theta(t))} \le \sqrt{(\theta(0))}\exp(\alpha t) + \int_0^t p(s)\exp(\alpha(t - s))\,\mathrm{d}s \qquad \forall t \in [0, T]. \tag{4.84}$$

PROOF. Let $A = \{t \in [0, T] \mid \theta(t) > 0\}$. Since θ is continuous we get that $A = \bigcup_{n \in N} I_n$, where I_n are disjoint intervals. Let $n \in \mathbb{N}$ be fixed. Then there exist $a_n, b_n \in \mathbb{R}$ such that $I_n = (a_n, b_n)$ or $I_n = [0, b_n)$. From (4.83) we deduce

$$\frac{\mathrm{d}}{\mathrm{d}t}(\sqrt{(\theta(t))}\exp(-\alpha t)) \le p(t)\exp(-\alpha t) \qquad \text{a.e.} \quad t \in I_n.$$

If we integrate this inequality we deduce

$$\sqrt{(\theta(t))}\exp(-\alpha t) - \sqrt{(\theta(a_n))}\exp(-\alpha a_n) \leq \int_0^t p(s)\exp(-\alpha s)\,\mathrm{d}s \qquad \forall t \in I_n.$$

If $I_n = (a_n, b_n)$ then $\theta(a_n) = 0$ and (4.84) follows for all $t \in I_n$. If $I_n = [0, b_n)$ then we can put in the above inequality $a_n = 0$ and (4.84) also follows for all $t \in I_n$. Hence (4.84) was proved for all $t \in A$ and it results that it holds for all $t \in [0, T]$.

Lemma 4.15 *Let* $\theta \in C^1([0, T], \mathbb{R})$, $\theta \geq 0$ *and* $\alpha > 0$, $\beta, \gamma \geq 0$. *If*

$$\dot{\theta}(t) \leq -2\alpha\theta(t) + 2\beta\sqrt{(\theta(t))} + 2\gamma \qquad \forall t \in [0, T] \tag{4.85}$$

then

$$\sqrt{(\theta(t))} \leq \sqrt{(\theta(0))}\exp(-\alpha t) + (\beta + \sqrt{(\beta^2 + 4\alpha\gamma)})/(2\alpha) \qquad \forall t \in [0, T]. \tag{4.86}$$

PROOF. Suppose $\gamma \neq 0$ (if $\gamma = 0$ Lemma 4.14 can be used). Let $\delta > 0$ and let M_δ be the set defined by

$$M_\delta = \{t \in [0, T] \mid \theta(t) \geq \delta + \sqrt{(\theta(0))}\exp(-\alpha t) + (\beta + \sqrt{(\beta^2 + 4\alpha\gamma)})/(2\alpha)\}.$$

If M_δ is not empty, since $0 \notin M_\delta$ and M_δ is a closed set we get $0 < t_\delta = \inf M_\delta$ and

$$\sqrt{(\theta(t_\delta))} = \delta + \sqrt{(\theta(0))}\exp(-\alpha t_\delta) + (\beta + \sqrt{(\beta^2 + 4\alpha\gamma)})/(2\alpha).$$

Using (4.85) and the above equality we get

$$\frac{\mathrm{d}}{\mathrm{d}t}(\sqrt{(\theta(t))} - \sqrt{(\theta(0))}\exp(-\alpha t))|_{t=t_\delta} \leq (-\alpha\delta^2)/\sqrt{(\theta(t_\delta))} < 0$$

hence at least in a neighbourhood of t_δ the function $\sqrt{(\theta(t))} - \sqrt{(\theta(0))}\exp(-\alpha t)$ is decreasing, i.e. $t_\delta > \inf M_\delta$, a contradiction. Hence $M_\delta = \varnothing$ for all $\delta > 0$ and (4.86) will follow.

Lemma 4.16 *Let* $r\colon \mathbb{R}_+ \to \mathbb{R}_+$ *be a continuous function such that* $\lim_{t\to+\infty} r(t) = 0$ *and* $p(t)\colon \mathbb{R}_+ \to \mathbb{R}_+$ *be given by* $p(t) = \int_0^t r(s)\exp(-c(t-s))\,\mathrm{d}s$ *with* $c > 0$. *Then* $\lim_{t\to+\infty} p(t) = 0$.

PROOF. Let us denote by $M = \max_{t\in\mathbb{R}_+} r(t)$. For all $\varepsilon > 0$ there exists $t_\varepsilon > 0$ large enough such that $\mathrm{e}^{-ct_\varepsilon} < c\varepsilon/(2M)$ and $r(t) < c\varepsilon/2$ for all $t > t_\varepsilon$. For all $t > 2t_\varepsilon$ we have

$$p(t) = \exp(-ct)\left[\int_0^{t_\varepsilon} r(s)\exp(cs)\,ds + \int_{t_\varepsilon}^{t} r(s)\exp(cs)\,ds\right]$$
$$\leq \exp(-ct)\left[\frac{M}{c}\exp(ct_\varepsilon) + \frac{\varepsilon}{2}\exp(ct)\right]$$
$$\leq \frac{M}{c}\exp(-ct_\varepsilon) + \frac{\varepsilon}{2} < \varepsilon;$$

hence $\lim_{t\to+\infty} b(t) = 0$.

The following simple algebraic lemma was used in Section 2 of Chapter 4.

Lemma 4.17 *Let X be a real Hilbert space and $A, B \in B(X)$ be two symmetric operators. Suppose that there exist $\alpha, \gamma > 0$ such that*

$$\langle Ax, x\rangle_X \geq \alpha\|x\|_X^2, \qquad \langle (B-A)x, x\rangle_X \geq \gamma\|x\|_X^2 \qquad \forall x \in X. \tag{4.87}$$

Let $\beta = \alpha\gamma/(\|B\|_{B(X)}\cdot\|A\|_{B(X)}^2)$; then we have

$$\langle (A^{-1} - B^{-1})x, x\rangle_X \geq \beta\|x\|_X^2 \qquad \forall x \in X. \tag{4.88}$$

PROOF. Let $\langle\, ,\rangle_A$, $\|\cdot\|_A$ and $\langle\, ,\rangle_B$, $\|\cdot\|_B$ be the inner products and norms generated by A and B, i.e.

$$\langle x, y\rangle_A = \langle Ax, y\rangle_X, \qquad \|x\|_A^2 = \langle x, x\rangle_A \tag{4.89}$$

$$\langle x, y\rangle_B = \langle Bx, x\rangle_X, \qquad \|x\|_B^2 = \langle x, x\rangle_B \tag{4.90}$$

for all $x, y \in X$. If we bear in mind that $\|x\|_B^2 \leq \|B\|_{B(X)}\|x\|_X^2$, then from (4.87) we get

$$\|x\|_A^2 \leq (1 - \gamma/\|B\|_{B(X)})\|x\|_B^2 \qquad \forall x \in X. \tag{4.91}$$

For all $y \in X$ we have $\langle B^{-1}Ay, Ay\rangle_X^2 = \langle B^{-1}Ay, Ay\rangle_A^2 \leq \|B^{-1}Ay\|_A^2\|y\|_A^2$ and if we use (4.91) we obtain

$$\langle B^{-1}Ay, Ay\rangle_X^2 \leq \|y\|_A^2\|B^{-1}Ay\|_B^2(1 - \gamma/\|B\|_{B(X)}).$$

If we remark that $\|B^{-1}Ay\|_B^2 = \langle B^{-1}Ay, Ay\rangle_X$ then from the above inequality we deduce

$$\gamma/\|B\|_{B(X)}\|y\|_A^2 \leq \|y\|_A^2 - \langle B^{-1}Ay, Ay\rangle_X.$$

If we put $y = A^{-1}x$ we get

$$(\gamma/\|B\|_{B(X)})\langle A^{-1}x, x\rangle_X \leq \langle (A^{-1} - B^{-1})x, x\rangle_X \qquad \forall x \in X$$

and if we bear in mind that $\langle A^{-1}x, x\rangle_X \geq \alpha(\|A\|_{B(X)}^2)\|x\|_X^2$, then we obtain (4.88).

Bibliographical notes

The material presented in Section 1 is standard and can be found in many books on functional analysis. For more information in the field we refer the reader to the works of Dunford and Schwartz (1958), Rudin (1966), and Yosida (1971).

For a complete treatment on general theory of convex functions as well as for proofs of the results exhibited in Section 2.1 the reader is referred to the works of Moreau (1966–67), Rockafellar (1968, 1970), Ekeland and Temam (1974), and Barbu and Precupanu (1978). Theorem 2.2 belongs to Mosco (1976) and extends a result of Ky-Fan (1972). The theory of elliptic variational inequalities has been the subject of much development during the last decades. For a more extensive study we refer the reader to the works of Lions and Stampacchia (1967), Stampacchia (1969), Lions (1969, Ch. 2), Mosco (1970, 1976), and Kikuchi and Oden (1980). Proofs of the results given in Section 2.3 can be found for instance in the books of Brezis (1973), Barbu (1976, 1984), and Pascali and Sburlan (1978). Theorems 2.3 and 2.4 were originally given by Minty (1962) and were extended by Browder (1965*a*,*b*) in the case of real reflexive Banach spaces. Theorem 2.5 due to Kerner and Vainberg can be found in the book of Vainberg (1967, p. 79).

Theorems 3.1 and 3.2 represent extensions in Banach spaces of the well-known Cauchy–Lipschitz existence theorem. Theorem 3.3 belongs to Lovelady and Martin (1972) and was also investigated by Pavel and Ursescu (1974). The most extensive treatise on abstract theory of semigroups of linear operators is the classical book of Hille and Phillips (1957); other general references on this topic are, for instance, the books of Dunford and Schwartz (1958), Kato (1966), Yosida (1971), and Pazy (1983). Theorem 3.4 gives a complete characterization of the infinitesimal generator of a C_0 semigroup and represents the starting point of the subsequent systematic development of the theory of semigroups of bounded linear operators. It was obtained independently by Hille (1948) and Yosida (1948).

Theorem 3.5 is due to Dascalu and Ionescu (1991). A different approach to the construction of the extended semigroup given in Theorem 3.5 was obtained by Da Prato and Grisvard (1984) (see also Cazenave and Haraux (1990, p. 41)). Theorem 3.6 is due to Phillips (1954a) and Theorem 3.7 is due to Hille (1952); see also Phillips (1954*b*). Lemma 3.6 is a particular case of Theorem II of Kato (1973). Theorem 3.8 belongs to Segal (1963) and, with minor adjustments, Theorem 3.9 can be found in Pazy (1983, p. 189). Theorem 3.10 is due to Ionescu (1990*b*) while Theorem 3.11 is a simple consequence of Theorem 1 in Barbu (1976, p. 164); Lemma 3.9 can also be obtained from a result due to Kato (1967). For other information concerning non-linear semigroups and applications to multivalued evolution equations we refer the reader to Brezis (1973) and Barbu (1976).

For more detailed information on approximation methods for elliptic variational inequalities we refer the reader to Glowinski *et al.* (1976), Ciarlet (1978). For example, Lemmas 4.1 and 4.2 can be found respectively in Ciarlet (1978, p. 186) and Glowinski *et al.* (1976, p. 43), and Lemma 4.3 can be found in Céa (1964, p. 93). Ilioi (1980, p. 82), or in Orthega and Rheinboldt (1970). With minor adjustments, Lemma 4.4 concerning the error estimate of Euler's method can be found in the works of Henrici (1962, p. 26) or Isaacson and Keller (1966, p. 368). Lemmas 4.5–7 and the results from Section 4.3 were obtained by Ionescu (1988*a*). A different method in the study of nonlinear evolution equations of the form (4.41), (4.42) was given by Sofonea (1990*b*), (1992).

REFERENCES

Adams, R. S. (1975). *Sobolev spaces*. Academic Press, New York.

Anzellotti, G. (1983). On the existence of the rates of stress and displacement for Prandtl–Reuss plasticity. *Quarterly of Applied Mathematics*, **61**, 181–208.

Anzellotti, G. and Gianquinta, M. (1980). Existence of the displacement field for an elastoplastic body subject to Hencky's law and Von Mises yield condition. *Manuscripta Mathematica*, **32**, 101–36.

Bai, Y. (1982). Thermo-plastic instability in simple shear. *Journal of the Mechanics and Physics of Solids*, **30**(4), 195–207.

Bai, Y. (1990). Adiabatic shear bounding. *Res. Mechanica*, **31**, 133–203.

Barbu, V. (1976). *Nonlinear semigroup and differential equations in Banach spaces*. Editura Academiei, Bucharest-Noordhoff, Leyden.

Barbu, V. (1984). *Optimal control of variational inequalities*. Pitman, Boston.

Barbu, V. and Precupanu, Th. (1978). *Convexity and optimization in Banach spaces*. Editura Academiei, Sijthoff & Noordhoff International Publishers, Bucharest.

Bochner, S. (1933). Integration von Funktionen, deren Wert die Elemente eines Vektoraumes sind. *Fundamenta Mathematica*, **20**, 262–76.

Brezis, H. (1968). Équations et inéquations non linéaires dans les espaces vectoriels en dualité. *Annales de l'Institut Fourier*, **18**(1), 115–75.

Brezis, H. (1973). *Opérateurs maximaux monotones et semigroups de contractions dans les espaces de Hilbert*. North-Holland, Amsterdam.

Browder, F. E. (1965*a*). Existence and uniqueness theorems for solutions of nonlinear boundary value problems. *Proc. AMS Symposium of Applied Mathematics*, **17**, 24–9.

Browder, F. E. (1965*b*). Nonlinear elliptic boundary value problems. *Transactions of the American Mathematical Society*, **117**, 530–50.

Burns, T. J. and Trucano, T. G. (1982). Instability in simple shear deformation of strain-softening materials. *Mechanics of Materials*, **1**, 313–24.

Cazenave, T. and Haraux, A. (1990). *Introduction aux problèmes d'évolutions sémi-linéaires*. Ellipse, Paris.

Céa, J. (1964). Approximation variationnelle des problèmes aux limites. *Annales de l'Institut Fourier*, **14**(2), 345–444.

Céa, J. (1971). *Optimization. Théorie et algorithmes*. Dunod, Paris.

Chen, H. Tz., Douglas, A. S. and Malek-Madani, R. (1989). An asymptotic stability condition for inhomogeneous simple shear. *Quarterly of Applied Mathematics*, **47**(2), 247–62.

Ciarlet, P. (1978). *The finite element method for elliptic problems*. North-Holland, Amsterdam.

Ciarlet, P. (1988). *Mathematical elasticity. Volume I: Three dimensional elasticity*. Studies in Mathematics and its Applications, 20, North-Holland, Amsterdam.

Cristescu, N. (1963). On the propagation of elasticplastic waves in metallic rods. *Bulletin de l'Académie Polonaise des Sciences*, **11**, 129–33.

Cristescu, N. (1964). Some problems on the mechanics of extensible strings. In *Anelastic solids* (ed. H. Kolstry and W. Prager), pp. 118–32, Springer-Verlag.

Cristescu, N. (1967). *Dynamic plasticity*. North-Holland, Amsterdam.

Cristescu, N. (1975). Plastic flow through conical converging dies, using a viscoplastic constitutive equations. *International Journal of Mechanical Sciences*, **17**, 425–33.

Cristescu, N. (1976). Drawing through conical dies—An analysis compared with experiments. *International Journal of Mechanical Sciences*, **18**(1), 45–9.

Cristescu, N. (1977). Speed influence in wire drawing. *Revue Roumaine de Sciences techniques—Mécanique Appliquée*, **22**(3), 391–9.

Cristescu, N. (1980). On the optimum die angle in fast wire drawing. *Journal of Mechanics and Working Technology*, **3**(3–4), 275–87.

Cristescu, N. (1985). Plasticity of compressible/dilatant rock-like materials. *International Journal of Engineering Sciences*, **23**(10), 1091–100.

Cristescu, N. (1987). Elastic/viscoplastic constitutive equations for rock. *International Journal of Rock Mechanics Mining Sciences & Geometrics Abstract*, **24**(5), 271–82.

Cristescu, N. (1989). *Rocky Rheology*. Kluwer Academic Publishers, Dordrecht.

Cristescu, N. and Suliciu, I. (1982). *Viscoplasticity*. Martinus Nijhoff, Editura Tehnică, Bucharest.

Cristescu, N., Ionescu, I. R. and Rosca, I. (1991). Creep of coal in long wall workings (submitted).

Da Prato, G. and Grisvard, P. (1984). Maximal regularity for evolution equations by interpolation and extrapolation. *Journal of Functional Analysis*, **58**, 107–24.

Dascalu, C. and Ionescu, I. R. (1991). Weak solutions for dynamic processes of rate-type viscoplastic materials. Preprint, Institute of Mathematics, Bucharest, no. 1 (to appear in *Proc. 4th International Conference on Hyperbolic Problems*, Taormina 1992, A. Donato, Ed.)

Djaoua, M. and Suquet, P. (1984). Évolution quasistatique des milieux viscoplastiques de Maxwell–Norton. *Mathematical Methods in Applied Sciences*, **6**, 192–205.

Douglas, A. S. and Chen, H. Tz. (1985). Adiabatic localisation of plastic strain in antiplane shear. *Scripta Metallurgica*, **19**, 1277–80.

Douglas, A. S., Malek-Madani, R. and Chen, H. Tz. (1987). Stability conditions for shearing in plates. *Proc. International Conference on Impact Loading and Dynamic Behaviour of Materials*, Bremen, F.R.G.

Dunford, N. and Schwarz, J. (1958). *Linear Operators, Part I, General Theory*. Interscience, New York.

Duvaut, G. and Lions, J. L. (1970). Écoulement d'un fluide rigide-viscoplastique incompressible. *Comptes Rendus de l'Académie des Sciences de Paris*, **270**, 58–61.

Duvaut, G. and Lions, J. L. (1972). *Les inéquations en mécanique et en physique*. Dunod, Paris.

Eirich, F. R. (ed.) (1956). *Rheology. Theory and applications*. Academic Press, New York.

Ekeland, I. and Temam, R. (1974). *Analyse convexe et problèmes variationels*. Dunod, Gauthiers-Villars, Paris.

Făciu, C. (1989). An energetical study of the solutions for initial and boundary value problems in viscoplasticity. D. Phil. Thesis, University of Bucharest (in Romanian).

Făciu, C. and Mihailescu-Suliciu, M. (1987). The energy in one-dimensional rate-type semilinear viscoelasticity. *International Journal of Solid Structures*, **23**(11), 1505–20.

Fichera, G. (1972). *Existence theorems in elasticity.* Handbuch der Physik VI/a2, Springer-Verlag, Berlin.

Fortin, M. (1972). Calcul numérique des écoulements des fluides de Bingham et des fluides newtoniens incompressibles par la méthode des elements finis. Thesis, University of Paris VI.

Fressengeas, C. (1988). Analyse dynamique elasto-viscoplastique de l'héterogénéité de la deformation plastique de cisaillement. *Journal de Physique*, Colloque C 3, Supl.au. no. **9**(49), 277–82.

Fressengeas, C. and Molinari, A. (1987). Instability of plastic flow in shear at high strain rates. *Journal of Mechanics and Physics of Solids*, **35**, 185–211.

Gagliardo, E. (1957). Caratterizzazioni della trace sulla frontiera relative alume classi di funzioni in *n* variabile. *Rendiconti dell Seminario Matematico, Universita di Padova*, **27**, 284–305.

Geiringer, H. and Freudenthal, A. M. (1958). The mathematical theories of the inelastic continuum. *Handbuch der Physik*, Springer-Verlag, Berlin.

Germain, P. and Muller, P. (1980). *Introduction à la mécanique des milieux continus.* Masson, Paris.

Geymonat, G. and Raous, M. (1977). Elements finis en viscoélasticité périodique. *Lecture Notes in Mathematics*, **606**, 150–66.

Geymonat, G. and Suquet, P. (1984). Functional spaces for Norton–Hoff materials. Preprint no. 84-6. *Laboratoire de Mécanique Générale des Milieux Continus.* Montpellier.

Glowinski, R. (1973). Sur l'écoulement d'un fluide de Bingham dans une conduite cylindrique. *Journal de Mécanique*, **13**(4), 601–21.

Glowinski, R., Lions, J. L. and Trémolières, R. (1976). *Analyse numérique des inéquations variationnelles.* Tome I, II. Dunod, Paris.

Gurtin, M. E., Williams, W. O. and Suliciu, I. (1984). On rate-type constitutive equations and the energy of viscoelastic and viscoplastic materials. *International Journal of Solid Structures*, **16**, 607–17.

Hahn, W. (1967). *Stability of motion.* Springer-Verlag, Berlin.

Halphen, B. (1978). Problèmes quasi-statiques en viscoplasticité, Thesis, Paris, 1978.

Haraux, A. (1981). *Nonlinear evolution equations—Global behaviour of solutions.* Springer-Verlag, Berlin.

Henrici, P. (1962). *Discrete variable methods in ordinary differential equations.* Wiley, New York.

Hille, E. (1948). *Functional analysis and semi-groups.* American Mathematical Society Colloquium Publishers, 31, New York.

Hille, E. (1952). Une généralisation du problème de Cauchy. *Annales de l'Institut Fourier*, **4**, 31–48.

Hille, E. and Phillips, R. S. (1957). *Functional analysis and semigroups.* American Mathematical Society Colloquium Publishers, 31, Providence, R.I.

Hlaváček, I. and Nečas, J. (1981). *Mathematical theory of elastic and elasto-plastic bodies: an introduction.* Elsevier, Amsterdam.

Ilioi, C. (1980). *Optimization problems and algorithms for approximations of the solutions.* Editura Academiei, Bucuresti (in Romanian).

Ionescu, I. R. (1985). A boundary value problem with a non-local viscoplastic friction law for the Bingham fluid. *Studii si Cercetări Matematice*, **37**(1), 60–5.

Ionescu, I. R. (1988*a*). Error estimates of a numerical method for a nonlinear evolution equation. *Analele Universităţii din Bucuresti*, Matematica-Informaticŏ, **2**, 64–74.

Ionescu, I. R. (1988*b*). Dynamic processes for a class of elastic-viscoplastic materials. *Preprint series in mathematics*, INCREST, Bucharest, 64, and *Studii si Cercetări Matematice*, **44**(12), 113–125 (1992).

Ionescu, I. R. (1990*a*). Error estimates of an Euler's method for a quasistatic elastic-viscoplastic problems. *Zeitschrift für Angewandte Mathematik und Mechanik* (ZAMM), **70**(3), 173–80.

Ionescu, I. R. (1990*b*). Functional and numerical methods in rate-type viscoplasticity. Thesis, University of Bucharest, and *Studii si Cercetări Matematice*, **42**(4), 309–99 (in Romanian).

Ionescu, I. R. (1991). Some existence results in one dimensional dynamic viscoplasticity with hardening. *IMA Journal of Applied Mathematics*, **47**, 217–28.

Ionescu, I. R. and Predeleanu, M. (1991). On the dynamic shearing problem for rate and temperature dependent media (to appear in *Quarterly Journal of Mechanics and Applied Mathematics*).

Ionescu, I. R. and Sofonea, M. (1984). A variational formulation of a boundary value problem in the study of the Bingham fluid. *Revue Roumaine des Sciences Techniques. Série de Mécanique Appliquée*, **30**(4), 357–63.

Ionescu, I. R. and Sofonea, M. (1986). The blocking property in the study of the Bingham fluid. *International Journal of Engineering Science*, **24**(3), 289–97.

Ionescu, I. R. and Sofonea, M. (1988). Quasistatic processes for elastic-visco-plastic materials. *Quarterly of Applied Mathematics*, **46**(2), 220–43.

Ionescu, I. R. and Vernescu, B. (1988). A numerical method for a viscoplastic problem. An application to wire drawing. *International Journal of Engineering Science*, **26**(6), 627–33.

Ionescu, I. R., Molnar, I. and Vernescu, B. (1985*a*). A finite element model of wire drawing. I. Variational formulation and numerical method. *Revenue Roumaine des Sciences Techniques. Série de Mécanique Appliquée*, **36**(6), 611–22.

Ionescu, I. R., Rosca, I. and Sofonea, M. (1985*b*). A variational method for nonlinear multivalued operators. *Nonlinear Analysis, Theory, Methods & Applications*, **9**(2), 259–73.

Ionescu-Tulcea, C. and Ionescu-Tulcea, A. (1972). *Topics in the theory of liftings*. Springer-Verlag, Berlin.

Isaacson, E. and Keller, H. B. (1966). *Analysis of numerical methods*. Wiley, New York.

John, O. (1974). On the solution of the displacement boundary value problem for elastic–inelastic materials. *Applikace Mathematiky*, **19**(2), 61–71.

Johnson, C. (1976). Existence theorems for plasticity problems. *Journal de Mathématiques Pures et Appliquées*, **55**, 431–44.

Kato, T. (1966). *Perturbation theory of linear operators*. Springer-Verlag, New York.

Kato, T. (1967). Nonlinear semigroups and evolution equations. *Journal of the Mathematical Society of Japan*, **19**(4), 508–20.

Kato, T. (1973). Linear evolution equations of 'hyperbolic' type, II. *Journal of the Mathematical Society of Japan*, **25**(4), 648–66.

Kato, T. (1975). Quasilinear equations of evolution with application to partial differential equations. *Lecture Notes in Mathematics*, 448, Springer-Verlag, pp. 25–70.

Kikuchi, N. and Oden, J. T. (1980). Theory of variational inequalities with applications to problems of flow through porous media. *International Journal of Engineering Sciences*, **18**, 1173–84.

Komura, J. (1967). Nonlinear semigroups in Hilbert spaces. *Journal of the Mathematical Society of Japan*, **19**, 493–507.

Kurtz, T. G. (1973). A limit theorem for perturbed operator semigroups with application to random evolution. *Journal of Functional Analysis*, **12**, 55–67.

Ky-Fan (1972). A minimax inequality and applications. In *Inequalities III* (ed. Shido), Academic Press, pp. 103–13.

Laborde, P. (1979). On visco-plasticity with hardening. *Numerical Functional Analysis and Optimization*, **1**(3), 315–39.

Léné, F. (1974). Sur les matériaux élastiques a énergie de déformation non quadratique. *Journal de Mécanique*, **13**(3), 499–534.

Le Tallec, P. (1990). *Numerical analysis of viscoelastic problems.* Masson, Paris.

Lions, J. L. (1969). *Quelques méthodes de résolution des problèmes aux limites non linéaries.* Dunod, Gauthiers-Villars, Paris.

Lions, J. L. and Magenes, E. (1968). *Problèmes aux limites non-homogènes.* I, Dunod, Paris.

Lions, J. L. and Stampacchia, G. (1967). Variational inequalities. *Communications on Pure and Applied Mathematics*, **XX**, 493–519.

Lovelady, D. L. and Martin, R. H. Jr. (1972). A global existence theorem for a nonautonomous differential equation in a Banach space. *Proceedings of the American Mathematical Society*, **35**, 445–9.

Malvern, L. E. (1969). *Introduction to the mechanics of a continuous medium.* Prentice-Hall, Englewood Cliffs.

Matthies, H., Strang, G. and Cristiansen, E. (1979). *The saddle point of a differential program in energy methods in finite element analysis* (ed. R. Glowinski, E. Rodin and O. C. Zienkiewicz), Wiley, New York.

Mazilu, P. and Sburlan, S. (1973). *Functional methods for the solving of the equations of the theory of elasticity.* Editura Academiei, Bucharest (in Romanian).

Mihăilescu-Suliciu, M. and Suliciu, I. (1979). Energy for hypoelastic constitutive equations. *Archive of Rational Mechanics and Analysis*, **71**, 327–44.

Mihăilescu-Suliciu, M. and Suliciu, I. (1985). On the method of characteristics in rate-type viscoelasticity. *ZAAM*, **65**(10), 479–86.

Mihăilescu-Suliciu, M., Suliciu, I. and Williams, W. (1989). On viscoplastic and elastic–plastic oscillators. *Quarterly of Applied Mathematics*, **47**(1), 105–16.

Minty, G. J. (1962). Monotone (nonlinear) operators in Hilbert spaces. *Duke Mathematical Journal*, **29**, 341–6.

Molinari, A. and Leroy, Y. M. (1990). Existence and stability of stationary shear bonds with mixed boundary conditions. *Comptes Rendus de l'Académie de Sciences de Paris*, **310**(II), 1017–23.

Moreau, J. J. (1965). Proximité et dualité dans un espace hilbertien. *Bulletin de la Société Mathématique de France*, **93**, 273–83.

Moreau, J. J. (1966–67). *Fonctionnelles convexes.* Séminaire sur les équations aux dérivées partielles, Collège de France.

Moreau, J. J. (1975). *Application of convex analysis to the treatment of elastic-plastic systems.* Lecture Notes in Mathematics, 503 (ed. P. Germain and B. Nayroles), Springer-Verlag.

Mosco, U. (1970). Perturbation of variational inequality. *Proceedings of the American Mathematical Society, Symposium of Pure Mathematics*, **28**, 182–94.

Mosco, U. (1976). Implicit variational problems and quasi-variational inequalities. In *Nonlinear operators and the calculus of variations*. Lectures Notes in Mathematics, 543, Springer-Verlag.

Nayroles, B. (1970). Essai de théorie fonctionnelle des structures rigides plastiques parfaites. *Journal of Mécanique*, **9**, 491–506.

Nečas, J. (1967). *Les méthodes directes en théorie des équations elliptiques*. Academia, Praha.

Nečas, J. and Kratochvil, J. (1973). On existence of solutions of boundary value problems for elastic–inelastic solids. *Commentationes Mathematical Universitaries Caroline*, **14**(4), 775–860.

Nicolescu, M. (1960). *Mathematical Analysis, Vol. III*. Editura Tehnica, Bucharest (in Romanian).

Oden, I. T. and Pires, E. (1981). Contact problems in elastostatics with non-local friction laws. *TICOM Report*, 81-82, The University of Texas at Austin.

Oldroyd, J. G. (1947*a*). A rational formulation of the equations of plastic for a Bingham solid. *Proceedings of the Cambridge Philosophical Society*, **43**(1), 100–17.

Oldroyd, J. G. (1947*b*). Two-dimensional plastic flow of a Bingham solid. A plastic boundary-value theory for slow motion. *Proceedings of the Cambridge Philosophical Society*, 43(3), 383–95.

Oldroyd, J. G. (1947*c*). Rectilinear plastic flow of a Bingham solid. Flow between eccentric circular cylinders in relative motion. *Proceedings of the Cambridge Philosophical Society*, **43**(3), 396–405.

Ornstein, P. (1962). A non-inequality for differential operators in the L^1 norm. *Archives of Rational Mechanics and Analysis*, **11**, 40–9.

Orthega, J. M. and Rheinbolt, W. C. (1970). *Iterative solutions of nonlinear equations in several variables*. Academic Press.

Pascali, D. and Sburlan, S. (1978). *Nonlinear mappings of monotone type*. Sijthoff Noordhoff International Publisher, Editura Academiei, Bucharest.

Pavel, N. (1977). *Differential equations attached to some nonlinear operators in Banach spaces*. Editura Academiei, Bucharest (in Romanian).

Pavel, N. and Ursescu, C. (1974). Existence and uniqueness for some nonlinear functional equations in a Banach space. *Analele Stiiţifice ale Universiţătii din Iasi*. f.I. Matematică, **20**, 53–8.

Pazy, A. (1983). *Semigroups of linear operators and applications to partial differential equations*. Springer-Verlag, New York.

Perzyna, P. (1963). The constitutive equations for rate sensitive plastic materials. *Quarterly of Applied Mathematics*, **20**(4), 321–32.

Pettis, B. J. (1938). On integration in vector space. *Transactions of the American Mathematical Society*, **44**, 277–304.

Phillips, R. S. (1954*a*). Perturbation theory for semi-groups of linear operators. *Transactions of the American Mathematical Society*, **74**, 198–221.

Phillips, R. S. (1954*b*). A note on the abstract Cauchy problem. *Proceedings of the National Academy of Sciences, USA*, **40**, 244–8.

Podio-Guidugli, P. and Suliciu, I. (1984). On rate-type viscoelasticity and the second law of thermodynamics. *International Journal of Non-Linear Mechanics*, **19**(6), 545–64.

Reiner, M. (1960). *Deformation, strain and flow*. H. C. Lewis, London.

Rockafellar, E. T. (1968). Convex functions, monotone operators and variational inequalities. *Proc. NATO Institute*, Venice.

Rockafellar, E. T. (1970). *Convex analysis*. Princeton University Press.

Rudin, W. (1966). *Real and complex analysis*. McGraw-Hill, New York.

Rudin, W. (1973). *Functional Analysis*. McGraw-Hill, New York.

Schwartz, L. (1957). Théorie des distributions a valeurs vectorielles. *Annales de l'Institut Fourier*, **7**, 1–141.

Schwartz, L. (1958). Théorie des distributions a valeurs vectorielles. *Annales de l'Institut Fourier*, **8**, 1–209.

Schwartz, L. (1967). *Théorie des distributions*. Hermann, Paris.

Segal, I. (1963). Non-linear semi-groups. *Annals of Mathematics*, **78**, 339–64.

Sofonea, M. (1982*a*). Sur l'écoulement de Poiseuille du fluide rigide-viscoplastique de Bingham. *Studii si cercetări matematice*, **34**(4), 388–94 (in Romanian).

Sofonea, M. (1982*b*). Variational inequalities with blocking property. *International Journal of Engineering Sciences*, **20**(9), 1001–7.

Sofonea, M. (1987). Evolution problems for a class of thermo-viscoplastic materials. *Preprint series in mathematics*, **32**, INCREST, Bucharest.

Sofonea, M. (1988). Functional methods in thermo-elastic-viscoplasticity. Thesis, University of Bucharest (in Romanian).

Sofonea, M. (1989*a*). A fixed point method in viscoplasticity with strain hardening. *Revue Roumaine de Mathématiques Pures et Appliquées*, **34**(6), 553–60.

Sofonea, M. (1989*b*). On existence and behaviour of the solution of two uncoupled thermo-elastic-viscoplastic problems. *Analele Universităţii Bucureşti, Matematică*, **38**(1), 56–65.

Sofonea, M. (1989*c*). Quasistatic processes for elastic-visco-plastic materials with internal state variables. *Annales Scientifiques de l'Université Blaise Pascal* (Clermont II). **25**, 47–60.

Sofonea, M. (1990*a*). Some remarks on the behaviour of the solution in dynamic processes for rate-type models. *Zeitschrift für Angewandte Mathematik und Physik* (ZAMP), **41**, 656–68.

Sofonea, M. (1990*b*). Some remarks concerning a class of nonlinear evolution equations in Hilbert spaces. *Annales Scientifiques de l'Université Blaise Pascal* (Clermont II), **25**, 13–20.

Sofonea, M. (1992). Error estimates of a numerical method for a class of nonlinear evolution equations (to appear in *Revista Columbiene di Matematicos*).

Sofonea, M. (1991). Problèmes mathématiques en elasticité et viscoplasticité. *Cours de DEA de Mathématiques Appliquées*, Université Blaise Pascal (Clermont II).

Soos, E. and Teodosiu, C. (1983). *Tensorial calculus with applications in solid mechanics*. Editura Ştiintifică si Enciclopedică, Bucureşti (in Romanian).

Stampacchia, G. (1969). *Variational inequalitites*. Publicazioni I.A.C., Serie III, N. 25, Roma.

Suliciu, I. (1984). Some energetic properties of smooth solutions in rate-type viscoelasticity. *International Journal of Non-Linear Mechanics*, **19**(6), 525–44.

Suliciu, I. and Sabac, M. (1988). Energy estimates in one dimensional rate-type viscoplasticity. *Journal of Mathematical Analysis and Applications*, **131**(2), 354–72.

Suquet, P. (1978*a*). Sur un nouveau cadre fonctionel pour les équations de la plasticité. *Comptes Rendus de l'Académie des Sciences de Paris*, serie A, **282**, 1129–32.

Suquet, P. (1978*b*). Un espace fonctionnel pour les équations de la plasticité. *Annales de la Faculte Scientifique de Toulouse*, **1**, 37–87.

Suquet, P. (1981*a*). Evolution problems for a class of dissipative materials. *Quarterly of Applied Mathematics*, 391–414.

Suquet, P. (1981*b*). Sur les équations de la plasticité existence et regularité des solutions. *Journal de Mécanique*, **20**(1), 3–39.

Suquet, P. (1982). Plasticité et homogènisation. Thesis, University of Paris VI.

Temam, R. (1979). *Navier–Stokes equations: theory and numerical analysis*. North-Holland.

Temam, R. (1983). *Problèmes mathématiques en plasticité*. Méthodes Mathématiques de l'informatique 12. Gauthiers-Villars, Paris.

Temam, R. and Strang, G. (1978). Existence de solutions relaxées pour les eqúations de la plasticité: étude d'un espace fonctionnel. *Comptes Rendus de l'Academie de Sciences de Paris*, **287**, serie A, 515–518.

Temam, R. and Strang, G. (1980). Functions of bounded variation. *Archive of Rational Mechanics and Analysis*, **75**, 7–21.

Truesdell, C. (1974). *Introduction à la mécanique rationnelle des milieux continus*, Masson, Paris.

Vainberg, M. M. (1967). *Variational methods for the study of nonlinear operators*. Holden-Day, San Francisco.

Westfreid, S. (1980). Étude du comportement asymptotique pour les modèles de viscoplasticité. Thèse de 3-eme cycle, Paris.

Wright, T. W. and Walter, J. W. (1987). On stress collapse in adiabatic shear bands. *Journal of Mechanics and Physics of Solids*, **35**(6), 701–20.

Yang, W. H. (1980). A generalized von Mises criterion for yield and fracture. *Journal of Applied Mechanics and Technical Physics*, **47**, 297–300.

Yosida, K. (1948). On the differentiability and representation of one parameter semigroups of linear operators. *Journal of Mathematical Society of Japan*, **1**, 15–21.

Yosida, K. (1971). *Functional Analysis*. Springer-Verlag, Berlin.

Zenner, C. and Holloman, J. H. (1944). Effect of strain rate upon plastic flow of steel. *Journal of Applied Physics*, **15**, 22–32.

INDEX